The Legacy of John von Neumann

PROCEEDINGS OF SYMPOSIA
IN PURE MATHEMATICS
Volume 50

The Legacy of John von Neumann

James Glimm, John Impagliazzo,
and Isadore Singer, Editors

AMERICAN MATHEMATICAL SOCIETY
PROVIDENCE, RHODE ISLAND

PROCEEDINGS OF THE SUMMER RESEARCH INSTITUTE
ON THE LEGACY OF JOHN VON NEUMANN
HELD AT HOFSTRA UNIVERSITY
HEMPSTEAD, NEW YORK
MAY 29–JUNE 4, 1988

with support from the National Science Foundation, Grant DMS-8722314;
the Alfred P. Sloan Foundation, Grant B 1987-65; and IBM Corporation

2000 *Mathematics Subject Classification.* Primary 28Dxx, 46Lxx, 65–XX, 76–XX, 81–XX, 00–XX; Secondary 01–XX.

Library of Congress Cataloging-in-Publication Data

The Legacy of John von Neumann / James Glimm, John Impagliazzo, and Isadore Singer, editors.
p. cm. — (Proceedings of symposia in pure mathematics; v. 50)
ISBN 0-8218-1487-7 (hardcover; alk. paper) ISBN 0-8218-4219-6 (softcover; alk. paper)
1. Von Neumann, John, 1903-1957—Congresses. 2. Mathematics—Congresses. I. Glimm, James. II. Impagliazzo, John. III. Singer, I. M. (Isadore Manuel), 1924– . IV. American Mathematical Society. V. Series.

QA29.v66L44 1990 90-33765
510—dc20 CIP

---

10 9 8 7 6 5 4 3 2 11 10 09 08 07 06

Contents

Preface

John von Neumann is one of the giants of modern mathematics. His career was marked by the founding and invention of a number of new fields of mathematics and by fundamental contributions to many others. His interests spanned pure mathematics, basic and applied science, mathematical analysis, the foundation of quantum mechanics, and scientific computing.

A conference, cosponsored by the American Mathematical Society and Hofstra University, was organized to study the legacy of John von Neumann. Experts in the many fields that were initiated and influenced by von Neumann met from 29 May to 4 June 1988, at Hofstra University to examine the ideas of von Neumann and to trace these ideas and their development and evolution to the present day. The present volume records some of these lectures and also includes views of von Neumann as a person, as he was known by people close to him. It documents, in a powerful and coherent fashion, a major current in the scientific and intellectual life of our times and the individual who gave it birth.

Von Neumann believed in the power of mathematical reasoning and he believed in its importance to modern civilization. This belief, expressed through his life work, is the legacy of John von Neumann.

The conference was conceived by Marshall Harvey Stone and was his final scientific activity. It is thus fitting to honor Marshall Stone by dedicating this volume to him.

James Glimm
John Impagliazzo
Isadore Singer

Proceedings of Symposia in Pure Mathematics
Volume **50** (1990)

John von Neumann: A Personal View

MARINA v.N. WHITMAN

I must confess, first of all, to considerable misgivings when I was first invited to participate in this symposium. Obviously, I have no competence to comment on the many specialized areas in which my father made noteworthy contributions, subjects that will be covered in varying degrees of detail by many expert specialists during the course of the conference. And as far as personal recollections are concerned, my uncle Nicholas clearly has an enormous advantage, partly because of his chronological proximity to my father and partly because of the intensive study he has made of my father's background and family environment, some of the fruits of which he has just shared with us.

I couldn't quite see what I had to add to these many varieties of expertise, but I finally agreed to participate in the spirit of Woody Allen's famous phrase that "ninety percent of life is just showing up". I didn't want to pass up the opportunity to make whatever contribution I can to this significant occasion, and I've already made ninety percent of it simply by standing here in front of you.

I will spend a few minutes commenting briefly on what, from my observation point, motivated my father. I think what I have to say on that will be very complementary to what Nicholas has already said. And then I will spend a few more minutes on his own view of his intellectual legacy and his concerns—his fears, in fact—about its longevity.

First, as regards motivation, from the vantage point of someone whose adult relationship with John von Neumann was relatively brief but close: Looking back over what has been said and written during the thirty-plus years since his death, it seems to me that the forces that drove him have often been poorly understood, even badly misunderstood. One purportedly serious evaluation, for example, combined pop sociology and Marxist ideology to conclude that von Neumann was driven by a desire for power and a

1980 *Mathematics Subject Classification* (1985 *Revision*). Primary 01A70.
This paper is in final form and no version of it will be submitted for publication elsewhere.

fascination with being close to the rich and powerful.

Now, there is no question that my father enjoyed the good life, liked to live well, and counted a number of celebrities among his friends and colleagues. But I believe that his wide-ranging activities, including his worldly involvements as well as his strictly intellectual pursuits, were motivated by two profound convictions. The first was the overriding responsibility that each of us has to make full use of whatever intellectual capabilities we were endowed with. He had the scientist's passion for learning and discovery for its own sake and the genius's ego-driven concern for the significance and durability of his own contributions. And the second was the critical importance of an environment of political freedom for the pursuit of the first, and for the welfare of mankind in general.

I'm convinced, in fact, that all his involvements with the halls of power were driven by his sense of the fragility of that freedom. By the beginning of the 1930s, if not even earlier, he became convinced that the lights of civilization would be snuffed out all over Europe by the spread of totalitarianism from the right. So he made an unequivocal commitment to his home in the new world and to fight to preserve and reestablish freedom from that new beachhead.

In the 1940s and 1950s, he was equally convinced that the threat to civilization now came from totalitarianism on the left, and his commitment was just as unequivocal to fighting it with whatever weapons lay at hand, scientific and economic as well as military. It was a matter of utter indifference to him, I believe, whether the threat came from the right or from the left. What motivated both his intense involvement in the issues of the day and his uncompromisingly hardline attitude was his belief in the overriding importance of political freedom, his strong sense of its continuing fragility, and his conviction that it was in the United States, and the passionate defense of the United States, that its best hope lay.

Now, a little bit about John von Neumann's legacy, from the vantage point of his daughter and only child: In particular, I will focus on his concerns during the last year or two of his life. Especially toward the end, but even before, he became deeply concerned about the question of his ongoing legacy. He was concerned partly with immortality in the personal or religious sense, but that is a topic for another day.

He was also profoundly concerned, however, with the nature of his legacy in this world, in two respects. One had to do with the durability of his work, his intellectual contributions; he was surprisingly insecure about that. And, interestingly enough, I don't think he was a very accurate prophet regarding what turns the practical applications of his pioneering work would take. For example, he clearly expected that the computer would have its impact primarily on scientific research and military work. He was particularly interested in its role in advancing the accuracy of weather forecasting and, ultimately, climate modification. I don't think progress in this area has been nearly as far or as fast as he hoped and expected. Similarly, I think he anticipated that the theory of games would have more impact, to borrow John MacDonald's book title, on poker, business, and war than it has turned out to have had so far, at least.

On the other hand, if anyone had ever told him that the company I work for, General Motors, would produce and utilize literally millions of computers every year (each of the roughly eight million vehicles we produce each year contains several, not to mention the ones in our plants and offices), I think he would have been startled. And the notion of adults fulminating against computers as corrupters of youth in the form of video games would have amused and perhaps secretly pleased the playful, childlike aspect of his personality.

In fact, my father foresaw the inadequacy not only of his own forecasts but of such forecasts in general. In the 1955 *Fortune* article my uncle Nicholas quoted a few minutes ago, he said: "All experience shows that technological changes profoundly transform political and social relationships. Experience also shows that these transformations are not a priori predictable, and that most contemporary first guesses concerning them are wrong."

The second focus of John von Neumann's concern about his earthly legacy was, to put it succinctly, me. I was his only offspring and, toward the end of his life, he became acutely conscious that all his eggs were in one basket, genetically speaking (if biological inaccuracy can be forgiven for the sake of metaphor). So he put tremendous pressure on me to perform up to the peak of my abilities, and made clear his displeasure with the path I appeared to be taking. I had married young, right out of college, and he thought that this was a bad beginning, not because he was not quite fond of my husband, but simply because he feared (and it was a reasonable fear, in the 1950s) that a woman who married young was very probably reducing her chances of making a significant intellectual or professional contribution. And, as Nicholas has pointed out, he tended to take a somewhat statistical view of the world.

Statistically he was right, of course, but I like to think that in this particular case he was wrong. I'm no John von Neumann, obviously, but I have had a reasonably successful and highly rewarding career as an academic economist, a presidential adviser, and now as a corporate executive.

And beyond me is the next generation, the grandchildren whose accomplishments he couldn't foresee because he died far too early, before they were born or even contemplated.

I'm deeply sorry that he never got to know the grandson who has translated a six-year-old's dream of "someday finding a cure for cancer" into a career as a molecular biologist doing research on the chemistry of intercellular message transmission. He would have found comforting continuity, I think, in the fact that Malcolm and his colleagues are exploring mysteries of the universe as fundamental as many of those that preoccupied my father, as well as because the work that they are doing is in some sense a fitting sequel to von Neumann's own explorations, culminating in his posthumously-published *The computer and the brain*, into relationships among mechanical, electrical, chemical, and biological processes.

I'm equally sorry that he could never know the granddaughter who, having avoided scientific subjects throughout her university career, realized after graduation that her newly-focused goal of radically improving health care would require medical training and that that, in turn, would require filling the substantial gap between ninth-grade biology and premedical requirements. She filled the gap and is now a medical student, firm in her belief in the chemical-biological unity of mind and body and determined to fashion a career which will harness her skills in the service of *mens sana in corpore sano* for those whose access to such services is limited by accidents of geography or socioeconomic situation.

John von Neumann would have felt reassured and gratified, I believe, by the choices his progeny and his progeny's progeny have made, to do what he considered most important, that is, to utilize our intellectual capacities right up to their limits, to try to achieve whatever potential we have. He would have been relieved and gratified that the forces of light appear to be holding their own, perhaps even gaining, against the forces of darkness in the sphere of political freedom. He would have been surprised and perhaps amused, as well as disappointed in some areas, at the twists and turns his contributions have taken in affecting our everyday lives. He would have felt reassured to know that we are all still here to talk about such matters in the year of our Lord 1988, given the fears he expressed in the article Nicholas quoted that mankind might not make it to 1980. And he would be tremendously gratified to know that the intellectual leaders gathered here are devoting their Memorial Day weekend, as well as the week that follows, to a discussion of the legacy of John von Neumann.

Vice President and Group Executive, General Motors Corporation

Proceedings of Symposia in Pure Mathematics
Volume **50** (1990)

Remembering John von Neumann

PETER D. LAX

I am just about the youngest member of the generation that had direct scientific contact with John von Neumann. I am here to give an account of some of the things von Neumann said, and some of the things said about him.

When von Neumann showed unusual mathematical talent at age 15 his parents arranged for him to be tutored by Gábor Szegö. Mrs. Szegö often recalled that Szegö came home with tears in his eyes from his first encounter with the young prodigy. The brilliant solutions to the problems posed by Szegö, written on the stationery of his father's bank, can be seen in the von Neumann Archives in Budapest. After Szegö left Budapest, von Neumann's mathematical education was taken over by Michael Fekete; this resulted in a joint publication in 1922 on the Transfinite Diameter, a subject to which Fekete subsequently devoted most of his scientific life.

The work which first brought von Neumann center stage was his theory of selfadjoint operators, about which Marshall Stone spoke at this conference. The following story illustrates well the originality of von Neumann's approach: In the thirties Erhardt Schmidt asked the young Franz Rellich to describe his recent research in Hilbert space theory. Rellich was happy to oblige and started in the now standard fashion: "Let H be a Hilbert space and L a linear operator..." at which Schmidt interrupted him: "Please, young man, say infinite matrix." The success of von Neumann's approach hinged on giving up the notion of operators as infinite matrices and substituting a geometric view. Von Neumann retained a lifelong interest in the theory of selfadjoint operators; I recall how pleased and excited he was in 1953 when he learned of Kato's proof of the selfadjointness of the Schrödinger operator for the helium atom.

Von Neumann combined, in a unique fashion, extreme quickness, very broad interests, and a fearsome technical prowess. The popular saying was:

1980 *Mathematics Subject Classification* (1985 *Revision*). Primary 01A70.
This paper is in final form and no version of it will be submitted for publication elsewhere.

"Most mathematicians prove what they can, von Neumann proves what he wants." Once in a discussion about the rapid growth of mathematics in modern times, von Neumann was heard to remark that whereas thirty years ago a mathematician could grasp all of mathematics, that is impossible today. Someone asked him: "What percentage of all mathematics might a person aspire to understand today?" Von Neumann went into one of his five-second thinking trances, and said: "About 28 percent."

It is hard to believe today that von Neumann's great versatility led some pure and puritanical mathematicians to charge that he had merely skimmed the cream off subjects, and left to others the arduous task of getting to the bottom. The story is told of the time when the mathematics faculty of the Institute for Advanced Study was discussing the appointment of Enrico Bombieri. His great contributions to function theory, calculus of variations, minimal surfaces, and algebraic number theory were described; everyone was favorably impressed, except one distinguished professor who commented: "This fellow dabbles in this, dabbles in that; he will end up like von Neumann."[1]

Von Neumann was addicted to thinking, and in particular to thinking about mathematics. Eisenhart surmised that his legendary quickness was enhanced by his having thought before about a great many questions. The obverse side of his compulsive thinking was a boredom with most mathematical discoveries except those that had something very surprising about them.

During and after the Second World War, von Neumann took on many official duties, time-consuming and bearing a heavy burden of responsibility. He did this not only because of a sense of obligation, but also out of a deep desire to leave his mark on the world, not only on the world of mathematics. About 15 years ago Eugene Wigner visited Budapest and received a celebrity treatment, including two long interviews on television. He was asked if it were true that in the late forties and early fifties the scientific policy of the U.S. Government was set by von Neumann. Wigner replied with his characteristic precision: "That is not quite so; but after von Neumann analyzed a problem, it was clear what had to be done."

His official obligations left less time for scientific work, if not for the doing of it, for recording it. He was willing to lecture on his work and ideas, but didn't have the leisure to write them up for publication. He didn't have much time for preparing them either; but so great was his power of concentration that his lectures were models of good organization and his sentences grammatically resplendent, with a slight tendency toward germanic length. After one of the rare occasions when he couldn't finish a proof he remarked ruefully: "I knew three different ways of proving this result, but unfortunately I chose a fourth way."

[1]It was recently pointed out by someone present that the second part of the statement belongs to the ben trovato category.

Problems of computing in science, the design of algorithms, the design of computing machines, and the structure of programming languages fascinated von Neumann endlessly. I heard him recall how, early in his forays into programming, he spent much time proving that every program that could be coded in one particular language could also be coded in another particular language. "Only later did it occur to me," he said, "that I was merely reproving a special case of Turing's theorem."

It is fashionable these days to talk about non-von Neumann computers, meaning ones that process in a parallel rather than serial manner. This ever so slightly pejorative designation is unfair to von Neumann; at the time he formalized the concept of a computer, the state of technology pointed to a serial design. There is little doubt that today, when sophisticated chips are abundantly available at low cost, von Neumann would have favored a parallel design. It is a backhanded compliment to his intellectual prestige that those who propose to change the paradigm from serial to parallel behave as if they were committing symbolic patricide.

The tragedy of von Neumann's early death robbed mathematics and the sciences of a natural leader and an eloquent spokesman, and deprived a whole younger generation of beholding the most scintillating intellect of this century.

COURANT INSTITUTE OF MATHEMATICAL SCIENCES, NEW YORK UNIVERSITY

Proceedings of Symposia in Pure Mathematics
Volume **50** (1990)

Nonmathematical Reminiscences about Johnny von Neumann

FRANÇOISE ULAM

As I am not a mathematician, it was as a sort of camp follower that I attended a few such gatherings in the past. It is a new experience for me to be asked to participate, so I shall begin these after dinner reminiscences with a few simple vignettes:

I never saw Johnny other than properly dressed in a business suit and tie, no matter what the occasion. I have at home, and I expect most of you have seen, the photo of Johnny and Klari at the Grand Canyon, posing in a group of riders, before the descent to the bottom. Johnny is the last one, hatless but fully dressed in coat and tie, on the only mule facing the wrong way!

For a nonathletic, nonoutdoorsy person he could surprise you sometimes, as when he joined the Lake Peak hike Edward described this afternoon—that was a great summer favorite of the Fermis and Bethes. That memorable day I noticed he wore his usual city clothes!

More seriously, in Los Alamos he also worked! I can still see him in the coffee room of T-division covering a blackboard with the first stirrings of flow-diagram coding, while casting unconscious sidelong glances at every feminine pair of legs that passed by! His lines and arrows, instructions and addresses boggled my nontechnical mind, so I chose to stay with one of the old mechanical machines on which Stan and Everett's simplified H-bomb calculations were made as others prepared the more detailed Von Neumann-Evans trial runs while they waited for one of the two Maniacs Edward described to become operational.

Speaking of the Maniac, my ten-year-old granddaughter was amazed when I told her that the ancestors of her pocket calculator used to fill a room, and that the programs it contains are derived from those first created by a friend of her grandfather—the very programs just mentioned, that were born of the need to calculate whether it was possible or not to build an H-bomb, on

1980 *Mathematics Subject Classification* (1985 *Revision*). Primary 01A70.
This paper is in final form and no version of it will be submitted for publication elsewhere.

the subject of which I felt Johnny was more in tune with Edward than with Oppenheimer, though he defended the latter eloquently during the so-called Affair.

I first met Johnny in 1941 when Stan introduced me to him, at a mathematical meeting in Chicago shortly after we were married. That morning the mathematicians had been offered, for distraction, a slaughterhouse tour. Of course, Johnny had gone along! (Stan and I more cowardly passed this up.) I say "of course" because it was not long before I became aware that he seemed to be always willing to go wherever the action was! He joined us after the tour, pale and shaken, are admitted that it had been a rather gory trip.

There were but few other times when I saw him visibly upset and not his usual composed self. One, when he came to us immediately upon returning from the Alamagordo bomb test. Another, on what was to be his last visit to Los Alamos. He already knew the nature of his illness, and did not look well at all. But at our house for lunch, he would not tell Stan what was wrong. He wanted, I think, to spare him.

At our first meeting I had been struck by the intensity of his dark, dark eyes and his cheerful rotundity. I told Stan his appearance reminded me of Riquet à la Houppe, a short, roly-poly knight in a French fairy tale. Stan must have repeated this to him for in a following math letter he took time to remark: "My respectful homage to Françoise... God knows that my similarity to Riquet à la Houppe... is rigorously exterior. I am a miserable sinner, and I never rescue maidens... But her high and utterly undeserved consideration thrills me to the core of my being." I was amused by the impeccably polite way in which he was making fun of me! and properly impressed that his encyclopedic knowledge included the chubby Perrault knight.

Our acquaintance began in earnest during his and Klari's visits to Los Alamos. It is in that context that I got to know him, more often than not in association with Klari, who became a close friend. As we lived a stone's throw from their quarters at the Lodge it was easy for them to walk over after work, for a meal, or a drink, or parties on our lawn, or other entertainment their visits always generated. For in addition to being a hard worker, Johnny seemed to be a bon vivant par excellence, always ready, for instance, for a wild drive to neighboring Spanish cafés in search of the peppery enchiladas he loved. Stan said they probably reminded him of Hungarian goulash!

I used to marvel how anyone could be so sociable and still do creative work, until I realized that you, mathematicians, live a double life. Even when you are here, with us, on the earth, your eyes have a look that shows you are cerebrating! "Thinking!" you call it. I suppose that Johnny's mind was always tuned to the far reaches of abstract cogitations and it did not matter whether he was at work or play.

Play was when on Christmas Eves he and Klari would come to help us put together our daughter's unassembled toys. Can you see us four on the floor, struggling with the parts of a dollhouse? Johnny and Stan could never figure out how to insert tab A into slot B!

Play was when we went with friends to the banks of the Rio Grande for candle-lit dinners at Edith Warner's adobe house, which had no running water or electricity! This Easterner in mocassins cooked on a wood stove, and her meals were served by her long-braided Indian friend, Tilano. Johnny appreciated, as we did, the contrast of their timeless Indian life with the 20th century reality "on the hill".

One year we were invited to the von Neumanns in Princeton for a family celebration. Johnny and Klari's mothers were there, Michael and Nikki, too, and a number of Princeton friends. Mrs. von Neumann, Sr. seemed definitely the respected matriarch, and Johnny, the head of the clan and a devoted family man. The same man who could solicitously inquire about Claire's measles in a letter full of math.

This brings me to what I want to mention next, namely Johnny's correspondence with Stan. After Johnny died, Stan asked me to gather all the Johnny letters I could find in his files. As he never threw a letter away there were quite a few, many in Johnny's clear long hand on an assortment of ocean liners, hotels, or university stationery that attest to his peripatetic existence. They span most of his American life from 1935 to his death in 1957.

Needless to say, they are first and foremost mathematical but they are also a wonderful echo of their non-ivory-tower, non-absent-minded-professor conversations. Rational, more intellectual than affective, laced with touches of irony, they alternate math with gossip, politics, and family news. I have a feeling, as when they talked, that this constant switching of subjects provided them with a necessary release from the tension of abstract thought.

I'd like to give you a few samples of the tone and tenor of their more mundane parts. They show Johnny better than I can.

The earliest prewar ones are full of gaiety and parties: "Do come and stay as long as possible... the Robertsons are yelling for caviar!" "Come in time to make the Harvard-Princeton game... I have already tickets for both of us." Can you imagine Stan and Johnny at a football game? Do you suppose they talked mathematics and exchanged unprintable jokes the whole time?

Then as the clouds of war break over Europe and finally engulf the world, the tone, the preoccupations, change: "I have been made a member of the War Department's 'Scientific Advisory Committee to the Ballistics Research Laboratory' at Aberdeen... I have been put on the 'War Preparedness Committee' of the AMS+AMA too. The functions of all these sets, and sets of sets are not very well defined as yet, but I suppose that they will be when 'The Day' comes around... " (of the U.S. entry into the war).

After America's entry into the conflict: "I think as before *qu'on les aura.* May take honorable 2–3–4 years. But it's going to be a peculiar little world... Life is getting complicated. I am writing a book on economics with Morgenstern... But I'm getting more and more snowed under by war work."

Not realizing the extent of his involvement, Stan and I in Madison couldn't understand why Johnny's letters were becoming scarce and Stan complained, which brought back a torrent of excuses on unmarked stationery: "I would be very sorry if you interpreted my behavior as discourtesy, it is most certainly not that,... I have to lead an absolutely irregular life, ... Since I came back from England I have been in 3–4 different places each week. I am now in the Southwest,... I really hope that you will understand ... "

This letter, dated November 1943, was obviously from Los Alamos, and shortly after, Stan received an invitation to join the Manhattan Project. He promptly informed Johnny, who replied, "I am very glad that Mr. Hughes 'and all that he stands for' has come through. (Mr. Hughes was the personnel manager at Los Alamos)... If you want war work, this is probably a quite exceptional opportunity ... " Prophetic words indeed!

From then on, and between his visits to "the hill", the letters are about computing, about what they called the "S" (for Super) problem, and about his escalating involvement with government and military science. In this manner, the Los Alamos connection, the development of Monte Carlo and other computing techniques, their interest in cellular automata and the brain, were their link to the last.

And now, since it is Stan who more appropriately should have been here tonight, but has unfortunately also 'left"—to use Erdösian language—I will conclude with a few of his remarks:

"... Von Neumann was a giant in the breadth of his knowledge, especially... now that the diversity and complexity of contemporary problems enormously surpass the situation confronting [people like] Poincaré and Hilbert... Yet, he admitted to me that he felt he did not know even a third of mathematics... So, at his suggestion and for amusement I concocted for him an oral doctoral examination in various fields in such a way that he would not be able to pass it... I found one problem in differential geometry, one in number theory, one in algebra and a couple of others. And he agreed that he could not have answered any of the questions and the exam would have been a complete failure...

"At Los Alamos, he was one of the originators... of implosion... He worked on the whole project, scientifically and politically, especially with the hydrogen work... In fact he was unable to play the role of senior scientist or advisor without being actively engaged... To my mind and to my taste, the most important work he did is what he did when he was getting older... namely his speculations on automata, on the brain, and his contributions to computing and to problems in hydrodynamics...

"His quickness was quite remarkable. He saw immediately the possibilities of Monte Carlo... It was very pleasant to discuss it with him. Too bad he did not live to see how computers have revolutionized everything... His role in their development was tremendous..."[1]

[1]From a conversation with Rota published in *From Cardinals to Chaos, Reflections on the Life and Legacy of Stanislaw Ulam*, Necia Grant Cooper, editor. Cambridge University Press, 1989, pp. 303–306.

Proceedings of Symposia in Pure Mathematics
Volume **50** (1990)

The Extraordinary Inspiration of John von Neumann

ISRAEL HALPERIN

When John von Neumann died in 1957, LIFE magazine (February 1957) published the following tribute: Nobel Laureate physicist Hans Bethe said of von Neumann, "I have sometimes wondered whether a brain like von Neumann's does not indicate a species superior to that of man."

Throughout the world mathematicians and others had marvelled at the lightning speed with which von Neumann analyzed and solved complex problems. In a memorial without precedent, the American Mathematical Society published a special 129 page Bulletin, May 1958, describing von Neumann's "brilliant mind, his warm personality, and his work which will live in mathematics and the other sciences to which he contributed so much".

Because of the rise of Hitlerism, Goettingen ceased to be the world center for mathematics and in the early 1930s Princeton became the new Mecca for mathematicians, Fine Hall at Princeton University began in 1933 to house not only the mathematicians of the university but also those of the new Institute for Advanced Study. During 1933–1936 there were at Princeton, among others: the professors Eisenhart, Lefschetz, Wedderburn, Church, H. P. Robertson, Wigner, Bochner, Wilks, Tucker, Bohnenblust, Veblen, von Neumann, Alexander, Weyl, Einstein; and a stream of visitors, among them Montgomery, Brauer, Coxeter, Gödel, Bernays, Ulam, Albert, Dirac, Pauli, Jessen, Myers, Zippin, Nakayama, Ted Martin, Levinson, Bergman, Infeld, Chabauty, and Bouckaert. There were students, doctoral and postdoctoral, among them Jacobson, Walker, Kleene, Rosser, Murray, Givens, Vanderslice, Steenrod, Clarkson, Malcolm Robertson, Dowker, and Taub.

In this galaxy of stars von Neumann, a professor at the Institute, simply radiated excitement. His lectures on Hilbert Space, measure theory, rings of operators (called now von Neumann algebras), and continuous geometry, fascinated a large audience. At the daily afternoon tea he engaged some

---

1980 *Mathematics Subject Classification* (1985 *Revision*). Primary 01A70; Secondary 06C20.
This paper is in final form and no version of it will be submitted for publication elsewhere.

group in a most lively and stimulating discussion. With obvious delight he explained, clarified, and analyzed problems on the spot and gave help to one and all. But sometimes he would stand apart, deep in thought, his brown eyes staring into space, his lips moving silently and rapidly, and at such times no one ventured to disturb him.

Von Neumann's researches usually attracted the attention of schools of followers. But strangely, only a few showed serious interest in von Neumann's work on complemented modular lattices (special case: continuous geometries) and regular rings (special case: continuous rings). Yet this work contains deep, interesting theorems, in particular: the coordinatization theorem, the arithmetics of continuous rings, and the identification of certain continuous geometries with the set of projections of the finite type rings of operators.

This seeming lack of interest may have been due to the fact that his manuscripts were published only after his death: two in mathematical journals, the book *Continuous geometry* by Princeton University Press in 1960, and the Memoir *Continuous geometries with a transition probability* by the American Mathematical Society in 1981. That much remains to be researched was indicated by Irving Kaplansky's brilliant discovery (Annals of Math., vol. 61, (1955), pages 524-541) that every complete orthocomplemented modular lattice is automatically a continuous geometry.

Professors at the university direct doctoral theses but those at the Institute do not. Unaware of this, in 1934 I asked von Neumann if he would direct my doctoral thesis. He replied Yes, and suggested the problem of identifying the Hilbert space closure and adjoint of nth-order linear differential operators. Marshall Stone, in his huge volume *Linear transformations in Hilbert Space*, had solved the case for first order and his methods generalized to higher orders. My not particularly outstanding thesis was accepted and I moved into an ardent study of continuous geometry. In 1936, as a postdoctoral Fellow at Yale, I found a partly new proof, with weaker axioms, for von Neumann's transitivity of perspectivity. Von Neumann invited me to visit Princeton and talk with him.

He gave me most cordial encouragement, let me have his unpublished manuscripts to study, and later took the initiative to recommend me to Marshall Stone for a B. P. Instructorship at Harvard. This warm, generous concern made a deep impression on me.

Von Neumann spent the summer of 1940 as visiting professor at the University of Washington at Seattle and I went there simply to be in his presence. One weekend, he and his wife Klari took me with them on a trip around the Olympic Peninsula. They took turns driving, both of them in a completely normal way. (I discount the stories of von Neumann's erratic driving which I suspect he encouraged to amuse his colleagues.) During the trip I asked what he anticipated would be the future trend in mathematics. He replied that some strong influence from outside mathematics was needed. In

1945 he addressed the Canadian Mathematical Congress in Montreal, dealing with partial differential equation problems for which he felt computers were needed; he predicted that future mathematicians would use computers in place of paper and pencil to try out ideas concerning complex problems. Forty-three years later Hale Trotter reported that every professor of mathematics at Princeton University was about to have a computer terminal in his or her office.

Looking at a list of geometers one day I suggested that his name should be there but he said no, he was not a geometer. To the obvious question, he replied he would probably call himself an algebraist. Though probably never in doubt about his own talent, von Neumann expressed profound admiration for certain other mathematicians such as Hilbert and Gödel. When others discovered simpler proofs for some of his theorems, von Neumann was quick to express his appreciation. But in fact his own proofs usually contain powerful new techniques and it would be a great pity if these were neglected.

UNIVERSITY OF TORONTO, CANADA

Proceedings of Symposia in Pure Mathematics
Volume **50** (1990)

The Philosophical Legacy of John von Neumann, in the Light of its Inception and Evolution in his Formative Years

NICHOLAS VONNEUMAN

Thank you for the introduction, and thanks also to the sponsors and the organizing committee for inviting me. I consider it a great privilege indeed to have this opportunity to participate in this Symposium in honor of my brother, John von Neumann.

To the question: which philosopher did John follow? the general answer is that he was a pragmatist, and that in itself is an indication of an underlying philosophy. But from my perspective he followed, or at least considered, one specific philosopher: Goethe. After all, Goethe was a philosopher, among many other professions. We studied Faust in school very thoroughly, both in the original and in Hungarian translation. We discussed it for years and reread it occasionally throughout our respective lifetimes.

There were three passages in particular which aroused our attention. The first passage is the concluding eight lines of the second part:

> "Alles vegängliche ist nur ein Gleichnis"—"All things transitory but as symbols are sent/Earth's insufficiency here grows to event/The indescribable here it is *done*/The Woman-Soul leadeth us upward and on" (Bayard Taylor's translation).

There are many interpretations for these eight lines. But relevant to our purposes was our speculation, right or wrong, that within Goethe's overall approach there is a unifying force behind all manifestations of nature, which we cannot fully comprehend here on this earth, but we can try to explain it by *means at our disposal*. From my perspective, it was in this spirit that John tried to satisfy his curiosity and to tackle the mysteries of nature with means at *his* disposal; namely, atomic and subatomic phenomena,

---

1980 *Mathematics Subject Classification* (1985 *Revision*). Primary 01A70.

This article is based in part on "John von Neumann, as seen by his brother" by N. A. Vonneuman (©1987 Nicholas A. Vonneuman).

through the means of quantum mechanics; weather and climate phenomena, through the means of numerical meterology; central nervous system mysteries, through analogies with the logic and architecture of computers; and genetics and inheritance, through the means of theory of self-reproducing automata.

As for the second passage, in Part I before departing on their journey, to Faust's question "where are we going" Mephistopheles replies: "The little world and then the great we'll see." Pursuant to our studies we associated the *little* world with Part I, dealing with basic human emotions and passions of man standing alone, and we associated the *great* world with Part II, dealing with man's more sophisticated intellectual ambitions, no longer standing alone but together with others as members of society. The discussion of this aspect was then leading to the professionals' responsibility to society within the scope of their respective profession, leading ultimately to John's views on the social responsibility of scientists.

The third passage is Faust's monologue at the opening of the First Part: "In the beginning was the Act", and the corresponding statement in Part II: "The deed is everything, the Glory naught." This we discussed in the context of the redeeming value of action, leading ultimately to John's views emphasizing the redeeming value of practical applications in his profession.

So much for Goethe. But we have many additional examples for the background of the evolution of John's philosophical, ethical, scientific, and related concepts in childhood and young adulthood, in the course of family discussions around the dinner table, to the extent that these concepts reappeared later at the other end of their evolution in the background of the work of the mature scientist, and to the extent of my observations as a witness, and by now one of the very few surviving witnesses of that early period.

I will now mention some of these, but by doing so I do not claim that there is necessarily a causal connection between exposure in the early environment and reappearance of concepts in mature years. I only claim that from my perspective there was a significant correlation.

Around the dinner table we listened frequently to my father's comments on his own profession and business activities, as a lawyer and banker engaged in financing, and his underlying theories, namely, financial statement analysis and criteria for credit qualifications; and the practical applications, namely, selecting investment opportunities or borrowers. This usually led to a general discussion of the underlying theories also of other professions in relation to their respective practical applications. Father also used to discuss the technological aspects which his industrial applications involved. If a newspaper enterprise was involved, he would have brought home samples of type font and the printing press was discussed. If financing involved a textile enterprise, such as for instance the Hungaria Jacquard Textile Weaving Factory, then the discussion centered around the automatic Jacquard loom and its modern adaptations. Again we have a strange coincidence that much later John became involved with IBM punch cards.

Discussions also involved many other subjects: politics, science, theatre, literature, and art, displaying already at this early stage John's diversified curiosity and interests in all directions. For instance, from high school studies we recalled, perhaps erroneously, that the light-sensitive receptors on the retina perform exactly the same function as the silver grains on the photonegative film, so that a complete and fully detailed picture is projected into the retina which is then recorded as such. John commented that maybe this is not quite so simple, because the retina presumably merely takes samples; and then he pointed to the *amazing* fact that whereas on man-made panels the wire bundles exit towards the rear, on the retina they move forward in a "transparent" medium before bending back and exiting as the optic nerve. Another discussion on the linear input to the ear contrasted with the multidimensional or area input to the eye. But, he warned, this may be misleading, since the cochlea is not necessarily recording merely a *sequence* of changing sound frequencies, but may record also other incidentals of the soundworld around us. These early interests are again recognizable in his subsequent speculations in the context of central nervous system circuitry.

And now, an early example for axiomatic deduction. A prize contest in an English language periodical. Who wrote this:

> "They know this well my baron and my men/Gascony, England, Normandy, Poitou/That I had never follower so low/ Whom I would leave in prison to my gain./ I say this not as a reproach to them/ But prisoner I am."

Since I did not know the answer, I asked John and he responded promptly without hesitation: Richard Coeur de Lion (King Richard the First)—Did you know this poem?—No—Then how did you identify the poet?—"Very simple (he said), Gascony, England, Normandy, and Poitou were in one feudal hand only during the early Plantagenets, and from there it was quite easy to associate with Richard's crusades and European captivity. But of course (he added) this is a translation, since quite obviously the early Plantagenets spoke Norman-Medieval French." I found out much later that the translation was that of Henry Adams, and that the "Prison Song" is only one of Richard's most perfect poems, usually referred to as gems of English literature!

And now, an example of an early display of a future statistician's approach. The catastrophy of the "Titanic" remained in the headlines for many years, particularly after details of the U.S. Senate and British Board of Trade inquiries became better known. I got quite excited about the argument that if only one of the several critical contributing factors had not occurred, then the tragedy could have been averted. John assured me that my concern was illogical. What I should have said was: if another *set of random circumstances had prevailed*...and then some of them could have been worse. Example: one of the significant contributing factors was First Officer Murdoch's order from the bridge: "hardastarboard–fullspeedastern," which in retrospect and with hindsight was a mistake. But trade it for a high wind or storm, and then none of the lifeboats could have been lowered!—Shouldn't this remind us of the language of a future statistician?—We then also discussed the broader *moral* issues involved, namely, responsibility for training persons in command position in how to behave in emergency situations under stress or panic, such as for instance: Who would have been responsible for training the ship's officers in high speed maneuvering? Who would have been responsible for equipping the ship with the additional lifeboats for which the davits were designed? Responsibility for the failure of Captain Lord of the nearby SS Californian to order his radio operator on the air at the first sighting of the Titanic's distress signals? Was J. P. Morgan directly or indirectly responsible, in his capacity as founder and financier of International Mercantile Marine, owners of both the Titanic's White Star Line and the Californian's Leyland Line? Again, all of this led to discussions of social responsibilities of professionals in their respective field of competence.

Around the dinner table or otherwise John loved to tell jokes in a pleasant raconteurian style, not necessarily claiming originality, nor for the sake of entertainment, but for the benefit of discussion and analysis of the hidden meaning or implied message involved in a joke. Two examples follow.

The first example shows Emperor Franz Joseph walking with an aide in the Prater, Vienna's famous public park, on a warm early spring afternoon. Couples were sitting on benches, in the lawn and shrubbery, in each others' arms with tender embraces. His majesty looks around with astonishment, and finally asks his aide what these young people are doing. The aid turns to his majesty's ear and in a low voice gives the answer ... such low voice that others could not hear it and history could not record it. His majesty, in turn, looks around once more with an expression of bewilderment, and says: "Also machen das die Leute noch"—"So it is still being done." In context, this illustrates political science and Franz Joseph's alleged senility at the outbreak of World War I.

The second example shows a street scene in Berlin during World War I. A man at a street-corner was yelling repeatedly: "The Kaiser is an idiot." Two police agents appear out of nowhere and arrest him for high treason.—"But I was referring to the Austrian Kaiser, not to our Kaiser."—"You *can't* fool *us*. We *know* who the idiot is." In context, this again illustrates political science, but now tied to semantic ambiguities.

Many more jokes are available. But my point is that this sense of humor reappeared later as very dry humor in the background of the most serious statements of the mature scientist. Four examples are now given.

(1) In the context of comparing living organisms with man-made devices.

The living organism when injured can restore itself. "But if we hit a man-made device with a sledgehammer, no such restoring tendency is apparent."

(2) In the context of weather forecasting with highspeed computers. It is not only a question of doing it faster but doing it at all. For example, computations for a 30 days global weather prediction may take one to thirty days. "If it takes more than 30 days, then we will probably not do it."

(3) During the famous, if not infamous, Oppenheimer hearings of the AEC Personnel Security Board, where John appeared as a witness, the counsel for the Board presented John with a question based on an incredibly complicated *hypothetical* situation in which someone acted badly (insinuating Oppenheimer) and asked John whether he would have acted in the same manner in that situation (insinuating him). He replied, "... you are telling me now to hypothesize that somebody else acted badly and ask me whether I would have acted in the same way. But isn't this a question of when did you stop beating your wife?"

(4) In the context of maintenance problems with vacuum tube devices. "It is desirable that on a day consisting of 24 hours, we should *not* spend more than 24 hours on maintenance."

So much for jokes and humor. And now, in conclusion, let me list some of the messages left behind by John as what I consider his philosophical legacy, even though my conclusions are not based only on the examples which I presented now in a rush, but also on other material from my book, my notes, and related sources.

(1) John warned that

> "as a mathematical discipline travels far from its empirical source, it *may* lead to abstract inbreeding and danger of degeneration of the mathematical subject, and that at that point the remedy may be the rejuvenating return to the source, the reinjection of more or less directly empirical ideas."

All of this is applicable by analogy also to other professions or disciplines, in the context of *their* respective pure and abstract practice, in relation to practical application.

(2) Social responsibility of scientists has been vastly increased since the 1942 Chicago atomic pile, especially in the field of international relations. Consequently, scientists are no longer free to carry on their research in isolated "ivory towers", completely relieved from accounting to society for the possible uses of their discoveries. Consequently, the education of future scientists is not complete if limited to technical professional subjects, but must include, also, "history, law, economics, government and public opinion ... but without endangering the foundations upon which the sciences themselves rest and thrive."

(3) The inclusion of traces of humor into the background of messages, not for the fun of it, but for easier transmission of the message, particularly of an unpleasant message.

(4) A message to new generations: include in your studies subjects of science and technology, because evolution is so fast that by the time of graduation or entering professions, entirely new fields will be open which could not have been anticipated.

(5) Last, but by no means least, quoting from the editorial introduction to John's 1955 Fortune article entitled, *Can we survive technology*, his "constructive pessimism which does not hamper creativity." And John's own concluding words of this article in the context of the fact that technology and science are neutral although potentially both beneficial and dangerous: "the problems of the future of humanity can not be resolved by a single perscription, but only in reliance on day-to-day *opportunistic* measures, and reliance on the human qualities required: (namely) *patience, flexibility* (and) *intelligence.*"

Well, that's it. But I repeat that all of this was merely from my perspective, but nevertheless part of an overall scenario which may become of interest to John's future biographers. What I said about Goethe was also only from my perspective. After all, a detailed discussion of Faust would call for at least one additional five-day symposium. Thank you for listening.

Proceedings of Symposia in Pure Mathematics
Volume **50** (1990)

Von Neumann and the Early Days of Ergodic Theory

GEORGE W. MACKEY

1. Background. Ergodic theory, as a self-conscious new branch of mathematics, began in 1931 with the mean and pointwise ergodic theorems of John von Neumann and G. D. Birkhoff respectively. However the central notion of "ergodicity", or "metric transitivity" as it was originally called, was introduced a bit earlier. It was defined in a paper of Birkhoff and Paul Smith entitled *On the structure of surface transformations* published in 1928. All three of these papers appeared in the Proceedings of the National Academy of Sciences of the U.S.A.

Of course the circle of problems to which the notion of ergodicity and the two ergodic theorems made such important contributions has a much longer history. The problems in celestial mechanics which inspired the paper of Birkhoff and Smith go back at least as far as G. W. Hill. It was Hill who first fully realized the necessity of studying the qualitative properties of the solutions of differential equations, directly from the equations themselves, without seeking to find explicit integrals and published a paper on the three-body problem in 1878 which stimulated the extensive work of Poincaré on the same subject. Birkhoff took up where Poincaré left off and his paper with Smith was a direct outgrowth of this continuation of the work of Poincaré. One could go back even further and implicate works of Lagrange in the 1770s on "small divisors" and their effects on perturbation theory.

The idea of an ergodic theorem goes back in another direction and originated in the work of Boltzmann on the "kinetic theory of gases". Continuing earlier work of Maxwell, Boltzmann wished to explain the thermodynamical properties of gases by applying Newton's laws and probability considerations to the motions of the gas molecules. Let Ω denote the phase space of the dynamical system describing the motion of the collection of all of the molecules of a gas. For each real number t let T_t denote the one-to-one transformation

1980 *Mathematics Subject Classification* (1985 *Revision*). Primary 28D99, 01A70.

This paper is in final form and no version of it will be submitted for publication elsewhere.

of Ω onto itself which is such that if $\omega \in \Omega$ is a point in phase space then a state of the system described by ω at some time t_0 will be described by $T_t(\omega)$ at time $t_0 + t$. If f is a real valued function on Ω describing some "observable" of the system one can study the changes in f with time and (if it exists) can ask for $\lim_{a \to \infty} \frac{1}{a} \int_0^a f(T_t(\omega))\, dt$. In order to carry out his program Boltzmann wanted a proof that this "time average" of f can be obtained, without knowing the T_t, by taking the "space average" of f over the set Ω_E of all points $\omega' \in \Omega$ with $H(\omega') = H(\omega)$. Here H is the real valued function on Ω whose value at each $\omega' \in \Omega$ is the total energy of the system in that state. One calls the Ω_E the constant energy hypersurfaces. It follows, of course, from the law of conservation of energy that the Ω_E are invariant under the T_t so that one can restrict oneself to a particular Ω_E in considering both kinds of average. There is a "natural" integration process on the Ω_E (and on Ω) which is also T_t invariant so that the average of f over Ω_E makes good mathematical sense whenever Ω_E is of finite "measure" with respect to this process. Boltzmann wished to deduce the equality of the space and time averages from the hypothesis (which he called the ergodic hypothesis) that the orbit of any point of ω (that is, the set of all $T_t(\omega)$) fills up the whole hypersurface Ω_E on which ω lies. When it was pointed out to him that this hypothesis is impossible on topological groups whenever the Ω_E are more than one dimensional he modified it to the "quasi-ergodic hypothesis" which asserts that the $T_t(\omega)$ are dense in Ω_E. Boltzmann began his work in the late 1860s and formulated the ergodic hypothesis in 1887. He died in 1906, without having found an acceptable argument for the equality of space and time averages, and with his program as a whole incomplete and in trouble because of lack of agreement with experiment.

2. The years 1931 and 1932. The immediate stimulus for the work of Birkhoff and von Neumann in 1931 seems to have been a paper of B. O. Koopman published in the same year and stimulated in turn by M. H. Stone's paper of 1930 on the spectral theory of one-parameter groups of unitary operators. Koopman's paper was based on a simple observation which was much less obvious in 1931 than it would be today. Let Ω be the phase space of a dynamical system with T_t as above and form the Hilbert space $\mathscr{L}^2(\Omega, \zeta)$, where ζ is the measure corresponding to the natural integration process in Ω alluded to above. Then for each t, the operator $f \to g$, where $g(\omega) = f(T_t(\omega))$, is a unitary operator V_t and $t \to V_t$ is a one-parameter group of such. Stone's theorem may be applied to this one-parameter unitary group and allows one to classify dynamical systems using the invariants supplied by spectral theory. Applying operator theory to dynamical systems in this way was a startlingly new idea at the time. André Weil reports in his collected works that he had the same idea independently of Koopman. Both Koopman and Weil communicated their observations to von Neumann in conversation and von Neumann, one of the leading experts in operator theory, was inspired to try operator theoretic methods on the ergodic problem. In a very short time he had conceived and proved his celebrated mean ergodic theorem. In its modern formulation this is a theorem about arbitrary one-parameter unitary groups $t \to V_t$ and asserts that for each vector ϕ in the Hilbert space, $\lim_{T\to\infty} \frac{1}{T}\int_0^T V_t(\phi)\,dt$ exists in the sense of the metric defined by the Hilbert space norm and is a vector ψ which is such that $V_t(\psi) = \psi$ for all t. Now when V_t is defined by the Koopman construction and Ω_E is of finite measure the limit function ψ will have the property that $V_t(\psi) = \psi$ for all t if and only if $\psi(T_t(\omega)) = \psi(\omega)$ almost everywhere in ω for all t. If ψ is not almost everywhere constant then it follows from elementary measure theory that Ω_E can be written as a union of two disjoint measurable sets of positive measure each of which is invariant under the transformations T_t. The impossibility of such a decomposition of Ω_E is precisely the definition of metric transitivity. Thus if the action of the real line on Ω_E given by the T_t satisfies the hypothesis of metric transitivity one can conclude that von Neumann's limit function ψ is a constant almost everywhere and it is easy to deduce that the constant is equal to the space average $(\mu(\Omega_E))^{-1}\int \phi(\omega)\,d\zeta_E(\omega)$ of ϕ. Thus the mean ergodic theorem implies that the time average exists and is equal to the space average whenever the hypothesis of metric transitivity holds on the Ω_E. This is true, of course, only if one interprets the time average as being defined in the sense of "convergence in the mean". However in a second paper, published in the same volume of the National Academy Proceedings as the first, von Neumann argued in detail that convergence in this sense is all that is needed in order to draw the physical conclusions that Boltzmann wanted.

The difficult point that remains is that of establishing the metric transitivity of the action of the T_t on the Ω_E. This is still a problem in 1988 in spite of many attempts to settle it. However it was an important conceptual advance to have isolated the concept of metric transitivity as the concept that is needed. The original ergodic hypothesis of Boltmann certainly implies metric transitivity but is much too strong. It is known *not* to hold in any physically interesting example. The quasi-ergodic hypothesis is too weak. The "right" concept is metric transitivity and von Neumann proposed renaming this important concept "ergodicity". Von Neumann's terminology has now become standard.

Of course it is natural to wonder whether or not the time averages $\frac{1}{T}\int_0^T \phi(T_t(\omega))\,dt$ converge to a limit, not only in the $\mathscr{L}^2$ mean, but also for almost all ω. Even if not needed for Boltzmann's program it is an interesting question mathematically and turns out to have important applications in probability theory. Von Neumann mentioned this question to Birkhoff and Koopman while communicating his mean ergodic theorem to them. Birkhoff rose to the challenge and, using methods quite different from those of von Neumann, found a proof in a rather short time. The theorem which resulted is now known as the pointwise ergodic theorem or as the Birkhoff ergodic theorem. Its proof is much more difficult than that of the mean ergodic theorem and finding it is one of Birkhoff's most celebrated achievements.

In addition to making the indicated contributions to Boltzmann's program and stimulating Birkhoff to prove the pointwise ergodic theorem, von Neumann wrote a long and very influential paper entitled *Zur Operatorenmethode in der klassischen Mechanik.* This paper was published in 1932 in the Annals of Mathematics and marks the beginning of the systematic study of the concept of ergodicity (metric transitivity) for its own sake. The paper begins with a fully detailed proof of the mean ergodic theorem and then presents the statement and proof of an important (and still inadequately appreciated) decomposition theorem establishing the fact that the ergodic measure preserving actions of the real line are the fundamental building blocks from which all measure preserving actions can be constructed. The nature of this theorem can perhaps be best appreciated by looking at an example where its truth is more or less obvious. Moreover for greater simplicity we shall consider an example of an analogue of the theorem in which the additive group of the integers replaces the additive group of the real line. Consider the unit disk in the x,y plane, $x^2 + y^2 \leq 1$, and denote it by the symbol D. Let T_θ be the transformation $x, y \to x\cos\theta + y\sin\theta, -x\sin\theta + y\cos\theta$. Then T_θ is a one-to-one transformation of D onto itself which preserves Lebesgue measure μ_D in D. Hence the iterates T_θ^n $(n = 0, \pm 1, \pm 2, \dots)$ define a measure preserving action of the group Z on D. This action of course is far from ergodic. Indeed given any two real numbers r_1 and r_2 with $0 < r_1 < r_2 < 1$ the annulus $r_1^2 \leq x^2 + y^2 \leq r_2^2$ is a Z-invariant measurable set of positive measure whose complement is also of positive measure. On the other hand consider the fibering of D by the concentric circles $S_r\colon x^2 + y^2 = r^2$ where $0 \leq r \leq 1$. These are all Z-invariant sets into which D can be considered as decomposed but are unfortunately of measure zero. To compensate for this we notice that there exists a unique measure ν in the interval $0 \leq r \leq 1$ such that if μ_r denotes the usual linear Lebesgue measure in S_r then for every measurable subset A of D, $\mu_D(A)$ may be computed from ν and the $\mu_r(A \cap D)$ by the formula

$$\mu_D(A) = \int_0^1 \mu_r(A \cap D)\, d\nu(r).$$

In other words, we may reconstruct the action of Z on D, μ_D from its actions on the S_r, μ_r by integrating the latter with respect to r as indicated. Moreover whenever θ is an irrational multiple of π the action of Z on S, μ_r defined by T_θ is just that given by Birkhoff and Smith as an example of a metrically transitive action and hence is ergodic. We have decomposed our action as a direct integral of ergodic actions. Theorem 2 of von Neumann's paper states that every measure preserving action of the real line satisfying a mild regularity hypothesis admits an analogous decomposition into ergodic constituents and moreover that modulo inessential changes on sets of measure zero, this decomposition is *unique.*

The truth of Theorem 2 makes it clear that the problem of classifying one-parameter groups of measure preserving transformations reduces essentially to that of classifying those that are ergodic. Theorem 5 is rather easy but quite striking contribution to the latter problem. Before describing it, it will be useful to say a few words about a division of ergodic actions into two categories which differ fundamentally from one another. Let S, μ be any measure space on which we are given a one-parameter group $t \to T_t$ of measure preserving transformations and suppose that the triple S, μ, T is subject to certain, not very restrictive, regularity hypotheses. Suppose that for some s_0, the set of all $T_t(s_0)$—the so-called *orbit* of s_0—has positive measure. Then, if the given action is ergodic, the complement of this orbit is necessarily of measure zero. In other words by disregarding an (inessential) invariant subset of measure zero we obtain an action which is *transitive* in the sense that one orbit fills up the whole of S. Naively one might have supposed that all ergodic actions are *essentially transitive* in this sense. However, as the example of Birkhoff and Smith shows, this naive conjecture is false. There exist ergodic actions in which every orbit is of measure zero. We shall call such actions *properly ergodic*. The fact that they exist is of the greatest importance. If they did not, the Birkhoff ergodic theorem would be an easy consequence of von Neumann's decomposition theorem and the subject of ergodic theory would be exhausted by two theorems.

While it is trivial to determine all possible transitive measure preserving actions of the real line, and the same is true for actions of the integers, the corresponding statement is very far from true for the properly ergodic actions of either group. To this day, almost sixty years after the beginning of ergodic theory described above, we are still very far from any sembalance of a complete description of all possible ergodic actions of the real line or of the integers. We have many examples and at least two classes of examples within which a complete classification is known but not much more.

Theorem 5 in von Neumann's 1932 Annals paper is concerned with one of these classes. Let $t \to T_t$ be an ergodic one-parameter group of measure preserving transformations in a measure space S, μ and let $t \to V_t$ be the one-parameter unitary group associated with $t \to T_t$ by the Koopman construction. It is natural to say that the ergodic action has *pure point spectrum* if $t \to V_t$ can be decomposed as a *discrete* direct sum of (necessarily one-dimensional) irreducible representations of the real line. Each of these irreducible constituents is, of course, of the form $t \to e^{it\lambda}I$, where λ is some real number and I is the identity operator. The numbers λ which occur constitute the spectrum. If $\mathscr{L}^2(S, \mu)$ is separable, this spectrum is a countable set. The content of Theorem 5 can then be summarized as follows.

Given that an ergodic action has a countable pure point spectrum $\lambda_1, \lambda_2, \ldots$:

(1) Every λ_j occurs with multiplicity one.

(2) $\lambda_i \pm \lambda_j$ is equal to some λ_k for every i and j and each choice of $+$ or $-$, so that the spectrum is a countable subgroup of the group of all real numbers.

(3) Every countably infinite subgroup occurs for some ergodic action with a pure point spectrum.
(4) If two such actions have the same spectrum and the measures μ_1 and μ_2 are normalized so that $\mu_1(S_1) = \mu_2(S_2) = 1$, then there is an isomorphism B between S_1, μ_1 and S_2, μ_2 as measure spaces such that $BT_t^1 B^{-1} = T_t^2$ for all t.

Since one sees easily that the above described action is properly ergodic if and only if the group of all λ_j is dense in the real line one has continuum many inequivalent examples of properly ergodic actions with pure point spectrum.

In Theorem 6 von Neumann introduced a method for constructing an ergodic action of the additive group of the line out of an ergodic action of the integers and a positive measurable function f on the space S of the action. The space of the action of the line is the set of all $s, y \in S \times P$ with $y \leq f(s)$. Here P is the positive real axis. One defines $T_t(s, y)$ to be $s, y+t$ as long as $t \leq f(s) - y$ and for $f(s) - y < t < f(s) - y - f(\alpha(s))$ one defines $T_t(s, y)$ to be $\alpha(s), t - f(s) + y$. Here α is the measure preserving transformation of S associated with 1 in the given integer action. The definition for other t values should now be clear. In pictorial terms each point s, y is pushed up the interval $s, [0, f(s)]$ until it reaches the "top" $s, f(s)$. Then it is identified with $\alpha(s), 0$ and pushed up the interval $\alpha(s), [0, f(d\alpha(s))]$, etc. One speaks of the flow *built over* α, under the function f. Using this construction von Neumann was able to produce examples of properly ergodic actions with no point spectrum at all.

3. Afterwards. The long-range consequences of the work described thus far go considerably beyond clarifying and somewhat advancing Boltzmann's program in statistical mechanics. We shall first list some of these and then add explanatory comments in each case.

(A) It provided a tool for analyzing dynamical systems that could be applied to other problems than those of Boltzmann—in particular the celebrated three-body problem of celestial mechanics.
(B) It provided a powerful new tool for use in probability theory.
(C) By way of more or less straightforward generalizations from the real line and the integers it provided a new tool for analyzing arbitrary groups of diffeomorphisms of manifolds.
(D) By raising the question of classifying ergodic actions, of deciding which measure preserving actions are ergodic, and of how to decompose nonergodic actions into ergodic parts it created an important new branch of pure mathematics; a branch which is still thriving half a century later.

(A) An immediate consequence of von Neumann's decomposition theory was that one had an illuminating new way of looking at the old problem of seeking a "complete" set of "integrals of the motion" for any classical Hamiltonian dynamical system. Let $F_1, F_2, \ldots, F_N$ be any finite set of such integrals and for each N-tuple $c_1, c_2, \ldots, c_N = c$ of real numbers let Ω_c denote the subset of phase space Ω on which $F_j(\omega) = c_j$ for all j. The Ω_c are then invariant subsets under the group $t \to T_t$ of measure preserving transformations describing the time evolution of the system. Moreover a slight adaptation of the proof of the von Neumann decomposition theorem shows that one can decompose the given system as an integral over the vectors $c_1, c_2, \ldots, c_N = c$ of the various Ω_c. As in von Neumann's theorem one will have an invariant measure in each Ω_c uniquely determined up to a multiplicative constant for almost all c. It is then easy to see that, either these ergodic actions or almost all ergodic, or else there exists at least one further integral F_{N+1} which is *not* almost everywhere equal to $G(F_1, F_2, \ldots, F_N)$ for any function G of N real variables. Thus the system $F_1, F_2, \ldots, F_N$ is a "maximal" system of integrals if and only if the decomposition defined by the Ω_c coincides almost everywhere with von Neumann's decomposition into ergodic pieces. Moreover if in the von Neumann decomposition one has a set of constituents of *positive* measure each of which is *properly ergodic* then it is impossible to find a set $F_1, \ldots, F_N$ of integrals so large that knowing the values of all these integrals determines a single orbit. The search for all possible integrals must be replaced by an attempt to find the nature of the decomposition of the group $t \to T_t$ into its ergodic constituents and the failure of sufficiently many integrals to exist must be traced to the occurrence of *properly* ergodic constituents in this decomposition.

The distinction between ergodic actions with pure point spectrum and more general actions is also important here. In the so-called completely integrable systems studied by Liouville in 1855 one does *not* have a sufficient number of *globally defined* integrals in the sense we have been using it. One has instead a situation in which almost all the components of the von Neumann decomposition are either essentially transitive or have a pure point spectrum. Classically the existence of Ω_c's with pure point spectrum was perceived as the existence of what are sometimes called quasiperiodic motions. As far as the existence of von Neumann constituents, which are properly ergodic and have continuous spectrum, are concerned it is interesting to read the early work of G. D. Birkhoff on qualitative dynamics and even earlier work of Hadamard on "geodesic flows" where phenomena present themselves suggesting the need for considering something more general than quasiperiodic motions. [See, for example, G. D. Birkhoff, *Quelques théorèmes sur le mouvement des systèmes dynamiques*, Bull. Soc. Math. France **40** (1912), 305–323, and the paper of Hadamard cited therein.]

For more on geodesic flows see the comments under (D) below. There has recently been a certain amount of discussion and comment about the occurrence of "chaotic" phenomena in presumably deterministic dynamical systems. This occurrence is closely connected with the occurrence of components with no point spectrum in the von Neumann decomposition. How this comes about should become clearer after reading (B) below.

(B) Ergodic theory appeared on the scene just in time to be integrated with the modern measure theoretic approach to probability theory. The latter began with a paper of E. Borel published in 1909 and reached maturity of formulation in 1933 with the publication of Kolmogoroff's book, *Grundbegriffe der Wahrsheinlichheitsrechnung*. Important steps along the way were taken in Wiener's papers on Brownian motion published between 1920 and 1923 and in a paper of Steinhaus published in 1923. It is interesting that Kolmogoroff's book makes no mention of the ergodic theorem. It was only in 1934 that Doob, E. Hopf, and Khinchine independently recognized that the pointwise ergodic theorem can be looked upon as completely equivalent to a significant generalization of one of the basic theorems of probability theory—the so-called strong law of large numbers.

In the measure-theoretic approach to probability as systematized in Kolmogoroff's book one starts with a suitably restricted measure space Ω, μ such that $\mu(\Omega) = 1$. One thinks of the points ω in Ω as "events" in the "universe" Ω of all possible events. If E is a measurable subset of Ω, then $\mu(E)$ is the probability that the ω that actually occurs is in the set E. The space Ω, μ is given and fixed once and for all, and probability theory concerns itself with measurable functions defined on Ω. These are given the suggestive name of "random variables" and are usually (but not necessarily) real valued. If g is a real-valued random variable, then the probability that g takes on a value in the Borel set F of real numbers is just $\mu(g^{-1}(F))$. Thus the behavior of the random variable g taken in isolation is completely determined by the measure α on the real line $F \to \mu(g^{-1}(F))$. This is a "probability measure" in the sense that $\alpha([-\infty, \infty]) = 1$ and is what is called the distribution of the random variable g. When it exists, $\int_\Omega g(x)\,d\mu(x) = \int_{-\infty}^{\infty} x\,d\alpha(x)$ is called the "expected value" or expectation of the random variable g and if this is denoted by $\bar{g}$ then $\int (g(x) - \bar{g})^2\,d\mu(x) = \int_{-\infty}^{\infty} (x - \bar{g})^2\,d\alpha(x)$ is called the "variance" of g. It is clear that g has an expectation and a finite variance if and only if g is in $\mathscr{L}^2(\Omega, \mu)$. Of course, as long as only one random variable is involved, there is little point in introducing the "universe" Ω; everything can be expressed in terms of the measure α. It is in dealing with relationships between different random variables that the usefulness of Ω becomes clear—especially when there are infinitely many. For example, let g_1 and g_2 be two real-valued random variables and let α_1 and α_2 be their distributions. Then $\omega \to g_1(\omega), g_2(\omega)$ is a measurable function ψ from Ω to $R \times R$ and setting $\beta(F) = \mu(\psi^{-1}(F))$ defines a probability measure β in $R \times R$ which is far from uniquely determined by α_1 and α_2. It is called the joint distribution of the random variables g_1 and g_2. If $g_2 = g_1^2$ then β is supported by the curve $y = x^2$. On the other hand if g_1 and g_2 are so-called "independent" random variables then β is the product measure $\alpha_1 \times \alpha_2$ in which $\alpha_1 \times \alpha_2(F_1 \times F_2) = \alpha_1(F_1)\alpha_2(F_2)$. In fact, this is the definition of independence, and this definition seems to capture quite completely the intuitive notion of what it means for random variables to be independent.

To see the connection of the measure theoretic picture of probability theory with ergodic theory choose a probability measure space Ω, μ and let $\ldots, f_{-2}, f_{-1}, f_0, f_1, \ldots$ be a doubly infinite sequence of random variables. In probabilistic language this sequence is a "discrete parameter stochastic process". For each ω the doubly infinite sequence of real numbers $\ldots, f_{-2}(\omega)$, $f_{-1}(\omega), f_0(\omega), f_1(\omega), \ldots$ is called a "sample sequence of the process". Thus the function of two variables $n, \omega \to f_n(\omega)$ may be looked upon either as a family of sample sequences one for each ω or a family of random variables one for each n. For each pair n, d of integers with $d > 0$ let $\psi_{n,d}$ denote the mapping $\omega \to f_n(\omega), f_{n+1}(\omega), \ldots, f_{n+d-1}(\omega)$ of Ω into R^d and let $\mu_{n,d}$ be the probability measure in R^d, $F \to \mu(\psi_{n,d}^{-1}(F))$. $\mu_{n,d}$ is then the joint distribution of $f_n, f_{n+1}, \ldots, f_{n+d-1}$. In the special case in which $\mu_{n,d}$ depends only on d and is independent of n one says that the stochastic process is "stationary". Now in considering a single stationary stochastic process there is no loss in generality in supposing that the mapping $\omega \to \{f_n(\omega)\}$ of Ω into sample sequences is one-to-one and then a simple argument shows that there is an essentially unique measure preserving map β of Ω onto Ω such that $f_n(\omega) = f_0(\beta^n(\omega))$ for all n, and almost all ω. Conversely if β is a measure preserving transformation of a probability measure space Ω, μ onto itself and f is any real valued measurable function on Ω, then setting $f_n(\omega) = f(\beta^n(\omega))$ for all integers n defines a *stationary* stochastic process. In short one has a natural one-to-one correspondence (modulo the obvious equivalences) between all possible stationary stochastic processes on the one hand and all pairs consisting of a measure preserving action of the integers and a measurable function on the other. Of course the measure must be 1 on the whole space.

In understanding the significance of this correspondence it is useful to examine the meaning of ergodicity for the associated stochastic process. If one applies the von Neumann decomposition theorem to the action of the β^n on the underlying universe Ω, one divides all sample functions of the process into disjoint sets each associated with a single ergodic constituent. However in applying the theory one is usually given a sample function (actually a large segment of one) and wants to understand it by guessing the process from which it comes. For this purpose only the ergodic component in which it lies can be of any relevance since what goes on elsewhere has no influence on the sample function. Moreover once this ergodic component is known one has all one needs for further analysis of the sample. For these reasons there is seldom any loss in generality in restricting attention to those stationary stochastic processes in which the associated measure preserving action is indeed ergodic. Assuming ergodicity and assuming that the f_j are all L^1 functions so that they have expected values (necessarily the same for all j) one verifies easily that the pointwise ergodic theorem may be restated as follows: For almost all $\omega \in \Omega$ the corresponding sample sequence $\{f_n(\omega)\}$ has the property that

$$\lim_{N\to\infty} \frac{f_n(\omega) + f_{n+1}(\omega) + \cdots + f_{n+N}(\omega)}{N+1}$$

exists and equals the common expected value of the f_j. This is the so-called law of large numbers. Before 1934 it was known to be true only when the f_j are independent.

Let $\ldots, f_{-2}, f_{-1}, f_0, f_1, \ldots$ be a stationary stochastic process in which the action of the β^n on Ω is ergodic and in which the f_j are all in $\mathscr{L}^2(\Omega, \mu)$. Let $p(j) = \int f_j(\omega) f_0(\omega)\, d\mu(\omega)$. It is easy to prove that p is a so-called "positive definite function" on the integers whence it follows from a well-known theorem of Herglotz that there exists a unique measure ν on the unit circle such that $\int e^{ij\theta}\, d\nu(\theta) \equiv p(j)$. This measure is an important invariant of the process called its *spectrum*. In another important application of the ergodic theorem one can use it to prove that the positive definition function p can be computed from almost any sample function by the simple formula

$$p(j) = \lim_{N\to\infty} \frac{a_{j-1}a_{-1} + a_{j-2}a_{-2} + \cdots + a_{j-N}a_{-N}}{N}.$$

This shows that p coincides with the well-known auto correlation function introduced earlier by students of the statistics of time series. Using the ergodic theorem in a somewhat more elaborate way one can prove the possibility of reconstructing the entire stochastic process (not just its spectrum) from almost any sample sequence.

There is a simple and interesting connection between the spectrum of a stationary stochastic process, as just defined, and the spectrum, in the sense of the spectral theorem of operator theory, for the Koopman unitary operator U_β associated with the underlying measure preserving transformation β. By the spectral theorem for unitary operators there exists a projection valued measure $E \to P_E$ defined for all Borel subsets of the unit circle in the complex plane such that for every f in $\mathscr{L}^2(\Omega, \mu)$, $(U_\beta(f) \cdot f) = \int_0^{2\pi} e^{i\theta}\, d(P_\theta(f) \cdot f)$. From this it follows easily that

$$p(j) \equiv \int_0^{2\pi} e^{ij\theta}\, d(P_\theta(f) \cdot f) \quad \text{for all } j.$$

In other words the spectrum of the process defined by β and f is just the measure $E \to (P_E(f) \cdot f)$. While this measure depends on both f and β it is necessarily absolutely continuous with respect to those measures whose sets of measure zero are just those sets E for which $P_E = 0$. In particular if β has a pure point spectrum then the spectrum of *any* stationary process based on β is a countable set contained in this point spectrum. Since the degree to which the sample functions of a process are "truly random" or have futures determined by their pasts is strongly influenced by the nature of the spectrum of the process, it must also be strongly influenced by the nature of the underlying ergodic actions. It is for this reason that we may speak of some ergodic actions as being more "random" or "chaotic" than others.

We have confined our attention here to discrete stochastic processes because they are technically easier to study and are sufficient to make clear the nature of the connection with ergodic theory. One can also consider families of random variables parametrized in other ways. When the integers are replaced by the real numbers one speaks of continuous parameter stochastic processes and these are connected with ergodic actions of the additive group of the real line in much the same way that discrete parameter stochastic process are related to ergodic actions of the integers. Of course sample sequences must be replaced by sample functions and the measure μ in the universe Ω defines a measure in the space of sample functions. The celebrated Wiener measure which occurs in Wiener's Brownian motion theory (mentioned above) is of this character. Indeed, Wiener's work can be interpreted as the first rigorously defined example of an important type of continuous parameter stochastic process and the underlying ergodic action turns out to be the simplest example of an important type of ergodic action. Another of Wiener's important papers is illuminated by the development of ergodic theory and its connection with probability. This is *Generalized harmonic analysis* published in 1930 in Acta Mathematica, the year before the two ergodic theorems were proved. Here Wiener is concerned with a class of real or complex valued functions on the real line which are like almost periodic functions in that neither ordinary Fourier series expansions nor Fourier transforms are applicable to them. However they are much more general and as was realized a decade or so later are best understood as sample functions of continuous parameter stochastic processes. It is only when the underlying ergodic action has a pure point spectrum that these sample functions can be almost periodic.

The connection between ergodic theory and probability theory outlined in the preceding pages is in large part due to work of Khinchine and Wold published in 1934 and 1938 respectively. For lengthier and more thorough expositions in the spirit of the foregoing the reader may consult the following articles by the present author: *Ergodic theory and its significance for statistical mechanics and probability theory*, Adv. in Math. **12** (1974), 178–268 (especially §§2, 5, 6, 7 and 8) and §18 of *Harmonic analysis as the exploitation of symmetry–a historical survey*, Bull. Amer. Math. Soc. **3** (1980), 543–698.

(C) From the point of view of pure mathematics there is no reason to confine attention to the study of measure preserving actions of the real line and integer groups. The basic concepts make sense and are interesting for a more or less arbitrary group. Moreover one can formulate generalizations of the ergodic theorem whenever the group is such that one has a notion of "mean value" for suitable functions defined on the group. A somewhat less obvious generalization lies in relaxing the requirement that the measure μ be invariant. One says that a measure μ is "quasi-invariant" if μ and its translates μ_x by group elements x have the same sets of measure zero for all x and one can generalize the Koopman construction to associate a unitary group representation $x \to U_x$ by making use of the Radon-Nikodým derivatives of the μ_x with respect to μ to compensate for the noninvariance of μ. One defines ergodicity in the obvious way and can prove a generalization of the von Neumann decomposition theorem when G is separable and locally compact and S satisfies mild regularity conditions. Rather surprisingly one can even formulate and prove a generalization of the pointwise ergodic theorem when the invariant measure is replaced by one which is only quasi-invariant (see Hurewicz, Ann. of Math. (1944)).

As far as applications of these two generalizations of ergodic theory are concerned the first two known to the author were made by von Neumann himself and by von Neumann and his collaborator Murray. In the first of the celebrated von Neumann-Murray papers on operator algebras (published in the Annals of Mathematics in 1936) the authors proved the existence of Type II factors by an explicit construction starting with a properly ergodic action of an arbitrary countably infinite discrete group G. (See elsewhere in this volume for the definition of factors of Type I, Type II, and Type III.) While it is true that choosing G to be the integers would still have given a Type II factor much more general examples could be gotten by making G general. The first example of a Type III factor was not found until after the completion of the second von Neumann-Murray paper. The third paper of the series, written by von Neumann alone, was published in the Annals of Mathematics in 1940. Its main result is that one can obtain Type III factors by generalizing the Type II construction mentioned above by allowing measures which are only quasi-invariant.

In dealing with measures which are quasi-invariant with respect to a group action it is important to realize that any measure with the same null sets as the given one is also quasi-invariant and that for most purposes it does not matter very much which measure one chooses as long as they all have the same null sets. Indeed it is sometimes clarifying to introduce the concept of a "measure class" and realize that one can apply the concepts of ergodic theory whenever one has a group acting on a space which preserves some naturally occurring measure class. A potentially far reaching class of examples is suggested by the fact that every smooth manifold satisfying mild regularity conditions has a canonical measure class—the one defined locally by the class of Lebesgue measure. This measure class is invariant under every subgroup of the group of all diffeomorphisms. It follows that one can use the generalized von Neumann decomposition theorem and other concepts from ergodic theory to analyze more or less general diffeomorphism groups in much the same spirit as explained above for classical dynamical systems.

Another more or less unexploited potential application of ergodic theory involving quasi-invariant measures lies in extending the connection between ergodic theory and stationary stochastic processes to one between ergodic theory and nonstationary stochastic processes.

(D) Finally we conclude this essay with some remarks about the development of ergodic theory as a branch of pure mathematics. This is a huge subject and we have time for only a few indications. Some of the earliest examples of a properly ergodic action of the real line after those in von Neumann's 1932 Annals paper were given in papers by Hedlund published in 1934 and 1935. Let $\mathscr{M}$ be a complete Riemann manifold and let $\mathscr{M}_T$ be the manifold of all tangent vectors to $\mathscr{M}$. If C is any pointed geodesic on $\mathscr{M}$ then each point m of C has a tangent vector $\mathbf{v}$ and the pair $m, \mathbf{v}$ defines a point in $\mathscr{M}_T$. Thus C is canonically associated to a curve C^T in $\mathscr{M}_T$. It can be shown that the curves C^T in $\mathscr{M}_T$ for the various geodesics in $\mathscr{M}$ are the orbits of a unique action of the real line on $\mathscr{M}_T$. This action is such that on each orbit the length of the vector $\mathbf{v}$ is a constant so that the subset $\mathscr{M}_T^0$ of $\mathscr{M}_T$ consisting of all $m, \mathbf{v}$ for which $\mathbf{v}$ is of unit length is an invariant subset. The above described flow restricted to this invariant subset is called the "geodesic flow" on $\mathscr{M}_T^0$ or the "geodesic flow" for the Riemannian manifold $\mathscr{M}$. Hedlund found two types of two-dimensional Riemannian spaces for which he could prove the proper ergodicity of the geodesic flow. His proofs were long and difficult. Since a "free" particle in a Riemannian space is supposed to move along the geodesics one can think of geodesic flows as the flows of certain dynamical systems restricted to constant energy hypersurfaces. Hedlund's results were proved in a unified manner by a different and somewhat simpler argument by E. Hopf in 1936. The study of the ergodic character of geodesic flows has been an important topic in ergodic theory ever since.

Another topic of enduring interest has been the classification of ergodic actions of "shift" type. Let F be a finite set and let Ω be the set of all functions from the group Z of all integers under addition to F. Let F be made into a measure space by assigning a nonnegative real number $p(f)$ to each f in F in such a way that $\sum_{f\in F} p(f) = 1$ and defining $\mu(E) = \sum_{f\in E} p(f)$. Then regarding Ω as an infinite product of replicas of F and calling the product measure μ^α we get a measure in Ω such that $\mu^\alpha(\Omega) = 1$. Regarding Ω once again as the set of all functions a from Z to F we may define an action β of Z in Ω by setting $\beta(a)(n) = a(n+1)$. It is obvious that β is measure preserving and hence that its iterates define a measure preserving action of Z on Ω. One sees easily that these actions are properly ergodic and that they are precisely the actions associated with those stationary stochastic processes whose random variables are independent and take on only a finite number of values. For all choices of F and p these actions have the same Koopman spectrum. On the other hand it is indicative of the difficulty of the problem of classifying ergodic actions that for twenty-five years no one was able to decide whether two actions of shift type based on distinct finite measure spaces F_1, p_1 and F_2, p_2 were isomorphic as ergodic actions are not. The problem was finally solved by work of Kolmogoroff and Sinai in 1958 and 1959. They introduced a new invariant of ergodic actions of Z called the entropy and proved that the entropy of the shift defined by F and p is $-\sum_{f \in p} p(f) \log p(f)$. Since shifts with different entropy cannot be isomorphic it follows that one can have nonisomorphic shifts. More than a decade elapsed before Ornstein proved a converse: two shifts with the same entropy are isomorphic.

In these comments to (A), (B), (C), and (D) there has been no attempt at completeness. Many more interesting and important things could have been said. Our aim rather has been to give a general idea in a few pages.

Acknowledgment. The author is indebted to G. Birkhoff and A. Gleason for helpful comments. Gleason in particular read the whole typescript with care.

HARVARD UNIVERSITY

Proceedings of Symposia in Pure Mathematics
Volume **50** (1990)

Von Neumann and Ergodic Theory

DONALD S. ORNSTEIN

It is fair to say that the von Neumann and Birkhoff ergodic theorems marked the beginning of ergodic theory and are so fundamental that it is hard to think of a result in ergodic theory that does not make use of one of these ergodic theorems. Dynamical systems arising from physics formed the original context of these theorems but the fact that their proofs used only the measure theoretic structure of the dynamical system focused attention on measure-preserving transformations or flows on an abstract measure space.

The problem of classifying measure-preserving transformations or flows became a central problem in ergodic theory and von Neumann and Halmos were responsible for the first significant steps in this direction.

THEOREM (von Neumann-Halmos). *Let f_1 and f_2 be measure-preserving transformations on finite measure spaces X_1 and X_2. Let U_1 and U_2 be the unitary operators induced on L_2 of X_1 and X_2. Assume that f_1 and f_2 have discrete spectrum, i.e., the eigenfunctions of U_1 and U_2 span L_2. Then f_1 and f_2 are isomorphic if and only if U_1 and U_2 are the same as unitary operators on Hilbert space.*

f_1 and f_2 are isomorphic if and only if, after ignoring invariant sets of measure zero in X_1 and X_2, there is an invertible measure-preserving φ that maps X_1 onto X_2 and $f_2(\varphi(x_1)) = \varphi(f_1(x_1))$ for all $x_1 \in X_1$. If we drop the requirement that φ be invertible—φ may then be many-to-one—we say that f_2 is a factor of f_1. In this language an eigenfunction is a factor isomorphic to a rotation of the circle by the eigenvalue.

A consequence of the work of Halmos and von Neumann is the following

THEOREM. *If f has discrete spectrum, then so does any factor of f.*

Another theorem of Halmos and von Neumann identifies discrete spectrum with translations on compact Abelian groups.

1980 *Mathematics Subject Classification* (1985 *Revision*). Primary 47A35.

This paper is in final form and will not be submitted for publication elsewhere.

Von Neumann and Halmos classified the transformations that are in some sense the least random or least chaotic. Today we can classify the transformations and flows that are the most random and chaotic.

It seems clear that the most random kind of stochastic process is the process of independent random variables, in its simplest form the tossing of a fair coin. The corresponding transformation, the Bernoulli shift $(\frac{1}{2}, \frac{1}{2})$, is defined as the shift on $A^{\mathbf{Z}}$ with product measure, where A has two points each of measure $\frac{1}{2}$. The general Bernoulli shift $B(\dots p_i \dots)$, $p_i > 0$, $\sum p_i = 1$, is defined similarly.

Here in contrast to the Halmos-von Neumann case all Bernoulli shifts induce the same unitary operator.

The first steps toward the classification of Bernoulli shifts were taken by Kolmogorov and Sinai in 1958 when they introduced the concept of entropy into ergodic theory as an isomorphism invariant and used it to show that not all Bernoulli shifts were isomorphic (the entropy of $B(\dots p_i \dots) = -\sum p_i \log p_i$).

In 1970 Ornstein showed that $B(\dots p_i \dots)$ is isomorphic to $B(\dots q_i \dots)$ if and only if $\sum p_i \log p_i = \sum q_i \log q_i$. The continuous time version of this theorem contains the discrete time version. It says that there is a unique flow which, when observed at discrete time intervals, is a Bernoulli shift, and gives, in this way, all Bernoulli shifts.

THEOREM (Ornstein). (A) *There is a flow B_t such that B_1 is isomorphic to the Bernoulli shift $(\frac{1}{2}, \frac{1}{2})$ and for any t_0, B_{t_0} is also a Bernoulli shift; in fact we get all Bernoulli shifts by varying t_0. B_t is unique up to a constant scaling of the time parameter (i.e., if $\widetilde{B}_t$ is another flow such that $\widetilde{B}_{t_0}$ is a Bernoulli shift for some t_0, then there is a constant c and $\widetilde{B}_t$ is isomorphic to B_{ct}).*

(B) *The only factors of B_t are B_{ct} for $0 < c \leq 1$. Any flow, f_t, of positive entropy has B_{ct} as a factor (for any c that makes the entropy of B_{ct} less than or equal to the entropy of f_t).*

Part (B) shows that B_t has a unique place among flows of positive entropy. Positive entropy is often used as a precise definition of "chaotic". The zero entropy transformation (or flows) includes the transformations with discrete spectrum and are in a certain sense completely predictable.

The method used to prove the above theorem gives a criterion for showing that certain concrete dynamical systems are isomorphic to B_t. The first result along these lines was (Ornstein-Weiss): geodesic flow on a surface of negative curvature is isomorphic to B_t. (This was the first nontrivial chaotic flow whose measure theoretic structure was determined exactly.) By a similar method a very large number of very different looking flows have been shown to be isomorphic to B_t.

Here is a very small sampling of the flows that have been shown to be isomorphic to B_t or $\overline{B}_t$, the direct product of B_t, and the flow that rotates the circle. Geodesic flow on a manifold of negative curvature of any dimension is B_t. Axiom A attractors are B_t or $\overline{B}_t$ (Ratner). The motion of a billiard ball on a table with a convex obstacle is B_t (Ornstein-Gallavotti), the Lorentz attractor is B_t (Ratner). The following theorem is especially striking.

THEOREM (Pesin). *Let f_t be a smooth flow on a three-dimensional compact manifold that preserves a smooth invariant measure. If we ignore an invariant set of measure 0, then any ergodic component on which f_t has positive entropy has positive measure and on each such component f_t is either B_t or $\overline{B}_t$.*

B_t can also arise from a variety of stationary processes. A few examples are: A semi-Markov flow (where the particle remains in each of a finite number of states s_i, for a fixed time t_i, and then jumps according to the spin of a roulette wheel), is B_t or $\overline{B}_t$. In particular, if the process has two states and changes states with probability $\frac{1}{2}$, and if t_1/t_2 is irrational, then the process is B_t (Ornstein). Poisson processes, continuous time Markov processes on a finite number of states, or Brownian motion with reflecting barriers (so it remains in a finite interval or square) are isomorphic to the unique Bernoulli flow of infinite entropy (Shields).

Here are a few results for discrete time. In an algebraic context we have the beautifil result of Katznelson, Lind, and Miles-Thomas: Any ergodic automorphism of a compact Abelian group is isomorphic to Bernoulli shift. Recall that Halmos and von Neumann classified the ergodic translations on compact Abelian groups. For one-dimensional maps we have that the natural extension of the continued fraction transformation is Bernoulli (Smorodinsky); the natural extension of the maps $ax(1-x)$, for those a for which there is an absolutely continuous invariant measure, is Bernoulli (Ladrappier).

Von Neumann was the first to discover the connection between the nonexistence of certain kinds of factors and mixing properties or predictability properties.

THEOREM (von Neumann). *T has no rotation factors if and only if for any sets E and F, T^nE becomes independent of F along a sequence of n of density one.*

The above theorem plays an important role in Furstenberg's application of ergodic theory to combinatorics. It is also the prototype for several later results.

THEOREM (Sinai-Rochlin). *T has no factors of zero entropy if and only if for any partition P and any E, T^nE becomes independent of the σ-algebra $\bigvee_{-\infty}^{0} T^iP$.*

Transformations (or flows) that have no zero entropy factors (K transformations or K flows) were once believed to be the most random possible (i.e., in discrete time they were believed to be Bernoulli shifts). However, we now know that there are K transformations (and K flows) that are not Bernoulli (Ornstein). It was also believed that every transformation (or flow) was the product of K and zero entropy (the Pinsker conjecture). This also turned out to be false (Ornstein) and the classification problem between the two extremes is much more complicated than it might have been.

Von Neumann introduced the idea of amenable groups and the actions of these groups is the natural context for ergodic theory. We can view the iterates of a single transformation as an action of the integers and a flow as an action of the reals. In both these cases the group action usually represents the passage of time. There are however other symmetries that can be studied by ergodic theory. For example, the ergodic theory of the Ising model (which is invariant under translation of the $\mathbf{Z}^2$ lattice) involves the action of $\mathbf{Z}^2$ and the ergodic theory of infinite particle systems (which are invariant under translation of the space and the passage of time) in three dimensions involves the action of $\mathbf{R}^4$.

Recently the extension of the isomorphism theorems for chaotic systems to actions of general amenable groups has been worked out (Feldman, Ornstein, Rudolph, Weiss).

References

1. G. D. Birkhoff, *Proof of a recurrence theorem for strongly transitive systems*, Proc. Nat. Acad. Sci. USA, **17** (1931), 650–655.
2. ——, *Proof of the ergodic theorem*, Proc. Nat. Acad. Sci. USA, **17** (1931), 656–660.
3. P. R. Halmos, *On automorphisms of compact groups*, Bull. Amer. Math. Soc., **49** (1943), 619–624.
4. P. R. Halmos and J. von Neumann, *Operator methods in classical mechanics*, II. Ann. of Math., **43** (1942), 332–350.
5. A. N. Kolmogorov, *A new metric invariant of transient dynamical systems and automorphisms in Lebesgue spaces*, Dokl. Akad. Nauk SSSR, **119** (1958), 861–864 (Russian), MR 21, #2035a.
6. ——, *Entropy per unit time as a metric invariant of automorphisms*, Dokl. Akad. Nauk SSSR, **124** (1959), 754–755 (Russian), MR 21, #2035b.
7. B. O. Koopman and J. von Neumann, *Dynamical systems of continuous spectra*, Proc. Nat. Acad. Sci. USA, **18** (1932), 255–263.
8. J. von Neumann, *Proof of the quasi-ergodic hypothesis*, Proc. Nat. Acad. Sci. USA, **18** (1932), 70–82.
9. ——, *Zur Operatorenmethode in der Klassischen Mechanik*. Ann. of Math., **33** (1932), 587–642.
10. D. S. Ornstein, *Ergodic theory, randomness, and dynamical systems*, Yale University Press, New Haven, 1974.
11. D. S. Ornstein and B. Weiss, *Entropy and isomorphism theorems for actions of amenable groups*, J. d'Analyse Math., **48** (1987), 1–141.
12. ——, *Statistical properties of chaotic systems*, To appear in Bull. Amer. Math. Soc.
13. Ya. B. Pesin, *Characteristic Lyapunov exponents and smooth ergodic theory*, Russian Math. Surveys, **32**:4 (1977), 55–114.
14. Ya. G. Sinai, *On the notion of entropy of a dynamical system*, Dokl. Akad. Nauk SSSR, **124** (1959), 768–771.
15. ——, *A weak isomorphism of transformations with an invariant measure*, Dokl. Akad. Nauk SSSR, **147** (1962), 797–800; Soviet Math. Dokl., **3** (1962), 1725–1729, MR 28 #5164a; 28 #1247.
16. B. Weiss and D. S. Ornstein, *Geodesic flows are Bernoulli*, Israel J. Math., **14** (1973), 184–198.

Stanford University

Proceedings of Symposia in Pure Mathematics
Volume **50** (1990)

Nonconventional Ergodic Averages

HILLEL FURSTENBERG

Introduction. John von Neumann was the founder of what can be called the *operator theoretic* approach to ergodic theory. Ergodic theory is the study of transformations, or systems of transformations, of a measure space. The operator theoretic approach studies such transformations via the induced transformations—or operators—on function spaces. Suppose $(X, \mathscr{B}, \mu)$ is a measure space and T is a measurable transformation of $(X, \mathscr{B}) \to (X, \mathscr{B})$. We denote by U_T the operator defined by

$$U_T f(x) = f(Tx).$$

If T preserves the measure μ, then U_T acts as an isometry on each Lebesgue space $L^p(X, \mathscr{B}, \mu)$. In particular, if T is invertible, U_T defines a unitary operator on the Hilbert space $L^2(X, \mathscr{B}, \mu)$, making available the theory of these operators for the study of dynamical systems. This observation, due to B. O. Koopman, served as the inspiration for von Neumann's work, and the operator theoretic approach has remained a fundamental tool in ergodic theory.

Von Neumann's first application of this approach was to the mean ergodic theorem. This is the assertion that for $f \in L^2(X, \mathscr{B}, \mu)$, and for a measure preserving T, the ergodic average

$$\tfrac{1}{N}\{f + U_T f + U_f^2 f + \cdots + U_T^{N-1} f\}$$

converges in the norm of $L^2(X, \mathscr{B}, \mu)$ to a function Pf, where the operator P is the orthogonal projection of $L^2(X, \mathscr{B}, \mu)$ to the subspace of U_T-invariant functions. Von Neumann's proof went by way of the spectral theory for U_T. A contemporary proof highlighting the Hilbert space technique can be given in a few lines. In fact, let A be a bounded linear operator on the Hilbert space $\mathscr{H}$ with $\|A\| \le 1$. It is easily seen that $\mathscr{H}$ decomposes orthogonally into the subspace $\mathscr{H}_1$ spanned by $\{Ay - y : y \in \mathscr{H}\}$ and $\mathscr{H}_2 = \{x : A^*x = x\}$.

1980 *Mathematics Subject Classification* (1985 *Revision*). Primary 28D05.
This paper is in final form and no version of it will be submitted for publication elsewhere.

Now for $\|A\| \le 1$ we also have $\|A^*\| \le 1$ and $Ax = x \Leftrightarrow A^*x = x$. Let P_2 denote projection to $\mathscr{H}_2$. For both $\mathscr{H}_1$ and $\mathscr{H}_2$ we quickly verify that

$$\frac{1}{N}\sum_{n=0}^{N-1} A^n x \to P_2 x,$$

and this includes the mean ergodic theorem.

In another application of spectral theory to ergodic theory, von Neumann demonstrated the dynamical significance of a continuous spectrum for U_T. Von Neumann introduced the motion known today as weak mixing. T is a *weakly mixing* transformation of $(X, \mathscr{B}, \mu)$ if for each $A, B \in \mathscr{B}$, there is a set of times $\mathbf{N}_0 \subset \mathbf{N}$ with density 0 so that

$$(A \cap T^{-n}B) \to \mu(A)\mu(b) \quad \text{as } n \to \infty,\ n \notin \mathbf{N}_0.$$

Von Neumann shows that this occurs if and only if U_T has no nontrivial point spectrum, i.e., the only solution to $U_T f = \lambda f$ is $\lambda = 1$, $f =$ constant. We should note that the expression $\mu(A \cap T^{-n}B)$ is given by the inner product $\langle 1_A, U_T^n 1_B\rangle$ in $L^2(X, \mathscr{B}, \mu)$, and so it is quite natural that its behavior is governed by spectral properties of U_T.

A third paper, written jointly with P. R. Halmos, treats the opposite situation, namely the pure point spectrum case. In this case, that is, when $L^2(X, \mathscr{B}, \mu)$ is spanned by eigenfunctions of U_T, and when T is ergodic, the system is isomorphic to a group rotation. This means that the underlying space can be identified with a compact abelian group, the measure with Haar measure, and the measure preserving transformation is multiplication by a group element generating a dense subgroup. We shall call such a system a *Kronecker system*.

The general picture that presents itself as a consequence of the spectral analysis carried out by von Neumann is this: To any ergodic "measure preserving system" $(X, \mathscr{B}, \mu, T)$ one can associate a Kronecker system which is a "factor" of the given system. Denoting the latter by $(Z, \mathscr{D}, \nu, T_{g_0})$, there will be a measure preserving map from X to Z intertwining the transformation T and T_{g_0} so that the characters of the graph Z will lift to eigenfunctions of U_T on X. The Kronecker factor of $(X, \mathscr{B}, \mu, T)$ is characterized by the fact that all eigenfunctions of U_T arise in this way. There also exists an operator E_Z taking $L^2(X, \mathscr{B}, \mu)$ to $L^2(Z, \mathscr{D}, \nu)$ by averaging functions on X over the preimage of each point $z \in Z$. One can than show that for $f, g \in L^2(X, \mathscr{B}, \mu)$ the sequences

$$(1) \qquad \langle f, U_T^n g\rangle \quad \text{and} \quad \langle E_Z f, U_{T_{g_0}}^n E_Z g\rangle$$

are equivalent, where we say $\{a_n\} \sim \{b_n\}$ if

$$\frac{1}{2N+1}\sum_{-N}^{N} |a_n - b_n|^2 \to 0.$$

In one extreme case Z is a point so that $E_Z f = \int f\,d\mu$. The equivalence of the two expressions is (1) then amounts to the property of weak mixing. In the opposite extreme case, $X = Z$ and then $\langle f, U_T^n g\rangle$ is itself an almost periodic sequence. In general, one can say that the asymptotic behavior of $\langle f, U_T^n g\rangle$ is determined by the Kronecker factor.

The notion of the ergodic decomposition of a general measure decreasing system is also due to von Neumann. By virtue of this, one can confine one's attention to the structure of ergodic systems, the most general system being built up of ergodic components.

In what follows we will consider ergodic theorems for "nonconventional" ergodic averages. Specifically, we consider the following examples:

$$\text{(a)}\quad \frac{1}{N}\sum_{n=0}^{N-1} U_T^{n^2} f,$$

$$\text{(b)}\quad \frac{1}{N}\sum_{n=0}^{N-1} U_T^{n} f U_T^{2n} g,$$

$$\text{(c)}\quad \frac{1}{N}\sum_{n=0}^{N-1} U_T^{n} f U_T^{2n} g U_T^{3n} h,$$

$$\text{(d)}\quad \frac{1}{N}\sum_{n=0}^{N-1} U_T^{n} f U_T^{n^2} g.$$

In the first three cases the limit as $N \to \infty$ can be shown to exist in $L^2(X, \mathscr{B}, \mu)$ for approximately bounded functions; in case (d) the existence of the limit has not yet been established. In all these cases one encounters a new phenomenon whereby even if T is ergodic, the limit function need not be constant. The nature of these nonconventional ergodic limits will be governed by factors similar to Kronecker factors, but unlike the latter, they need not be determined by the spectrum of U_T. This will not be surprising since expressions such as

$$\int f_0 U_T^n f_1 U_T^{2n} f_2 U_T^{3n} f_3\, d\mu$$

cannot be identified within the context of the Hilbert space theory of U_T. The asymptotic behavior of such expressions leads to a more detailed structure theory for ergodic systems, one that extends naturally the picture based on the spectral considerations of von Neumann.

One application of nonconventional ergodic theorems is to questions of recurrence and multiple recurrence. For example, the information we can give regarding limits of expressions of the form (b) enables us to prove a "double recurrence" statement: In a measure preserving system $(X, \mathscr{B}, \mu, T)$, if $\mu(A) > 0$ then for some n, $\mu(A \cap T^{-n}A \cap T^{-2n}A) > 0$.

As we have shown in [F], multiple recurrence theorems of this type have implications for combinatorial theory. For example the aforementioned double recurrence property leads to Roth's theorem: If $S \subset \mathbf{Z}$ has positive upper density, then it must contain a three element arithmetic progression. Similarly, the analysis of case (d) leads to the assertions that a set S of positive upper density contains a solution to the nonlinear equation $x - y = (y - z)^2$.

The applications of nonconventional ergodic theorems to recurrence theory depend on convergence of the averages in L^2, or even weak convergence. The question of almost everywhere convergence is much more difficult. We remark that recently J. Bourgain has shown that the averages of the form (a) and (b) converge a.e. [**B1**, **B2**].

1. Measure preserving systems and their factors. A measure preserving system $(X, \mathscr{B}, \mu, T)$ consists of a measure space triple $(X, \mathscr{B}, \mu)$, where we assume μ is a probability measure, and an invertible measurable, measure preserving map $T \colon X \to X$. The system is *ergodic* if the only sets $A \in \mathscr{B}$ with $T^{-1}A = A$ satisfy $\mu(A) = 0$ or 1. Introducing the operator U_T as in the Introduction, $U_T f(x) = f(Tx)$, we see that ergodicity is equivalent to the condition that in $L^p(X, \mathscr{B}, \mu)$ the only functions with $U_T f = f$ are constants.

If $(Y, \mathscr{D}, \nu)$ is another measure space and $\pi \colon X \to Y$ is a measure preserving (not necessarily one-to-one) map, i.e., $\mu(\pi^{-1}(A)) = \nu(A)$, then assuming certain regularity conditions on $(X, \mathscr{B}, \mu)$ we have a "disintegration" of μ to probability measures μ_y defined for a.e. y with μ_y concentrated on the preimage $\pi^{-1}\{y\}$, and such that

$$\int f \, d\mu = \int \left\{ \int f \, d\mu_y \right\} d\nu(y). \tag{1.1}$$

This disintegration defines an operator $f \to E_Y(f)$ defined by

$$E_Y(f)(y) = \int f \, d\mu_y. \tag{1.2}$$

If $f = g \cdot \pi$, i.e., if f is lifted from a function on Y, then (1.2) implies that $E_Y(f) = g$.

The measurability of π implies that $\mathscr{D}^1 = \pi^{-1}(\mathscr{D}) \subset \mathscr{B}$. If we lift $E_Y f$ to X by composing with π we obtain a $\mathscr{D}^1$-measurable function. $E_Y(f) \circ \pi$ will be recognized as the *conditional expectation* $E(f|\mathscr{D}^1)$. We could also start with a sub-σ-algebra $\mathscr{D}^1$ of $\mathscr{B}$ and define Y as an appropriate identification space $Y = X/(\sim)$, and then we can retrieve the previous situation.

Now suppose that $(Y, \mathscr{D}, \nu, T)$ is another measure preserving system (we use the same letter T for the transformation), that $\pi \colon X \to Y$ is as before, and that, in addition, for $x \in X$

$$T\pi(x) = \pi(Tx). \tag{1.3}$$

This is equivalent to the statement that $\pi^{-1}(\mathscr{D})$ is a T-invariant sub-σ-algebra of $\mathscr{B}$. In this case we say that $(Y, \mathscr{D}, \nu, T)$ is a *factor* of $(X, \mathscr{B}, \mu, T)$. (Note that the morphism π is essentially onto since if $B \in \mathscr{B}$ and $\pi^{-1}(B) = \varnothing$, then $\mu(B) = \nu(\varphi) = 0$.) If $\mathscr{D}^1 \subset \mathscr{B}$ is a T-invariant sub-σ-algebra, then

$(X, \mathscr{D}^1, \mu, T)$ can also be regarded as a factor of $(X, \mathscr{B}, \mu, T)$ with $\pi =$ identity. Finally we note that in the case of a factor $X \xrightarrow{\pi} Y$, the operator E_Y commutes with U_T:

$$U_T E_Y = E_Y U_T. \tag{1.4}$$

2. A van der Corput lemma for vectors in Hilbert space. All our nonconventional ergodic theorems depend on a lemma which is a variant of a well-known result of van der Corput for equidistributed sequences. The formulation given here is due to V. Bergelson [**Be**].

LEMMA 2.1. *If $\{u_n\}$ is a bounded sequence of vectors in a Hilbert space $\mathscr{H}$, and for each $m \in \mathbf{Z}$ we set*

$$\gamma_m = \lim_{N\to\infty} \frac{1}{N} \sum_{n=0}^{N-1} \langle u_n, u_{n+m} \rangle, \tag{2.1}$$

then if

$$\frac{1}{M} \sum_{m=0}^{M-1} \gamma_m \to 0 \tag{2.2}$$

as $M \to \infty$, we will have

$$\left\| \frac{1}{N} \sum_{n=0}^{N-1} u_n \right\| \to 0.$$

PROOF. If N is much larger than M, then

$$\frac{1}{N} \sum_0^{N-1} u_n \sim \frac{1}{N} \frac{1}{M} \sum_0^{N-1} \sum_0^{M-1} u_{n+m}.$$

Use the fact that for any vectors in $\mathscr{H}$, $\|\frac{1}{N} \sum y_n\|^2 \le \frac{1}{N} \sum \|y_n\|^2$ to write

$$\left\| \frac{1}{N} \sum_0^{N-1} \frac{1}{M} \sum_0^{M-1} u_{n+m} \right\|^2 \le \frac{1}{N} \sum \left\| \frac{1}{M} \sum_0^{M-1} u_{n+m} \right\|^2$$
$$= \frac{1}{N} \frac{1}{M^2} \sum_{n=0}^{N-1} \sum_{m_1, m_2 = 0}^{M-1} \langle u_{n+m_1}, u_{n+m_2} \rangle.$$

As $N \to \infty$ this converges to

$$\frac{1}{M^2} \sum_{m_1, m_2 = 0}^{M-1} \gamma_{m_1 - m_2}.$$

But it is easily seen that (2.2) implies that the latter expression converges to 0 as $M \to \infty$. □

As a first application consider case (a) of the Introduction. We will prove

THEOREM 2.2. *Let $(X, \mathscr{B}, \mu, T)$ be a measure preserving system and let $\mathscr{B}_{\text{per}}$ denote the sub-σ-algebra of $\mathscr{B}$ generated by sets $B \in \mathscr{B}$ for which $T^m B = B$ for some m. Then*

$$\lim_{N\to\infty} \frac{1}{N} \sum_{n=0}^{N-1} U_T^{n^2} f = \lim_{N\to\infty} \frac{1}{N} \sum_{n=0}^{N-1} U_T^{n^2} E(f|\mathscr{B}_{\text{per}}). \tag{2.3}$$

Note that the limit of the right-hand side of (2.3) exists since functions in $L^2(X, \mathscr{B}_{\text{per}}, \mu)$ can be approximated by functions g with $T^m g = g$ for some m. In proving the theorem we will prove that the limit of the difference of the two expressions converges to 0.

PROOF. We assume $E(f|\mathscr{B}_{\text{per}}) = 0$, and we want to show

$$\frac{1}{N} \sum_{0}^{N-1} U_T^{n^2} f \to 0. \tag{2.4}$$

Note that for all $l \in N$, $\frac{1}{N}\sum_{n=0}^{N-1} U_T^{ln} f \to P_l f$, where P_l is the projection onto functions satisfying $U_T^l g = g$. But these functions are measurable with respect to $\mathscr{B}_{\text{per}}$ and so $P_l f = 0$. Now apply the lemma with $\mathscr{H} = L^2(X, \mathscr{B}, \mu)$ and $u_n = U_T^{n^2} f$. We have

$$\langle u_n, u_{n+m} \rangle = \int U_T^{n^2} f U_t^{(n+m)^2} \bar{f}\, d\mu = \int f U_T^{m^2} \{U_T^{2mn} \bar{f}\}\, d\mu.$$

Averaging over n we obtain

$$\gamma_m = \lim_{N\to\infty} \int f U_T^{m^2} \left\{ \frac{1}{N} \sum_{0} U_T^{2mn} \bar{f} \right\} d\mu = 0.$$

By the lemma we obtain (2.4). □

The system $(X, \mathscr{B}_{\text{per}}, \mu, T)$ is isomorphic to a Kronecker system for which the underlying group is a finite or profinite group. Theorem 2.2 states that the asymptotic behavior of $\frac{1}{N}\sum U_T^{n^2} f$ is determined by the projection onto the "maximal profinite factor" of $(X, \mathscr{B}, \mu, T)$.

3. The case of weakly mixing transformations. We can reformulate the notion of weak mixing as follows:

DEFINITION. A system $(X, \mathscr{B}, \mu, T)$ is *weakly mixing* if for $f, g \in L^2(X, \mathscr{B}, \mu)$

$$\frac{1}{N} \sum_{0}^{N-1} \left| \int f U_T^n g\, d\mu - \int f\, d\mu \int g\, d\mu \right| \to 0. \tag{3.1}$$

This can be seen to be equivalent to the characterization in the Introduction. It is also equivalent to the weaker statement that for each $\varepsilon > 0$ the set

$$\left\{ n \colon \left| \int f U_T^n g\, d\mu - \int f\, d\mu \int g\, d\mu \right| > \varepsilon \right\}$$

has density 0.

Consider now the average

$$\frac{1}{N}\sum_{0}^{N-1} U_T^n f U_T^{2n} g \tag{3.2}$$

for T weakly mixing and $f, g \in L^\infty(X, \mathscr{B}, \mu)$. We shall show that (3.2) converges in $L^2(X, \mathscr{B}, \mu)$ to $\int f\,d\mu \int g\,d\mu$. This will be a special case of a result we prove in the next section.

Note first that weak mixing implies ergodicity; for if $U_T f = f$ then by (3.1)

$$\int |f|^2\,d\mu = \left|\int f\,d\mu\right|^2,$$

which implies $f = $ constant a.e. It follows that if $g = $ constant in (3.2) then the average in question converges by the ergodic theorem to $\int f\,d\mu \int g\,d\mu$. Thus it suffices to take g with $\int g\,d\mu = 0$. We apply the van der Corput criterion: Let $u_n = U_T^n f U_T^{2n} g$.

$$\langle u_n, u_{n+m}\rangle = \int U_T^n f U_T^{2n} g U_T^{n+m} \overline{f} U_T^{2n+2m} \overline{g}\,d\mu = \int (f U_T^m \overline{f}) U_T^n (g U_T^{2m} \overline{g})\,d\mu.$$

Averaging over n we obtain by the ergodic theorem

$$\gamma_m = \int f U_T^m \overline{f}\,d\mu \int g U_T^{2m} \overline{g}\,d\mu.$$

Now use the weak mixing property for the second integral. We have that

$$\left|\int g U_T^m \overline{g}\,d\mu\right| > \varepsilon$$

for a set m of density 0 (the even subset of a set of density 0) and so

$$\frac{1}{M}\sum_{0}^{M-1} |\gamma_m| \to 0.$$

This proves our assertions.

The foregoing is a special case of the following remarkable result by V. Bergelson [**B**].

THEOREM 3.1. *Let $p_1(t), p_2(t), \ldots, p_k(t)$ be polynomials with integer coefficients such that each $p_i(t)$ is nonconstant and each $p_i(t) - p_j(t)$, $i \neq j$, is nonconstant. Then if $(X, \mathscr{B}, \mu, T)$ is weakly mixing and $f_1, f_2, \ldots, f_k \in L^\infty(X, \mathscr{B}, \mu)$*

$$\frac{1}{N}\sum_{n=0}^{N-1} U_T^{P_1(n)} f_1 U_T^{P_2(n)} f_2 \cdots U_T^{P_k(n)} f_k \to \int f_1\,d\mu \int f_2\,d\mu \cdots \int f_k\,d\mu.$$

The proof of this theorem is carried out by an intricate induction procedure, successively using Lemma 2.1 to deduce each case from a less complex one.

4. Some averages for T not weakly mixing. It is easily seen that

$$\lim_{N\to\infty}\frac{1}{N}\sum_{n=0}^{N-1}U_T^n f U_T^{2n} g$$

need not always be a constant for T ergodic. Suppose, for example, that $\varphi(x)$ is an eigenfunction of U_T: $U_T\varphi = \lambda\varphi$. The fact that U_T is an isometry on $L^2(X, \mathscr{B}, \mu)$ implies that $|\lambda| = 1$, and so $U_T\overline{\varphi} = \lambda^{-1}\overline{\varphi}$. If we set $f = \varphi^2$ and $g = \overline{\varphi}$ then

$$U_T^n f U_T^{2n} g = \lambda^{2n}\varphi^2\lambda^{-2n}\overline{\varphi} = \varphi|\varphi|^2.$$

Since T is ergodic and $U_T|\varphi| = |\varphi|$, we conclude that $|\varphi|^2 =$ constant and we will have

$$U_T^n f U_T^{2n} g = c\varphi,$$

whence

$$\frac{1}{N}\sum_{0}^{N-1}u_T^n f U_T^{2n} g = c\varphi.$$

We find in this case that the limit function is defined on the Kronecker factor. This is typical for the average in question. Namely, let $\mathscr{B}_{\mathrm{Kron}}$ denote the σ-algebra in $\mathscr{B}$ generated by eigenfunctions of U_T. If $(Z, \mathscr{D}, \nu, T)$ denotes the Kronecker factor of $(X, \mathscr{B}, \mu, T)$ and $\pi\colon X \to Z$ is the corresponding map, then $\mathscr{B}_{\mathrm{Kron}} = \pi^{-1}(\mathscr{D})$. We now have, by the methods of §3:

THEOREM 4.1. *For any ergodic system $(X, \mathscr{B}, \mu, T)$, if $f, g \in L^\infty(X, \mathscr{B}, \mu)$, then*

$$\lim_{N\to\infty}\frac{1}{N}\sum_{n=0}^{N-1}U_T^n f U_T^{2n} g = \lim_{N\to\infty}\frac{1}{N}\sum_{n=0}^{N-1}U_T^n E(f|\mathscr{B}_{\mathrm{Kron}})U_T^{2n}E(g|\mathscr{B}_{\mathrm{Kron}}). \tag{4.1}$$

From (4.1) we can deduce the "ergodic Roth theorem":

THEOREM 4.2. *If $(X, \mathscr{B}, \mu, T)$ is an ergodic measure preserving system and $A \in \mathscr{B}$ with $\mu(A) > 0$, then $\exists n$ with $\mu(A \cap T^{-n}A \cap T^{-2n}A) > 0$.*

PROOF. Let $f = 1_A$. Then

$$\mu(A \cap T^{-n}A \cap T^{-2n}A) = \int f U_T^n f U_T^{2n} f\, d\mu.$$

It will suffice to prove that

$$\lim_{N\to\infty}\frac{1}{N}\sum_{n=0}^{N-1}\int f U_T^n f U_T^{2n} f\, d\mu > 0. \tag{4.2}$$

By (4.1) this is the same as

$$\lim_{N\to\infty}\frac{1}{N}\sum_{n=0}^{N-1}\int f U_T^n E(f|\mathscr{B}_{\mathrm{Kron}})U_T^{2n}E(f|\mathscr{B}_{\mathrm{Kron}})\, d\mu. \tag{4.3}$$

Since $\mathscr{B}_{\mathrm{Kron}}$ is T-invariant, the product $U_T^n E(f|\mathscr{B}_{\mathrm{Kron}})U_T^{2n}E(f|\mathscr{B}_{\mathrm{Kron}})$ is measurable over $\mathscr{B}_{\mathrm{Kron}}$ and so the first factor in the integrand of (4.3) can be replaced by $E(f|\mathscr{B}_{\mathrm{Kron}})$. Thus we can assume that f in (4.2) is a nonnegative function measurable over $\mathscr{B}_{\mathrm{Kron}}$ or, what amounts to the same thing, we can assume that $(X,\mathscr{B},\mu,T) = (Z,\mathscr{D},\nu,T)$, a Kronecker system. Writing group operation additively, and letting α be the generating group element; $T\theta = \theta + \alpha$, we rewrite the expression in (4.2) as

$$\frac{1}{N}\sum_{n=0}^{N-1}\int_Z f(\theta)f(\theta+n\alpha)f(\theta+2n\alpha)\,d\theta. \tag{4.4}$$

by equidistribution of $n\alpha$ in Z we can write the limit of the average in (4.4) as

$$\iint f(\theta)f(\theta+\varphi)f(\theta+2\varphi)\,d\theta\,d\varphi. \tag{4.5}$$

Since

$$\lim_{\varphi\to 0}\int f(\theta)f(\theta+\varphi)f(\theta+2\varphi)\,d\theta \to \int f(\theta)^3\,d\theta > 0,$$

it follows that (4.5) is positive. This proves the theorem. □

By the general principle alluded to in the Introduction, the "ergodic Roth theorem" implies Roth's theorem which asserts that a subset of the integers of positive upper density contains a three-term arithmetic progression.

We shall say that a family of systems is *characteristic* for a type of average (e.g., (a), (b), (c), or (d) of the Introduction) if for an arbitrary system $(X,\mathscr{B},\mu,T)$ we can find a factor $(Y,\mathscr{D},\nu,T)$ isomorphic to a member of the given family and such that the average of the expression form functions $f, g, \ldots$ on X behaves asymptotically as the average of $E_Y f$, $E_Y g, \ldots$ on Y. Theorem 4.1 asserts this terminology that Kronecker systems form a characteristic family of systems for averages of type (b).

We turn next to averages of the form (c):

$$\frac{1}{N}\sum_{n=0}^{N-1} U_T^n f U_T^{2n} g U_T^{3n} h. \tag{4.6}$$

We shall show with two examples that Kronecker systems are not a characteristic family for these averages.

Let X be the 2-torus $X = \mathbf{R}^2/\mathbf{Z}^2$ whose elements we denote $x = (\theta,\varphi)$, $\theta,\varphi\in\mathbf{R}$ modulo 1. Let $T\colon X\to X$ be the transformation

$$T(\theta,\varphi) = (\theta+\alpha, \varphi+\theta),$$

where α is irrational. We can calculate T^n explicitly:

$$T^n(\theta,\varphi) = \left(\theta+n\alpha, \varphi+n\theta+\frac{n(n-1)}{2}\alpha\right).$$

If we write $p(x) = e^{2\pi i\theta}$ and $q(x) = e^{2\pi i\varphi}$, then

$$U_T^n p = e^{2\pi i n\alpha}p, \qquad U_T^n q = e^{2\pi i(n(n-1)/2)\alpha}p^n q.$$

This gives

$$U_T^n q^3 U_T^{2n} q^{-3} U_T^{3n} q = q$$

so that

$$\lim_{N\to\infty} \frac{1}{N} \sum_{n=0}^{N-1} U_T^n q^3 U_T^{2n} q^{-3} U_T^{3n} q = q. \tag{4.7}$$

We claim that we cannot replace q by $E(q|\mathscr{B}_{\text{Kron}})$ on the left side of (4.7). For $E(q|\mathscr{B}_{\text{Kron}}) = 0$. To see this, let φ be any eigenfunction: $U_T\varphi = \lambda\varphi$. Then

$$\int q\overline{\varphi}\, d\mu = \int U_T^n q U_T^n \overline{\varphi}\, d\mu = e^{2\pi i(n(n-1)/2)\alpha} \lambda^n \int q p^n \overline{\varphi}\, d\mu.$$

But the functions p^n are mutually orthogonal and, for any $f \in L^2(X, \mathscr{B}, \mu)$, $\int f p^n\, d\mu \to 0$. This shows that $\int q\overline{\varphi}\, d\mu = 0$ and so q is orthogonal to $L^2(X, \mathscr{B}_{\text{Kron}}, \mu)$.

In the foregoing example, $L^2(X, \mathscr{B}, \mu)$ was spanned by eigenfunctions of U_T and "generalized" eigenfunctions, where the latter refers to functions $g(x)$ for which $U_T g = hg$ and h is an ordinary eigenfunction of U_T. Systems of this type were studied by Abramov and we call them *Abramov* systems. The next example shows that Abramov systems are not sufficiently general to be characteristic for the average in (4.6).

Let $H(u, v)$ be an antisymmetric bilinear form on $\mathbf{R}^d$ taking integer values for $u, v \in \mathbf{Z}^d$. Let S^1 be the circle group, $S^1 = \{\zeta : \zeta \in \mathbf{C}, |\zeta| = 1\}$, and let N be the nilpotent group $N = \mathbf{R}^d \times S^1$ with multiplication

$$(\theta, \zeta)(\theta^1, \zeta^1) = (\theta + \theta^1, e^{2\pi i H(\theta, \theta^1)} \zeta\zeta^1).$$

If $\Gamma = \mathbf{Z}^d \times \{1\}$, then Γ is a discrete subgroup of N and N/Γ is a compact nilmanifold which will be the underlying space for the system we construct. Let $\alpha \in \mathbf{R}^d$ be such that $\mathbf{Z}\alpha$ is dense in $\mathbf{R}^d/\mathbf{Z}^d$ and let $\tau \in N$ be defined by $\tau = (\alpha, 1)$. We now define $T: N/\Gamma \to N/\Gamma$ as left multiplication by τ: $T(g\Gamma) = \tau g\Gamma$.

The functions $e^{2\pi i n\cdot\theta}$, $n \in \mathbf{Z}^d$, $\theta \in \mathbf{R}^d$, are defined in N/Γ and are eigenfunctions of U_T. We claim that up to constant multiples these are the only eigenfunctions and there are no generalized eigenfunctions. Thus the Abramov factor of this system is the d-torus.

The nilmanifold N/Γ is a circle bundle over the torus $\mathbf{R}^d/\mathbf{Z}^d$, and we choose a measurable section. To do this let $K: \mathbf{R}^d \to S^1$ satisfy

$$K(\theta + n) = K(\theta) e^{2\pi i H(\theta, n)}$$

for $\theta \in \mathbf{R}^d$, $n \in \mathbf{Z}^d$. Since

$$(\theta, K(\theta))(n, 1) = (\theta + n, e^{2\pi i H(\theta, n)} K(\theta)),$$

we find that the section $\{(\theta, K(\theta)), \theta \in \mathbf{R}^d\}$ of N over $\mathbf{R}^d$ is invariant under multiplication by Γ to the right. Hence it defines a section of N/Γ over $\mathbf{R}^d/\mathbf{Z}^d$. The function $q(\theta, \zeta) = \zeta/K(\theta)$ is invariant under right multiplication by Γ

and is defined on N/Γ. Moreover the functions $\psi(\theta)q^n$ span $L^2(N/\Gamma)$. We have

$$(4.8)\qquad Tq = e^{2\pi i H(\alpha,\theta)}\frac{K(\theta)}{K(\theta+\alpha)}q$$

and if $f = \sum C_n(\theta)q^n$ is a generalized eigenfunction for T we see from (4.8) that $C_l(\theta)q^l$ is also a generalized eigenfunction for the same eigenvalue, provided C_l doesn't vanish. Suppose the (generalized) eigenvalue is $\lambda e^{2\pi i m\cdot\theta}$ (m may be 0). Then

$$C_l(\theta+\alpha)e^{2\pi i l H(\alpha,\theta)}\frac{K(\theta)^l}{K(\theta+\alpha)^l} = \lambda e^{2\pi i m\cdot\theta}C_l(\theta).$$

Taking absolute values we see that $|C_l(\theta)|$ is constant and we can assume $|C_l(\theta)| = 1$ and replace K by K/C_l. We then have

$$(4.9)\qquad e^{2\pi i(lH(\alpha,\theta)-m\cdot\theta)} = \left(\frac{K(\theta+\alpha)}{K(\theta)}\right)^l.$$

Take $r \in \mathbf{N}$ with $r\alpha$ close to $n \in \mathbf{Z}^d$. Iterate (4.9) for $\theta, \theta+\alpha, \dots, \theta+(r-1)\alpha$. Then

$$e^{2\pi i(rlH(\alpha,\theta)-m\cdot r\theta-\frac{1}{2}r(r-1)\cdot\alpha)} = \left(\frac{K(\theta+r\alpha)}{K(\theta)}\right)^l.$$

This is close in measure to

$$\left(\frac{K(\theta+n)}{K(\theta)}\right)^l = e^{2\pi i l H(\theta,n)} \sim e^{2\pi i l H(\theta,r\alpha)}$$

and we find that

$$(4.10)\qquad e^{4\pi i r l H(\alpha,\theta)} \sim e^{2\pi i(r(r-1)/2)m\cdot\alpha}e^{2\pi i m\cdot r\theta}.$$

Replace θ by 2θ and divide by the square of (4.10) and we find that

$$(4.11)\qquad r\alpha - n \sim 0 \Rightarrow e^{2\pi i(r(r-1)/2)m\cdot\alpha} \sim 1.$$

Unless $m = 0$, $m\cdot\alpha$ is irrational and (4.11) would imply that $e^{2\pi i(r(r-1)/2)m\cdot\alpha}$ is an almost periodic function of r. Since this cannot be so, $m = 0$. This leads to an impossible situation in (4.10) and it follows that there are no further generalized eigenfunctions.

We shall give an informal proof of the fact that for the system under discussion, the Abramov factor which coincides with the Kronecker factor is not adequate to determine the asymptotic behavior of (4.6). In effect to assert that a factor Y is adequate, i.e., that

$$\frac{1}{N}\sum_{n=0}^{N-1} U_T^n f U_T^{2n} g U_T^{3n} h \sim \frac{1}{N}\sum_{n=0}^{N-1} U_T^n E_Y f U_T^{2n} E_Y g U_T^{3n} E_Y h$$

(where the right-hand side is to be identified with a function on X) which in particular implies that

$$\frac{1}{N}\sum_{n=0}^{N-1}\int j(x)f(T^n x)g(^{2n}x)h(T^{3n}x)\,d\mu(x)$$
$$\sim \frac{1}{N}\sum_{n=0}^{N-1}\int E_Y j(y)E_Y f(T^n y)E_Y g(T^{2n}y), Eyh(T^{3n}y)\,dV(y),$$

is to say that as n ranges over $\mathbf{N}$ and x ranges over X, the only constraints on the 4-tuple x, $T^n x$, $T^{2n}x$, $T^{3n}x$ are those imposed by constraints on $y, T^n y, T^{2n} y T^{3n} y$. (A formal argument would be based on the notion of relative independence; the components of the 4-tuple $x, T^n x, T^{2n}x, T^{3n}x$ are relatively independent given the distribution of $y, T^n y, T^{2n}y, T^{3n}y$.) Now in the Kronecker factor, the constraint imposed on $(\theta, \theta+n\alpha, \theta+2n\alpha, \theta+3n\alpha)$ is that of belonging to the subgroup $\{(\theta, \theta+\varphi, \theta+2\varphi, \theta+2\varphi)\}$, i.e., the subgroup of 4-tuples in arithmetic progression. Denote this subgroup by $W \subset \mathbf{R}^{4d}$. Now there is a simple algebraic reason for the fact that the 4-tuple $(x, T^n x, T^{2n}x, T^{3n}x)$ is subject to further constraints. Namely, let N^4_W denote the set of 4-tuples in N^4 given by

$$N^4_W = \{(\theta_1, \zeta_1; \theta_2, \zeta_2; \theta_3, \zeta_3; \theta_4, \zeta_4) | (\theta_1, \theta_2, \theta_3, \theta_4) \in W\}.$$

To assert that $(x, T^n x, T^{2n}x, T^{3n}x)$ is independent relative to $(\theta, \theta+n\alpha, \theta+2n\alpha, \theta+2n\alpha)$ is tantamount to saying that as we vary n and x the former 4-tuple ranges over the image of N^4_W in $(N/\Gamma)^4$. This image is $N^4_W/N^4_W \cap \Gamma^4$. However, if we set

$$L = \{(\theta_1, \zeta_1; \theta_2, \zeta_2; \theta_3, \zeta_3; \theta_4, \zeta_4) | \zeta_1\zeta_2^{-3}\zeta_3^3\zeta_4^1 = 1, (\theta_1, \theta_2, \theta_3, \theta_4) \in W\},$$

then L is a closed subgroup of N^4_W (as can be verified using the fact that if $(\theta_1, \theta_2, \theta_3, \theta_4)$ and $(\theta_1', \theta_2', \theta_3', \theta_4')$ belong to W, then $H(\theta_1, \theta_1') - 3H(\theta_2, \theta_2') + 3H(\theta_3, \theta_3') - H(\theta_4, \theta_4') = 0$) containing $N^4_W \cap \Gamma^4$. Since L contains the diagonal of N^4 and also $(1, \tau, \tau^2, \tau^3) = (0, 1; \alpha, 1; 2\alpha, 1; 3\alpha, 1)$, it follows that all the 4-tuples $(x, T^n x, T^{2n}x, T^{3n}x)$ are in the submanifold

$$L/N^4_W \cap \Gamma \subset N^4_W/N^4_W \cap \Gamma.$$

The nilmanifold systems such as the one described in the foregoing example turn out to form a characteristic family for the average (4.6) as well as the more general expression

$$\frac{1}{N}\sum U_T^{an} f U_T^{bn} g U_T^{cn} h \tag{4.12}$$

$(a, b, c \in \mathbf{Z})$. Details of this result will appear in joint work of B. Weiss and the author. Related results appear in work by J.-P. Conze and E. Lesigne (**[C,L]**, **[L]**). Using this result, it is possible to prove the existence of the limits of (4.12) in L^2.

5. Nonlinear expressions

THEOREM 5.1. *Let* $(X, \mathscr{B}, \mu, T)$ *be a measure preserving system and let* $\mathscr{B}' \subset B$ *be a* T*-invariant* σ*-algebra so that for all* $a, b, c \in \mathbf{Z}$ *and* $f, g, h \in L^\infty(X, \mathscr{B}, \mu)$

(5.1)

$$\frac{1}{N}\sum_0^{N-1} U_T^{an} f U_T^{bn} g U_T^{cn} h \sim \frac{1}{N}\sum_0^{N-1} U_T^{an} E(f|\mathscr{B}') U_T^{bn} E(g|\mathscr{B}') U_T^{cn} E(h|\mathscr{B}').$$

Then

(5.2)

$$\frac{1}{N}\sum_0^{N-1} \int f U_T^n g U_T^{n^2} h \, d\mu \sim \frac{1}{N}\sum_0^{N-1} \int E(f|\mathscr{B}') U_T^n E(g|\mathscr{B}') U_T^{n^2} E(h|\mathscr{B}') \, d\mu.$$

PROOF. Since (5.1) has the same content for U_T and $U_{T^{-1}}$ we can replace U_T in (5.2) by U_T^{-1}. Having done so we multiply the integrands in (5.2) by $U_T^{n^2}$ and we see that it will suffice to prove that

$$\frac{1}{N}\sum_0^{N-1} U_T^{n^2} f U_T^{n^2-n} g \sim \frac{1}{N}\sum_0^{N-1} U_T^{n^2} E(f|\mathscr{B}') U_T^{n^2-n} E(g|\mathscr{B}'). \tag{5.3}$$

It will suffice to prove (5.3) in the case that either $E(f|\mathscr{B}')$ or $E(g|\mathscr{B}')$ vanishes. We apply the van der Corput criterion to prove that in this case the left side of (5.3) converges to 0. Set $u_n = U_T^{n^2} f U_T^{n^2-n} g$;

$$\begin{aligned}\langle u_n, u_{n+m}\rangle &= \int U_T^{n^2} f U_T^{n^2-n} g U_T^{n^2+2mn+m^2} \bar{f} U_T^{n^2-n+2mn+m^2-m} \bar{g} \, d\mu \\ &= \int f U_T^{-n} g U_T^{2mn} \left(U_T^{m^2} \bar{f}\right) U_T^{(2m-1)n} \left(U_T^{m^2-m} \bar{g}\right) d\mu.\end{aligned}$$

Averaging over n we obtain in the integrand and expression of the form (5.1) with $a = -1$, $b = 2m$, and $c = 2m - 1$. But then (5.1) implies that

$$\gamma_m = \lim \frac{1}{N}\sum_{n=0}^{N-1} \langle u_n, u_{m+m}\rangle = 0.$$

By Lemma 2.1 our theorem follows. □

Theorem 5.1 can be used to prove that for any measure preserving system $(X, \mathscr{B}, \mu, T)$, if $A \in \mathscr{B}$ with $\mu(A) > 0$, then

$$\mu(A \cap T^{-n} A \cap T^{-n^2} A) > 0$$

for some n. This in turn implies the diophantine result alluded to in the Introduction.

REFERENCES

[Be] V. Bergelson, *Polynomial ergodic theorems*, Ergodic Theory Dynamical Systems **7** (1987), 337–349.

[B1] J. Bourgain, *Pointwise ergodic theorems on arithmetic sets*, Inst. Hautes Études Sci. Publ. Math. (to appear).

[B2] ——, *Double recurrence and almost sure convergence*, Preprint.

[C,L] J.-P. Conze and E. Lesigne, *Sur un théorème ergodique pour des mesures diagonales*, C. R. Acad. Sci. Paris, t. 306 Série I (1988), pp. 491–493.

[F] H. Furstenberg, *Recurrence in ergodic theory and combinatorial number theory*, Princeton Univ. Press, Princeton, N.J., 1981.

[K,vN] B. O. Koopmans and J. von Neumann, *Dynamical systems of continuous spectra*, Proc. Nat. Acad. Sci. **18** (1932), pp. 255–266.

[L] E. Lesigne *Théorèmes ergodiques pour un translation sur une nilvariété*, Ergodic Theory and Dynamical Systems, **9** (1989), pp. 115–126.

[vN] J. von Neumann, *Proof of the quasi-ergodic hypothesis*, Proc. Nat. Acad. Sci. **18** (1932), 70–82.

[vNH] J. von Neumann and P. R. Halmos, *Operator methods in classical mechanics*, Ann. of Math. (2) **43** (1942), 332–350.

HEBREW UNIVERSITY, JERUSALEM

Proceedings of Symposia in Pure Mathematics
Volume **50** (1990)

The Rings of Operators Papers

F. J. MURRAY

This note is an expanded and I, hope, more coherent version of a talk given on the evening of May 31. There was considerable interest in the early R.O. papers and this might be considered a type of introduction to R.O. and R.O.II which would never survive the Reaganomics practiced by journal editors [**3**], [**4**].

As I begin I am acutely aware of what happens on occasion on my favorite television program, "The Peoples Court". The given testimony seems to favor one of the litigants and as Judge Wapner begins to present his decision one expects that it will be in favor of this party. But then the litigant interrupts the judge with additional remarks and inadvertently gives further information which upsets his case. Today has gone pretty well for me and I should probably stop here and not present my own version of snatching Defeat from the Jaws of Victory.

I had "completed the requirements for the Ph.D." from Columbia in the spring session of 1934, with B. O. Koopman as my advisor. Professor Koopman had given a course in Hilbert space with M. H. Stone's book as text. My other main interest had been Banach's book, especially the "closed graph" type of theorem. The geometric formulation of analysis was fascinating and was a major interest of Professor Koopman. I constructed various Hilbert spaces for partial differential operators and obtained what I thought would be useful results. But Professor Stone on a visit to Columbia pointed out to me that the Hilbert spaces results I had obtained were readily available from the polar decomposition of a closed transformation with dense domain, given in [**2**]. This required adjustments, but it did mean that I could handle polar decomposition effectively.

When I began as a National Research Fellow in the fall of 1934, Professor von Neumann suggested that I read the reference [**1**]. This paper contained the result that $\mathbf{M}'' = \mathbf{M}$ for von Neumann algebras and results, which are

1980 *Mathematics Subject Classification* (1985 *Revision*). Primary 01A70, 16-03, 16A30.
Key words and phrases. von Neumann rings, factors, finite analogies.
This paper is in final form and no version of it will be submitted for publication elsewhere.

somewhat complicated to state, but what is used is the corollary: If **M** is a factor and $E_1 \in \mathbf{M}$ and $E_2 \in \mathbf{M}'$, then $E_1E_2 \neq 0$ if $E_1, E_2 \neq 0$.

When I returned, Professor von Neumann proposed the problem of factors. I do not recall this talk in detail but I do remember certain diagrams which I made to grasp the situation. In the finite-dimensional case, a typical $\mathbf{M}, \mathbf{M}'$ pair can be described as follows. Consider a rectangle of lattice points (i, j), say, with $i = 1, \ldots, 5$ and $j = 1, \ldots, 4$ (Figure 1). Suppose then we have a twenty-dimensional linear vector space with a unit coordinate vector associated with each lattice point, orthogonal to all the others. Then **M** would consist of the linear transformations on this space which are independent of j and $\mathbf{M}'$, the transformations which are independent of i.

More generally a ring of operators, **M**, closed under the adjoint operator, in the finite case, would be associated with a string of such boxes, each to the right and above the previous box (Figure 2). **M** consists of the transformations which do not depend on the j index, and $\mathbf{M}'$ the corresponding set which do not depend on the i index. The center of **M**, $\mathbf{M} \cap \mathbf{M}'$ consists of the transformations which are constant on each box.

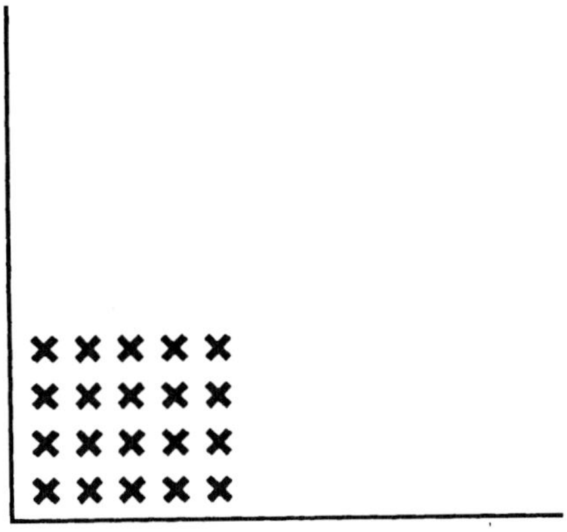

FIGURE 1

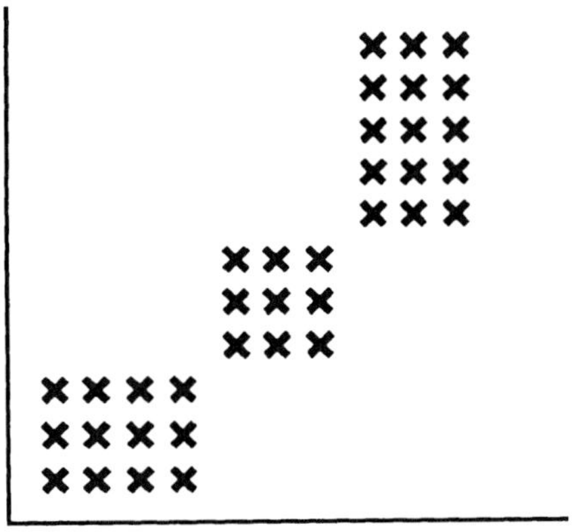

FIGURE 2

Professor von Neumann conjectured that if a factor pair, $\mathbf{M}, \mathbf{M}'$, in Hilbert space were both infinite, then $\mathfrak{H}$ could be expressed as an L_2 on the unit square interval, **M** would be the operators on x alone, and $\mathbf{M}'$ those on y alone.

All the diagrams have algebraic equivalents, which von Neumann preferred. He was tolerant of my diagram procedures but certainly not enthusiastic.

But there is an obvious lattice point diagram, corresponding to $l_1 \otimes l_2$, equivalent to his conjecture. (One can even let $\mathbf{M}'$ be finite dimensional.) One can show, using polar decomposition and the $E_1E_2 \neq 0$ corollary, that a factor **M** has the corresponding form if and only if it has a minimal projection.

This was quite revealing. It suggested that there was a direct analogy between the ranges of projections $\in \mathbf{M}$ and the partial isometric $W \in \mathbf{M}$ on one hand, and the cardinal number theory of infinite sets and one-to-one correspondence on the other hand. The critical point in the analogy is Bernstein's Theorem. For cardinal numbers, this says that if $C_1 \leq C_2$, i.e., a C_1 set can be mapped into a C_2 set, and if $C_2 \leq C_1$, then $C_1 = C_2$, i.e., a C_1 set and a C_2 set are in one-to-one correspondence. We both saw that we could obtain a similar result for the projection ranges by an analogous argument in which orthogonal ranges replace disjoint sets. When this result was written down, von Neumann described the five types of ranges for the dimension function in a discussion of less than half an hour.

The immediate problem was to obtain an example of a case other than type I (preferably a case of type II_1). The diagram procedure suggested one possibility (Figure 3).

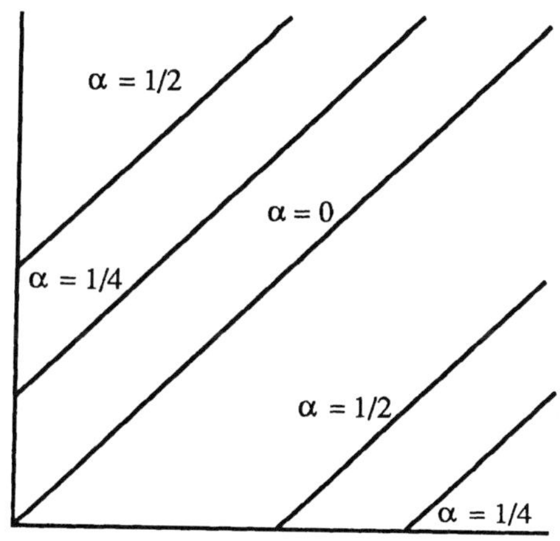

FIGURE 3

Consider the unit square and place a linear Lebesgue measure on the lines $y = x + \alpha$, mod 1, where α is a binary number. The corresponding inner product yields a Hilbert space in which a displacement by a binary fraction, mod 1, in either the x direction or the y direction yields a unitary transformation. We take $\mathbf{M}$ to be the ring determined by the x binary shifts and multiplication by the bounded Lebesgue functions of x. Then one must show that $\mathbf{M}'$ is the corresponding ring for y. For $f = \delta(y, x)$, $\mathrm{Tr}_{\mathbf{M}}(A) = (Af, f)$. The diagram immediately suggest these results, but of course, they have to be proved.

For type II, at least there is a general way to construct a diagram which yields certain results, but after these something much deeper is required.

Let $H \in \mathbf{M}$ be s.a., positive definite with bound 1, and suppose that the abelian von Neumann algebra determined by H is maximal in $\mathbf{M}$. Suppose for convenience that the dimension ratio for $\mathbf{M}, \mathbf{M}'$ is one. Then the procedures of R.O.II show how one can construct a corresponding $H' \in \boldsymbol{M}'$ with analogous properties and the resolutions of the identity for H and H' yield a diagram with uniform element along the diagonal (Figure 4).

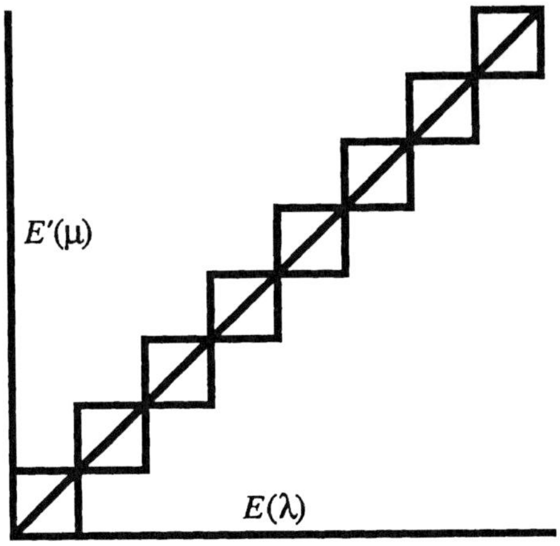

FIGURE 4

REFERENCES

1. J. von Neumann, *Zur Algebra der Funktionaloperatoren*, Math. Ann. **102** (1929), 370–427.
2. ——, *Über adjungierte Funktionaloperatoren*, Ann. of Math. (2) **33** (1932), 294–310.
3. F. J. Murray and J. von Neumann, *On rings of operators*, Ann. of Math. (2) **37** (1936), 116–229.
4. ——, *On rings of operators* II, Trans. Amer. Math. Soc. **41** (1937), 208–248.

DUKE UNIVERSITY

Proceedings of Symposia in Pure Mathematics
Volume **50** (1990)

Operator Algebras – An Overview

RICHARD V. KADISON

§**1. Introduction.** The subject popularly known as "operator algebras" (a shortened form of "algebras of Hilbert space operators") made its first appearance in a paper **[31]** published in the 1929–1930 Mathematische Annalen. In that article, a construct which von Neumann calls a "ring of operators" is defined, and a theorem, the Double Commutant Theorem, that is, arguably, the most basic theorem of the subject is proved. Somewhat later (the late 1940s), "rings of operators" was used interchangeably with "W^*-algebras," the 'W^*' referring to "weak-operator closed" and '$*$' to the central role of the adjoint operation, denoted by '$*$,' in the definition of this construct. In his classic text **[D]**, the first in the subject, Dixmier referred to these algebras as 'von Neumann algebras' following a suggestion made to him by J. Dieudonné.

Let $\mathscr{H}$ be a Hilbert space over the complex numbers **C** and let $< x, y >$ denote the (positive-definite) inner product of two vectors x and y in $\mathscr{H}$. The "length" or "norm" of a vector x is $< x, x >^{\frac{1}{2}}$ (denoted by $\|x\|$). The mapping $x \to \|x\|$ provides $\mathscr{H}$ with a norm. By definition, $\mathscr{H}$ is *complete* relative to the metric that assigns $\|x - y\|$ as the distance between the vectors x and y.

If $\mathscr{K}$ is another such Hilbert space and T is a linear transformation of $\mathscr{H}$ into $\mathscr{K}$, then T is continuous relative to the metric topologies, just described, if and only if

$$\sup\{\|Tx\| : x \in \mathscr{H}, \|x\| \leq 1\} < \infty.$$

This supremum, denoted by $\|T\|$, is referred to as the "norm" or "bound" of T. We say that T is *bounded* in this case. (The terms 'bounded' and 'continuous' are used interchangeably for linear transformations.)

The family $\mathscr{B}(\mathscr{H}, \mathscr{K})$ of all bounded linear transformations of $\mathscr{H}$ into $\mathscr{K}$ is itself a linear space, where $(aT + S)(x)$ is defined as $a(Tx) + Sx$. When $\mathscr{K} = \mathscr{H}$, we write '$\mathscr{B}(\mathscr{H})$' in place of '$B(\mathscr{H}, \mathscr{H})$.' In this case, $\mathscr{B}(\mathscr{H})$ is

1980 *Mathematics Subject Classification* (1985 *Revision*). Primary 46L05, 46L10.
This paper is in final form and no version of it will be submitted for publication elsewhere.

an algebra, where $(TS)x$ is defined as $T(Sx)$. This algebra has the operator I (the identity transform, $Ix = x$ for all x in $\mathscr{H}$) as a unit element. The mapping $T \to \|T\|$ endows $\mathscr{B}(\mathscr{H},\mathscr{K})$ with a norm relative to which it is a Banach apace; $\mathscr{B}(\mathscr{H})$ is a Banach algebra ($\|TS\| \leq \|T\|\,\|S\|$ and $\|I\| = 1$). The metric on $\mathscr{B}(\mathscr{H},\mathscr{K})$ that assigns $\|T - S\|$ as the distance between T and S gives rise to the *norm* or *uniform* topology on $\mathscr{B}(\mathscr{H},\mathscr{K})$.

The linear transformations of $\mathscr{H}$ into $\mathbf{C}$ (a one-dimensional Hilbert space) are called *(linear) functionals.*

The bounded linear functionals on $\mathscr{H}$ (elements of $\mathscr{B}(\mathscr{H},\mathbf{C})$) are described by the celebrated Riesz Representation Theorem. If φ is in $\mathscr{B}(\mathscr{H},\mathbf{C})$, there is a unique vector x in $\mathscr{H}$ such that $\varphi(y) = \langle y, x\rangle$ for each y in $\mathscr{H}$. The mapping $\varphi \to x$ is a conjugate-linear (that is, $a\varphi$ corresponds to $\bar{a}x$), isometric (that is, $\|\varphi\| = \|x\|$), isomorphism of $\mathscr{B}(\mathscr{H},\mathbf{C})$ onto $\mathscr{H}$. If T is in $\mathscr{B}(\mathscr{H},\mathscr{K})$ and z is in $\mathscr{K}$, then $x \to \langle Tx, z\rangle$ is in $\mathscr{B}(\mathscr{H},\mathbf{C})$. It follows that there is a unique vector T^*z in $\mathscr{H}$ such that, for all x in $\mathscr{H}$, $\langle Tx, z\rangle = \langle x, T^*z\rangle$. The mapping T^* is in $\mathscr{B}(\mathscr{K},\mathscr{H})$. The linear transformation T^* of $\mathscr{K}$ into $\mathscr{H}$ is called *the adjoint* of T. The equalities, $(a\ T + S)^* = \bar{a}\ T^* + S^*, (T^*)^* = T, \|T\| = \|T^*\|, \|T^*T\| = \|T^*\|\,\|T\|$, are easily proved. When $\mathscr{H} = \mathscr{K}$, we have that $(TS)^* = S^*T^*$. In this same case, we say that T is *selfadjoint* when $T = T^*$.

With $\mathscr{S}$ a subset of $\mathscr{B}(\mathscr{H})$, we define the *weak-operator closure* $\mathscr{S}^-$ of $\mathscr{S}$ to be the set of those T in $\mathscr{B}(\mathscr{H})$ such that, given a positive ε and vectors $x_1, \cdots, x_n; y_1, \cdots, y_n$ in the Hilbert space $\mathscr{H}$, there is an S in $\mathscr{S}$ for which $|\langle (T - S)x_j, y_j\rangle| < \varepsilon$, $j = 1, \cdots, n$. This closure operation defines a topology on $\mathscr{B}(\mathscr{H})$ called the *weak-operator topology.* A subalgebra $\mathscr{R}$ of $\mathscr{B}(\mathscr{H})$ that is weak-operator closed, contains I, and contains T^* when it contains T, is called a *von Neumann algebra.*

§2. **Motivation.** The reducibility and irreducibility results of I. Schur and the Peter-Weyl theory that figured so prominently during the period of von Neumann's university training led him to consider the analogous questions for families of operators (linear transformations) on a Hilbert space. This, in turn, led him to his Double Commutant Theorem, the heart of his 1929–1930 Mathematische Annalen paper, where "rings of operators" are first introduced.

In a sense, then, von Neumann algebras grew out of the early period of group representations. It is popular and natural, in view of von Neumann's involvement with the more rigorous mathematical formulation of the basics of quantum mechanics, to ascribe an important place to the "new physics" in motivating the introduction of "rings of operators," but that interpretation does not stand up to scrutiny. It seems equally clear that von Neumann's commitment to pursuing the study of his "rings of operators" was not, at first, large; he did not return to it for five years.

In 1935, F. J. Murray, a gifted young mathematician, who had just received his Ph.D. at Columbia University with a thesis on linear operators and spectral theory, arrived in Princeton as a National Research Fellow intent on studying under von Neumann. In the spirit of a postdoctoral research project, von Neumann proposed to Murray that he examine "rings of operators" from the point of view of the Wedderburn structure theory for matrix algebras. In particular, Murray was to concentrate on those von Neumann algebras whose centers consist of scalar multiples of I, the *factors*. Their most primitive guess, at the outset, was that a factor is isomorphic to $\mathscr{B}(\mathscr{H})$ for some Hilbert space $\mathscr{H}$; that would be a worthy result and the end of the project. Before the first year of the investigation had passed, a rich and intriguing world of mathematical phenomena had opened before Murray and von Neumann.

At an early stage, von Neumann joined Murray in the investigation on a full-time basis in what became a seven-year project, one of this country's most successful mathematical collaborations. As their research proceeded, other layers of motivation developed. From the examples of factors they constructed, the very clear applicability of the subject to the study of transformations on measure spaces (ergodic theory) and to representations of groups (through a generalized group algebra) was assured. Factors of type II_1, a class of factors they discovered, have associated algebras of unbounded operators that permit all the formal algebraic manipulations being used by the founders of quantum mechanics in their mathematical formulation. Von Neumann felt that the factors of type II_1 might provide the natural framework for quantum mechanics. That hope was not to be realized, but the factors of type II_1 are basic to the analysis of the class, factors of type III, that are essential to a mathematical presentation of quantum statistical mechanics and quantum field theory.

§3. **Functions and operators.** At the same time as the Murray-von Neumann collaboration began, M. H. Stone recognized the possibility of extracting topological spaces from algebraic structures [**42**] and representing the algebraic structures by functions on the spaces. The earlier extension of the spectral theorem to include unbounded selfadjoint operators and the resulting function calculus of such operators developed by both Stone and von Neumann provided a major impetus to Stone's ideas and to his analysis of $C(X)$, the algebra, under pointwise operations, of continuous complex-valued functions on a locally compact Hausdorff space X. Stone proved [**43**] that an algebra with a partial ordering, satisfying some natural conditions, is isomorphic to the algebra of real-valued functions in $C(X)$ and applied this to yield such an isomorphism of a commutative algebra of bounded selfadjoint operators.

Combining Stone's ideas with his work on Banach algebras and his study of the first three articles in the series of Murray and von Neumann, I. M.

Gelfand undertook the characterization of those norm-closed subalgebras of $\mathscr{B}(\mathscr{H})$ that contain T^* when they contain T. He and M. Neumark **[15]** proved that the Banach algebras in question are those with an involution $(A \to A^*)$ having the properties of the adjoint operation on $\mathscr{B}(\mathscr{H})$. The Banach algebras of Gelfand and Neumark have a unit and the subalgebras of $\mathscr{B}(\mathscr{H})$ contain I. The norm-closed subalgebras of $\mathscr{B}(\mathscr{H})$ stable under the adjoint operation have come to be known as C^*-algebras. They occupy a key position in the subject. Each von Neumann algebra is, in particular, a C^*-algebra. The Gelfand-Neumark article **[15]** recast Stone's result **[43]** in Banach algebra form and used it as a key lemma.

THEOREM. *Let $\mathscr{A}$ be a commutative Banach algebra with a unit element I and an involution $A \to A^*$ satisfying $(aA+B)^* = \bar{a}A^* + B^*, (AB)^* = B^*A^*, A^{**} = A, \|A^*A\| = \|A^*\|\,\|A\|$. Then there is a compact Hausdorff space X and an algebraic isomorphism φ of $\mathscr{A}$ onto $C(X)$ such that*

$$\varphi(A^*) = \overline{\varphi(A)}, \quad \|A\| = \|\varphi(A)\|(= \sup\{|\varphi(A)(x)| : x \in X\}) \quad (A \in \mathscr{A}).$$

If 'commutative' is omitted from the hypothesis of the theorem, there is a C^*-algebra $\mathscr{A}_0$ and an isometric isomorphism ψ of $\mathscr{A}$ onto $\mathscr{A}_0$ such that $\psi(A^*) = \psi(A)^*$. We refer to ψ as 'a * isomorphism.' It follows, in particular, that each $C(X)$ is * isomorphic to a commutative (abelian) C^*-algebra (with complex conjugation of functions as the involution on $C(X)$).

The identification of the family of commutative C^*-algebras with the family of function algebras $C(X)$ underlies the interpretation of the general study of C^*-algebras as 'noncommutative (real) analysis,' the point of view that has dominated the subject since the late 1940s. Analysis is very largely a study of $C(X)$, its associated structures, and operations (differentiation and integration) on these. If $C(X)$ is replaced, in such considerations, by a "noncommutative" $C(X)$, that is, a noncommutative C^*-algebra, then the definitions of the related structures and operations are usually forthcoming. The results of classical analysis can then be reformulated as questions and conjectures involving a C^*-algebra. The answers to these questions and the proofs of the conjectures (or counterexamples) tend not to be so forthcoming.

The view of the study of C^*-algebras as noncommutative (real) analysis guides the research and provides a large template for the motivation of the subject. When noncommutative analysis is the appropriate analysis, as in quantum theory or classical (commutative) analysis situations with non-abelian groups acting, operator algebras provide the mathematical framework.

As noted, each von Neumann algebra is a C^*-algebra. An abelian von Neumann algebra is isomorphic to some $C(X)$. In this case, the X has very special properties; each open set in X has a closure that is open. The space X is said to be *extremely disconnected*. The converse is not true; there are extremely disconnected spaces X for which $C(X)$ is not isomorphic to a von Neumann algebra. There are subtle "measure-theoretic" requirements that X must fulfill for there to be such an isomorphism. The function representation of an abelian von Neumann algebra can be formulated cogently in terms of measure theory.

Let S be a set and μ a (σ-finite) measure on S. The linear space $L_2(S, \mu)$ of functions (absolutely) square integrable with respect to μ, provided with the inner product, $\langle f, g\rangle = \int f\bar{g}d\mu$, is a Hilbert space $\mathscr{H}$. With f an essentially bounded measurable function on S and g in $\mathscr{H}$, we define $M_f(g)$ to be the product $f \cdot g$. Then $fg \in \mathscr{H}$ and M_f is a bounded linear operator on $\mathscr{H}$. The family $\mathscr{A}$ of these *multiplication operators* is an abelian von Neumann algebra on $\mathscr{H}$, with the property that each T commuting with all the multiplication operators is a multiplication operator. We say that $\mathscr{A}$ is *maximal abelian.*

If we specialize the construction, taking for S and μ one of $[0, 1]$ with Lebesgue measure, a finite or countably infinite set of points each with positive measure, or the union of two such measure spaces, we arrive at what seem to be some special abelian von Neumann algebras. Surprisingly, each abelian von Neumann algebra on a separable Hilbert space is isomorphic to one of the few just constructed. If the von Neumann algebra is maximal abelian, then there is even an isomorphism (unitary transformation) of the separable Hilbert space onto $L_2(S, \mu)$ that transforms the algebra onto the multiplication algebra. The family of essentially bounded measurable functions on S is an algebra, the *multiplication algebra* of (S, μ), under the pointwise operations of addition and multiplication of functions. The mapping that assigns M_f to f is an isomorphism of this algebra onto the maximal abelian algebra of multiplication operators.

From this discussion, we see that the theory of abelian von Neumann algebras is a version of measure theory. As with general C^*-algebras and continuous function theory, we recognize that the theory of the general von Neumann algebra is *noncommutative measure theory.* This point of view leads us to identify the selfadjoint idempotent operators ($E = E^*$ and $E = E^2$), the projections, in a von Neumann algebra $\mathscr{R}$ with the characteristic functions of measurable sets in a measure space. In effect, the projections are the "measurable sets" in our "noncommutative measure space." It is a basic technical fact that there are many projections in a von Neumann algebra. In fact, the linear span of the set of projections in a von Neumann algebra is norm dense in that algebra.

§**4. Factors.** In studying the most noncommutative von Neumann algebras, the factors, an examination of the "measurable sets," the projections provide us with a first glimpse of the underlying structure. By analogy with the "atoms" of a measure space, sets of positive measure with no proper subsets of smaller positive measure, we define a *minimal projection* in a von Neumann algebra $\mathscr{R}$ to be a nonzero projection E in $\mathscr{R}$ such that if F is a nonzero projection in $\mathscr{R}$ whose range is contained in the range of E (equivalently, $FE = F$), we say that F is a subprojection of E in this case and write ($F \leq E$), then $F = E$. We study our factors for minimal projections.

THEOREM. *If the factor $\mathscr{M}$ contains a minimal projection, then $\mathscr{M}$ is isomorphic to $\mathscr{B}(\mathscr{H})$ for some Hilbert space $\mathscr{H}$.*

Projections E and F such that $EF = 0$ are called *(mutually) orthogonal*. The dimension of $\mathscr{H}$ in the preceding theorem is the cardinality of a maximal orthogonal family of minimal projections in $\mathscr{M}$. This theorem tells us most of what we can expect to learn about factors with a minimal projection concerning isomorphisms (about the factors themselves, not the von Neumann subalgebras, of course). The problem that Murray and von Neumann faced at this (early) point in their investigation was whether or not all factors had minimal projections. Certainly, $\mathscr{B}(\mathscr{H})$ has minimal projections, the projections with one-dimensional range, but does the fact of being central (that is, having center consisting of scalar's multiples of I) force the factor to have a minimal projection? They solved the problem in the negative during their first year of work by constructing an important class of examples through the use of ergodic theory techniques.

With hindsight, we can see that there should be factors without minimal projections, just as there are measure spaces without atoms. But, of course, had this not been the case, the concept of a noncommutative measure space would have little validity.

Murray and von Neumann conceived the idea of comparing the "sizes" of projections in a von Neumann algebra. Two projections E and F in a von Neumann algebra are said to be *equivalent* (*modulo* $\mathscr{R}$) when some A in $\mathscr{R}$ maps the range of E isometrically onto the range of F. In this case, $AE (= V \in \mathscr{R})$ has the same properties and $V^*V = E$, $VV^* = F$. Murray and von Neumann indicate this equivalence by writing $E \sim F$. When $E \sim F_0$ and $F_0 \leq F$, they write $E \precsim F$(and $E \prec F$ when E is, in addition, not equivalent to F).

Each T in $\mathscr{B}(\mathscr{H})$ has a unique decomposition as VH, the *polar decomposition* of T, where H is a positive selfadjoint operator (that is, $\langle Hx, x\rangle \geq 0$ for each x in $\mathscr{H}$) with square equal to T^*T and V maps the closure of the range of H (we write '$r(H)$' for this closure) isometrically onto $r(T)$ and maps the orthogonal complement of $r(H)$ to 0. We say that V is a *partial isometry* with *initial space* $r(H)$ and *final space* $r(T)$. If $R(H)$ is the projection with range $r(H)$ (we call $R(H)$ the *range projection* of H), then $V^*V = R(H)$ and $VV^* = R(T)$. We note that the components V and H of this polar decomposition lie in $\mathscr{R}$ when T does, from which we conclude that $R(H) \sim R(T)$, as a consequence of the uniqueness of those components and the main theorem of von Neumann's original paper [**31**].

THEOREM (DOUBLE COMMUTANT). *If $\mathscr{R}$ is a von Neumann algebra acting on a Hilbert space $\mathscr{H}$ and $\mathscr{R}'$ is the family of bounded operators T that* *commute with all the operators in* $\mathscr{R}$, *then* $\mathscr{R}'$ *is a von Neumann algebra and* $(\mathscr{R}')' = \mathscr{R}$.

The notation $\mathscr{F}'$ for the set of bounded operators T commuting with all the operators in the family $\mathscr{F}$ of bounded operators will be used in the sequel; $\mathscr{F}'$ is called the *commutant* of $\mathscr{F}$. When $\mathscr{F}$ is selfadjoint (that is, $\mathscr{F} = \mathscr{F}^*$), $\mathscr{F}'$ is a von Neumann algebra.

With T in $\mathscr{R}$ and VH the polar decomposition of T, it is easy to see that $R(H) = R(T^*)$. It follows that $R(T) \sim R(T^*)$, a very useful fact for dealing with the relations $\sim$ and $\precsim$. Using this fact, the range projections of EAF and FA^*E are equivalent for each A in a factor $\mathscr{M}$, where E and F are projections in $\mathscr{M}$. Thus nonzero subprojections of E and F are equivalent unless EAF is 0 for each A in $\mathscr{M}$. The fact that $\mathscr{M}$ has center the scalars allows us to conclude that EAF is 0 for each A in $\mathscr{M}$ if and only if one of E or F is 0. It follows that each pair of nonzero projections in a factor $\mathscr{M}$ have equivalent nonzero subprojections. A measure-theoretic-like argument now shows that either $E \precsim F$ or $F \prec E$. An argument patterned on the Cantor-Schröeder-Bernstein argument from the theory of cardinals shows that $E \sim F$ when $E \precsim F$ and $F \precsim E$ for projections E and F in a general von Neumann algebra. It follows that $\precsim$ is a partial ordering of the equivalence classes of projections in a von Neumann algebra and a total ordering of these classes in a factor.

Murray and von Neumann define infinite and finite projections in this framework modeled on the set-theoretic approach. The projection E in $\mathscr{R}$ is *infinite* (relative to $\mathscr{R}$) when $E \sim E_0 < E$ (that is, $E_0 \leq E$ and $E_0 \neq E$), for some subprojection E_0 of E in $\mathscr{R}$, and finite otherwise. It is a difficult theorem (proved by Murray and von Neumann) that the union of two finite projections (this union is the projection with range spanned by the ranges of both projections) is a finite projection.

At this stage, the possibility of several types of factors presented itself to Murray and von Neumann. If the factor $\mathscr{M}$ has a minimal projection (and is isomorphic, therefore, to some $\mathscr{B}(\mathscr{H})$), they refer to it as a *factor of type I* (of *type* I_n if $\mathscr{H}$ is n-dimensional, of *type* I_∞ when $\mathscr{H}$ is infinite dimensional). If $\mathscr{M}$ has no minimal projection but contains a (nonzero) finite projection, they refer to $\mathscr{M}$ as a *factor of type II* (of type II_1 if I is finite relative to the factor, of type II_∞ otherwise). The remaining possibility is the case where all nonzero projections in the factor are infinite. They refer to these as factors of type III. Of course, no two factors of different types could be isomorphic. But the paramount question was the existence of factors of the various types.

§5. **Examples.** Murray and von Neumann were able to construct factors without minimal projections and identify them as factors of types II_1 and II_∞ by means of examples built from groups acting "freely" by measure (and measurability) - preserving transformations on a measure space. These examples appear in the first paper [27] resulting from their collaboration.

Let G be a (discrete) group with unit e, and $\mathscr{A}$ be a maximal abelian (selfadjoint) subalgebra of $\mathscr{B}(\mathscr{H})$. (We use $\mathscr{A}$ to replace an explicit measure space, as we know we can from our earlier discussion.) Let $\mathscr{K}$ be the direct sum of copies of $\mathscr{H}$ indexed by elements of G. (In effect, $\mathscr{K}$ is the Hilbert space of functions x on G with values in $\mathscr{H}$, such that $\sum_{g\in G}\|x(g)\|^2 < \infty$, provided with the obvious linear structure and the inner product $\langle x,y\rangle = \sum_{g\in G}\langle x(g),y(g)\rangle$.) With S in $\mathscr{B}(\mathscr{H})$, there is a naturally associated operator $\Phi(S)$ in $\mathscr{B}(\mathscr{K})$ defined by $(\Phi(S)x)(g) = S(x(g))$. Let $U : g \to U(g)$ be a unitary representation of G on $\mathscr{H}$ (that is, a homomorphism of G into the group of unitary transformations of $\mathscr{H}$ into itself) and let $(V(g)x)(g')$ be $U(g)x(g^{-1}g')$. We assume that $U(g)AU(g)^*$ is in $\mathscr{A}$ for each A in $\mathscr{A}$ and g in G (that is, each $U(g)$ implements an automorphism of $\mathscr{A}$ and U gives rise to a representation of G by automorphisms of $\mathscr{A}$) and that $\mathscr{A}\cap(U(g)\mathscr{A}) = \{0\}$ for each $g(\neq e)$ in G (that is, G acts *freely*, by automorphisms, on $\mathscr{A}$). The mapping Φ "copies" $\mathscr{B}(\mathscr{H})$ isomorphically into $\mathscr{B}(\mathscr{K})$. It is easily checked that $V(gg') = V(g)V(g')$ and $V(g)\Phi(S)V(g)^* = \Phi(U(g)SU(g)^*)$ for each S in $\mathscr{B}(\mathscr{H})$. Thus the representation V gives rise to the "same" representation of G by automorphisms of the copy $\Phi(\mathscr{A})$ of $\mathscr{A}$ as U does. We say that G acts ergodically on $\mathscr{A}$ (through the representation U) when the only elements A in $\mathscr{A}$ such that $U(g)AU(g)^* = A$, for all g in G, are the scalars.

THEOREM. *The von Neumann algebra $\mathscr{R}$ generated by $\Phi(\mathscr{A})$ and the group $\{V(g)\}_{g\in G}$ is a factor if and only if G acts ergodically on $\mathscr{R}$. In this case, it is a factor of type I if and only if $\mathscr{A}$ has a minimal projection. In any event, $\mathscr{A}$ is a maximal abelian subalgebra of $\mathscr{R}$.*

Specific examples of the structures described above are obtained from a measure space $(S,\mathscr{S},m)$ that is countably separated ($\mathscr{S}$, the family of measurable sets, contains a countable family of $E_1,E_2,\dots$ of non-null sets of finite measure such that if s and t are distinct points of S, then $t\in E_j$ and $s\notin E_j$ for some j) and a group G of one-to-one mappings of S onto S that preserves measurability and measure 0 subsets and acts freely on S (that is, $m(\{s\in S : g(s)=s\}) = 0$ when g is not the unit element e of G). In this case, $\mathscr{A}$ is the multiplication algebra of the measure space (acting on $L_2(S,m)$). The Radon-Nikodým theorem yields, for each g, a non-negative, real-valued, measurable function φ_g on S such that $\int x(g(s))\,dm(s) = \int x(s)\varphi_g(s)\,dm(s)$ for each x in $L_1(S,m)$. If U_g is defined by $(U_gx)(s) = [\varphi_g(s)]^{\frac{1}{2}}x(g^{-1}(s))$, for each x in $L_2(S,m)$, then $g\to U_g$ is a unitary representation of G that gives rise to automorphisms of $\mathscr{A}$ satisfying $\mathscr{A}\cap(U_g\mathscr{A}) = \{0\}$ for each $g(\neq e)$ in G. We say that G acts *ergodically* on S when $m(g(S_0)\backslash S_0) > 0$ for some g in G unless $m(S_0) = 0$ or $m(S\backslash S_0) = 0$. The representation $g\to U_g$ acts ergodically on $\mathscr{A}$ if and only if G acts ergodically on S.

With U_g in place of $U(g)$, $L_2(S,m)$ for $\mathscr{H}$, and the multiplication algebra of (S,m) for $\mathscr{A}$, the conditions for the construction of $\mathscr{R}$ described earlier are satisfied.

THEOREM. *If G acts ergodically on S, then $\mathscr{R}$ is a factor and*

(i) *$\mathscr{R}$ is of the type I if and only if some point in S has positive measure; in this case, $\mathscr{R}$ is of type I_n where n is the number of points in S.*
(ii) *$\mathscr{R}$ is of type II when S admits a G-invariant measure m_0 such that $m_0(S_0) = 0$ if $m(S_0) = 0$. In this case, $\mathscr{R}$ is of type II_1 when $m_0(S) < \infty$ and of type II_∞ when $m_0(S) = \infty$.*
(iii) *$\mathscr{R}$ is of type III when there is no m_0 as described in* (ii).

Explicit examples can be displayed by choosing explicit measure spaces and group actions. We describe some.

(a) With S the group of integers, each integer assigned measure 1, let G be the group of all translations of S (under addition). This example results in a factor of type I_∞. If S is replaced by a finite cyclic group of order n, the resulting von Neumann algebra is a factor of type I_n.
(b) Let S be the circle (in **C**) with Haar-Lebesgue measure and let G be the group of rotations of S through angles that are rational multiples of π. The action of G is ergodic and the resulting factor is of type II_1. The same is true if we replace G by the group generated by a single rotation through an angle that is an irrational multiple of π.
(c) Let S be **R**, the set of real numbers, with Lebesgue measure and G be the (countable) group of translations by rationals. Again, G acts ergodically, and the resulting factor is of type II_∞.

In each of the examples (a), (b) and (c), the measure is invariant under the group action. In the example that follows, we augment the group of (c) by transformations that do not preserve the measure. Since the group of rational translations acts ergodically, Lebesgue measure is the only invariant measure (up to positive scalar multiples). Thus the augmented group admits no invariant measure.

(d) Let S be **R** with Lebesgue measure and let G be the (countable) group of all "rational" affine transformations $s \to as + b$ with a and b rational. The resulting factor is of type III.

It is not easy to show that the factor in Example (d) is of type III (or to prove (iii) of the more general theorem on types). Some of the techniques and constructs needed will appear when we discuss traces and weights. Several years elapsed before the factors of type III were constructed. They appear in the third paper of the Murray-von Neumann series **[30]**, a paper authored by von Neumann alone.

Another, more simply described, class of examples was found later by Murray and von Neumann. These appear in **[29]**. They provide one of the possible extensions of the notion of group algebra (over **C**) from finite to infinite (discrete) groups. Let G be an infinite (discrete) group and $\mathscr{H}$ be the family of complex-valued functions φ on G such that $\sum_{g\in G}|\varphi(g)|^2 < \infty$. With the inner product, $\langle\varphi,\psi\rangle = \sum_{g\in G}\varphi(g)\overline{\psi(g)}$, $\mathscr{H}$ becomes a Hilbert space (referred to as $l_2(G)$). Let $(L_{g_0}\varphi)(g)$ be $\varphi(g_0^{-1}g)$ for each g in G. Then L_g is a unitary operator on the Hilbert space $\mathscr{H}$ (for $L_{g^{-1}}$ is its inverse and $< L_g\varphi, L_g\psi > = < \varphi,\psi >$ for all φ and ψ in $\mathscr{H}$). Moreover, $L_gL_{g'} = L_{gg'}$; the mapping $g \to L_g$ is a (group) isomorphism of G into the group of unitary operators on $\mathscr{H}$. In the same way, we can define the unitary operators R_{g_0} by $(R_{g_0}\varphi)(g) = \varphi(gg_0)$. Let $\mathscr{L}_G$ and $\mathscr{R}_G$ be the weak-operator closures of the algebras of finite, complex linear combinations of the operators $\{L_g : g \in G\}$ and $\{R_g : g \in G\}$, respectively. Then $\mathscr{L}_G$ and $\mathscr{R}_G$ are von Neumann algebras. In addition, each of $\mathscr{L}_G$ and $\mathscr{R}_G$ is the commutant of the other. (Thus $(\mathscr{L}_G')' = \mathscr{L}_G$, an illustration of the Double Commutant Theorem.)

THEOREM. *The von Neumann algebra $\mathscr{L}_G$ is a factor if and only if each conjugacy class (other than the class of the group identity) is infinite. In this case, $\mathscr{L}_G$ is a factor of type II_1.*

The groups G satisfying the infinite conjugacy class condition are called *i.c.c. groups*. Some examples of such groups are $\mathscr{F}_n$, the free (nonabelian) groups on $n(\geq 2)$ generators, and Π, the group of those permutations of the integers that move at most a finite number of integers. In **[29]**, with the aid of these examples, Murray and von Neumann answer the most important question raised in their earlier work: Has the type classification of factors settled the problem of algebraic isomorphism of factors? More specifically, are all factors of type II_1 isomorphic? They answer this last question in the negative.

THEOREM. *$\mathscr{L}_\Pi$ is not isomorphic to $\mathscr{L}_{\mathscr{F}_n}$.*

It was abundantly clear from this result of Murray and von Neumann that there are an infinite number of mutually nonisomorphic factors of type II_1 and that infinite groups, distinguished by subtle commutativity properties, would provide such examples. It was equally clear that the implementation of that program would not be easy. Twenty-five years after the Murray-von Neumann result, that program was completed **[25, 26, 39]**, following basic advances in the program **[5, 14, 38, 40, 51]**. In another direction, a more innovative addition to the II_1 factor archives comes from an automorphism-group invariant invented by Connes **[9]** that produces an infinite number of nonisomorphic II_1 factors.

As this is written, we do not know the answer to the question of whether $\mathscr{L}_{\mathscr{F}_n}$ and $\mathscr{L}_{\mathscr{F}_m}$ are isomorphic when $n \neq m$.

§6. **States, weights, and traces.** A factor $\mathscr{M}$ of type I_n, with n finite, is isomorphic to the algebra $\mathscr{M}_n(\mathbf{C})$ of $n \times n$ complex matrices over $\mathbf{C}$. A key element of structure for $\mathscr{M}_n(\mathbf{C})$ (and $\mathscr{M}$) is the linear functional τ with the properties

$$\tau(AB) = \tau(BA) \quad (A, B \in \mathscr{M})$$
$$\tau(I) = 1.$$

We refer to τ as the *normalized trace* on $\mathscr{M}$. With the properties noted, τ is unique. In addition, τ takes non-negative real values at positive matrices. If we denote by $[a_{jk}]$ a matrix in $\mathscr{M}_n(\mathbf{C})$, where a_{jk} is the entry in row j and column k, then $\tau([a_{jk}])$ is $n^{-1}(\sum_{j=1}^{n} a_{jj})$.

A discovery that intrigued Murray and von Neumann greatly was the existence of a functional on a factor $\mathscr{M}$ of type II_1 with the main properties exhibited by the trace on $\mathscr{M}_n(\mathbf{C})$. They referred to this functional on $\mathscr{M}$ as the (normalized) trace. It was relatively easy to prove that τ is unique, and it was not difficult to determine the value that τ must assume at each element of $\mathscr{M}$. Proving that τ, so determined, is additive ($\tau(A + B) = \tau(A) + \tau(B)$) was quite another matter. The additivity of τ had Murray and von Neumann stopped throughout the development of their first paper [**27**]. They surmounted their difficulties and proved the additivity of τ in [**28**], their second paper. It was a very hard argument. (In conversation with this author, von Neumann remarked that "that was Murray.")

To define their trace, Murray and von Neumann proceeded in a measure-theoretic manner. With $\mathscr{M}$ a factor of type II_1, it can be shown that for each positive integer n and each projection E in $\mathscr{M}$ there are n equivalent mutually orthogonal projections in $\mathscr{M}$ with sum E. If we assign to I the measure (or "normalized dimension") 1 and use I in place of E, then each of the n equivalent projections should be assigned measure n^{-1}. Each projection in $\mathscr{M}$ is a (possibly infinite) sum of such (rational) projections, which provides it with a measure. There are, of course, obvious technical problems, but these can be overcome without great trouble. Murray and von Neumann arrived at a "dimension function" d that assigns to each projection E in $\mathscr{M}$ a number in $[0, 1]$. They noted that the range of d is precisely $[0, 1]$, and recognized that they were dealing with "continuous dimensionality." These matrix-like algebras in which the associated subspaces (projections) could assume a continuous range of dimensions fascinated Murray and von Neumann, as well they might.

From another point of view, the function d is a (noncommutative) measure on the "measurable" sets of (that is, projections in) $\mathscr{M}$ – and the unique such measure compatible with the equivalence relation on projections. It enjoys a Haar-measure-like status relative to the noncommutative measure space underlying $\mathscr{M}$. Examining $M_n(\mathbf{C})$, we see that $\tau(E)$ is the dimension of the subspace on which the projection E projects, normalized so that the full space has dimension 1. Thus in $\mathscr{M}$, $d(E)$ must be the value assigned as $\tau(E)$. The spectral theorem, which tells us, in effect, that each selfadjoint operator is a (limit of) linear combinations of mutually orthogonal projections ($A = \int \lambda \, dE_\lambda$), determines that value of τ at each selfadjoint operator in $\mathscr{M}$: ($\tau(A) = \int \lambda \, d\tau(E_\lambda)$). Finally, each T in $\mathscr{M}$ is a sum $A + iB$ where $A(= (T + T^*)/2)$ and $B(= (T - T^*)/2i)$ are selfadjoint. If τ is to be linear, we must define $\tau(T)$ as $\tau(A) + i\tau(B)$. As remarked, proving that $\tau(R + S) = \tau(R) + \tau(S)$ for each pair of operators R and S in $\mathscr{M}$ was a considerable challenge.

Murray and von Neumann defined a dimension function d on the projections of all factors by devices similar to those described for a factor of type II_1. They note that the type of the factor is completely determined by the range of the dimension function. In the case of a factor $\mathscr{M}$ of type I, with a (necessarily finite) minimal projection E they normalize the dimension function d so that $d(E) = 1$. In this case, d has range $\{0, 1, \dots, n\}$ when $\mathscr{M}$ is of type I_n, and range $\{0, 1, 2, 3, \dots\}$ when $\mathscr{M}$ is of type I_∞. If $\mathscr{M}$ is of type II_∞, there is no preferred finite projection on which to normalize the dimension function, whichever (equivalence class of a) finite projection is chosen to have dimension 1, the range of the dimension function is $[0, \infty]$. With $\mathscr{M}$ of type III, the range of the dimension function consists of just 0 and ∞.

Inspired by the construction of the Gelfand-Neumark article **[15]**, Segal **[41]** singled out special linear functionals on a C^*-algebra from which one can construct Hilbert spaces and adjoint-preserving homomorphisms (*representations*) of the C^*-algebra into the algebras of all bounded operators on the Hilbert spaces. (It was through such representations that Gelfand and Neumark built the * isomorphism of their Banach algebra with a C^*-algebra.) The functionals on which Segal focused have the property that they assume non-negative real values on positive operators on the C^*-algebra and assign the value 1 to I. He called such functionals *states*. (They correspond to the expectation functionals associated with states of quantum mechanical systems.) The states turn out to be norm continuous without further assumptions. As noted, the Murray-von Neumann trace τ on a II_1 factor and the normalized trace on $\mathscr{M}_n(\mathbf{C})$ are states.

If ρ is a state of the C^*-algebra $\mathscr{A}$ and we let $\langle A, B\rangle_\rho$ be $\rho(B^*A)$, then $\langle , \rangle_\rho$ is a positive (semidefinite) inner product on $\mathscr{A}$. As such, $<, >_\rho$ satisfies the Cauchy-Schwarz inequality. It follows from this that

$$\{A \in \mathscr{A} : \rho(A^*A) = 0\}(= \mathscr{L}_\rho)$$

is a (norm-closed) left ideal in $\mathscr{A}$ (the *left kernel* of ρ), for $B^*B \le \|B\|^2 I$ whence $A^*B^*BA \le \|B\|^2 A^*A$. The quotient vector space $\mathscr{A}/\mathscr{L}_\rho$ inherits the positive-definite inner product

$$\langle T + \mathscr{L}_\rho, S + \mathscr{L}_\rho\rangle = \langle T, S\rangle_\rho = \rho(S^*T) \quad (T, S \in \mathscr{A})$$

relative to which its completion $\mathscr{H}_\rho$ is a Hilbert space. If $\varphi(A)(T + \mathscr{L}_\rho)$ is defined to be $AT + \mathscr{L}_\rho$, the resulting linear operator on $\mathscr{A}/\mathscr{L}_\rho$ has norm not exceeding $\|A\|$ (relative to the Hilbert space norm on $\mathscr{A}/\mathscr{L}_\rho$); this operator extends, uniquely, to a bounded linear operator $\varphi(A)$ on $\mathscr{H}_\rho$. It is not difficult to verify that $\varphi(aA+B) = a\varphi(A)+\varphi(B)$, $\varphi(AB) = \varphi(A)\varphi(B)$, $\varphi(A^*) = \varphi(A)^*$, and $\varphi(I) = I$ (φ is a *representation* of $\mathscr{A}$ on $\mathscr{H}_\rho$). In addition, $\|\varphi(A)\| \le \|A\|$ and the image $\varphi(\mathscr{A})$ of $\mathscr{A}$ is norm-closed in $\mathscr{B}(\mathscr{H}_\rho)$ (so that $\varphi(\mathscr{A})$ is a C^*-algebra).

Through this construction, the state ρ, itself, has been "represented" in a special form:

$$\rho(A) = \langle \varphi(A)(I + \mathscr{L}_\rho), I + \mathscr{L}_\rho \rangle.$$

In this way, ρ "has become" $\omega_{x_\rho}|\varphi(\mathscr{A})$, the *vector state* ω_{x_ρ} of $\mathscr{B}(\mathscr{H}_\rho)$ restricted to the C^*-algebra $\varphi(\mathscr{A})$, where x_ρ is the vector $I + \mathscr{L}_\rho$ in $\mathscr{H}_\rho$ and $\omega_{x_\rho}(T) = \langle Tx_\rho, x_\rho \rangle$ for each T in $\mathscr{B}(\mathscr{H}_\rho)$. Note, too, that $\varphi(\mathscr{A})x_\rho$ is $\mathscr{A}/\mathscr{L}_\rho$, a dense subspace of $\mathscr{H}_\rho$. We say that x_ρ is a *generating* (or *cyclic*) *vector* for $\varphi(\mathscr{A})$. The vector states have a key place in the study of von Neumann algebras, a place whose importance became clear throughout the 1950s.

The construction associating a representation of a C^*-algebra with a state of that algebra, is basic to the study of C^*-algebras. It is known as the GNS construction (for Gelfand, Neumark, and Segal). It was not available, as such, to Murray and von Neumann though they had found a vector representation for the normalized trace τ on a factor $\mathscr{M}$ of type II_1. (See **[28**; Theorem II].) In the case of τ, the left kernel is (0). We speak of τ as a *faithful* state of $\mathscr{M}$; the corresponding representation is an isomorphism (is *faithful*) and τ is "represented" by a vector x_τ. From the basic property of the trace, we see that $\langle ABx_\tau, x_\tau \rangle = \langle BAx_\tau, x_\tau \rangle$ for all A and B in $\mathscr{M}$; we call such a vector x_τ a *trace vector* for $\mathscr{M}$. In any case, with $\mathscr{M}$ acting on a Hilbert space $\mathscr{H}$, Murray and von Neumann show that τ is a convex combination $\sum_{j=1}^n a_j\omega_j$ of vector states ω_j.

A von Neumann algebra $\mathscr{R}$ is more closely tied to a representation on a Hilbert space than a C^*-algebra need be. As a consequence, the states most relevant for the analysis of the structure of $\mathscr{R}$ have a certain continuity property related to the weak-operator topology on $\mathscr{R}$, which is associated with the action of $\mathscr{R}$ on its underlying Hilbert space. These are the states of $\mathscr{R}$, termed *normal*, that are weak-operator continuous on the unit ball of $\mathscr{R}$. It is fairly difficult to prove the following characterization of normal states.

THEOREM. *With $\mathscr{R}$ a von Neumann algebra acting on a Hilbert space $\mathscr{H}$, each normal state ω of $\mathscr{R}$ is a sum $\sum \omega_j$ of linear functionals $\omega_j(= \omega_{x_j}|\mathscr{R})$, where $\sum \|x_j\|^2 = 1$. (This sum may be (countably) infinite, in which case, $\sum_{j=1}^\infty \omega_j(A)$ converges to $\omega(A)$ for each A in $\mathscr{R}$.) If $\mathscr{R}$ has at least one faithful vector state, then each normal state of $\mathscr{R}$ is a vector state.*

If $\{E_n : n = 1,2,\dots\}$ is an orthogonal family of projections in the von Neumann algebra $\mathscr{R}$, then $\sum E_n x$ converges to a vector Ex for each x in the underlying Hilbert space $\mathscr{H}$. In this case, E is a projection in $\mathscr{R}$; we write $E = \sum_{n=1}^\infty E_n$. If ω is a normal state of $\mathscr{R}$, then $\omega(\sum E_n) = \sum \omega(E_n)$. Conversely, if the state ω satisfies the preceding equality for each orthogonal family of projections in $\mathscr{R}$, then ω is a normal state of $\mathscr{R}$. Thinking, again, of $\mathscr{R}$ as a noncommutative measure algebra, we see that the normal states of $\mathscr{R}$ must be viewed as the noncommutative integration processes on $\mathscr{R}$. A normal state of $\mathscr{R}$ restricted to the projections in $\mathscr{R}$ serves as a noncommutative (probability) measure on the (noncommutative) measurable sets of $\mathscr{R}$.

The inverse situation (given a "measure" on the projections of a von Neumann algebra, is there a corresponding integral—that is, normal state) was discussed by Mackey, in the case where the von Neumann algebra is $\mathscr{B}(\mathscr{H})$, at the end of the 1940s in connection with his rigorous development of the foundations of quantum mechanics [**24**]. It was answered in the affirmative (for $\mathscr{B}(\mathscr{H})$) when $\mathscr{H}$ has dimension greater than 2 by Gleason [**16**]. Many years elapsed before this fundamental question received a positive answer (largely by Christensen [**6**] and Yeadon [**50**]) for general von Neumann algebras (with no two-dimensional representations). When we note that this answer includes, in particular, passing from the dimension function on the projections of a factor of type II_1 to the normalized trace on the factor, some of the difficulty involved can be appreciated.

Of course, the noncommutative measure-theoretic interpretation of the theory of von Neumann algebras cannot rest with an analogue of finite measures; the infinite measure spaces must be understood as well. The replacement for states has been developed in that context. A *weight* ρ of a C^*-algebra $\mathscr{A}$ is a function on the positive elements in $\mathscr{A}$ taking non-negative real values and the value $+\infty$ and satisfying $\rho(A+B) = \rho(A)+\rho(B)$ and $\rho(tA) = t\rho(A)$ when A and B are positive operators in $\mathscr{A}$ and $t > 0$. By analogy with infinite measures, the analogues of the sets of finite measure, the positive integrable functions, and the absolutely square integrable functions have noncommutative analogues of special importance. Various left ideals arise from these and a construction akin to the GNS construction can be made giving rise to a representation corresponding to the given weight.

As with states, we can speak of *faithful weights* (if $\rho(A^*A) = 0$, then $A = 0$). If $\mathscr{A}$ is a von Neumann algebra $\mathscr{R}$ and the linear space consisting of the positive operators in $\mathscr{R}$ on which ρ assumes finite values is weak-operator dense in $\mathscr{R}$, we say that ρ is *semifinite*. For a von Neumann algebra $\mathscr{R}$, we are interested, as with states, in weights related to the action of $\mathscr{R}$ on the underlying Hilbert space. These weights ρ, called *normal*, have the property that there is a family $\{\rho_a\}$ of normal states ρ_a of $\mathscr{R}$ such that $\rho(A) = \sum_a \rho_a(A)$ for each positive A in $\mathscr{R}$.

Again, a trace-like functional on $\mathscr{B}(\mathscr{H})$ provides us with our most prominent example of a weight. Let $\{x_a\}$ be an orthonormal basis for $\mathscr{H}$ and define $\tau(A)$ to be $\sum_a \langle Ax_a, x_a\rangle$ for a positive A in $\mathscr{B}(\mathscr{H})$. In this case, τ is a faithful, normal, semifinite weight on $\mathscr{B}(\mathscr{H})$. In addition, $\tau(AA^*) = \tau(A^*A)$ for each A in $\mathscr{B}(\mathscr{H})$. We say that τ, with this property, is a *tracial weight* (on $\mathscr{B}(\mathscr{H})$).

The existence of a semifinite, faithful, normal, tracial weight on a factor $\mathscr{M}$ entails the existence of a nonzero, finite projection in $\mathscr{M}$. Thus $\mathscr{M}$ admits no such tracial weight if it is of type III. If $\mathcal{M}$ is of type II_∞, there is an orthogonal family $\{E_a\}$ of (equivalent) finite projections in $\mathcal{M}$ with sum I. The family $E_a\mathcal{M}E_a$ of operators of the form E_aTE_a $(T \in \mathcal{M})$ is a von Neumann algebra. It is not difficult to show that $E_a\mathcal{M}E_a$ (acting on $E_a(\mathcal{H})$) is a factor $\mathcal{M}_a$ of type II_1. With τ_a the normalized trace on $\mathcal{M}_a$, the function τ on $\mathcal{M}$ that assigns $\sum_a \tau_a(A)$ to each positive A in $\mathcal{M}$ is a faithful, normal, semifinite, tracial weight on $\mathcal{M}$ (the unique such weight up to positive multiples).

The fact that a factor that admits no normal, semifinite tracial weight must be of type III is the key to von Neumann's argument that some of the factors described in Section 5 are of type III. In the notation of the general construction at the beginning of Section 5, we can view the elements of $\mathcal{R}$ as matrices, indexed by elements of G, with entries from $\mathcal{B}(\mathcal{H})$. The entry at the g, h position is $U(gh^{-1})A(gh^{-1})$, where $g' \to A(g')$ is a mapping from G into $\mathcal{A}$; the matrix should represent an element of $\mathcal{B}(\mathcal{K})$. Let η be the mapping that assigns to this element (matrix) of $\mathcal{R}$ the element with matrix all of whose entries are 0 except for the diagonal entries and these are all equal to $A(e)$. (In other words, η is the mapping that changes all off-diagonal entries to 0.) Several properties of η are noteworthy: $\eta(I) = I, \eta(H) \geq 0$ if $H \geq 0$, and $\eta(ATB) = A\eta(T)B$ when $A, B \in \mathcal{A}$. The mapping η is the noncommutative analogue of a conditional expectation (from $\mathcal{R}$ onto $\Phi(\mathcal{A})$) and is referred to as a *conditional expectation.* If ρ_0 is a normal, semifinite weight on $\mathcal{A}$ that satisfies

$$\rho_0(A) = \rho_0(U(g)AU(g)^*) \quad (A \in \mathcal{A}),$$

then $\rho(T)$ defined as $\rho_0(\Phi^{-1}(\eta(T)))$ yields the normal, semifinite, tracial weight ρ on $\mathcal{R}$. Moreover, each such tracial weight on $\mathcal{R}$ arises in this way. The more difficult part of the proof of this assertion is the argument establishing that $\rho_0(A)$, defined as $\rho(\Phi(A))$ for each A in $\mathcal{A}$, gives rise to a *semifinite* weight ρ_0 on $\mathcal{A}$ when the normal, semifinite, tracial weight ρ on $\mathcal{R}$ is given.

A normal, semifinite weight on $\mathcal{A}$ with the given group-invariance property corresponds to a group-invariant measure on the underlying measure space, absolutely continuous with respect to the underlying measure, in the situation of the specific examples of Section 5 constructed from a measure space and a group acting on it. In some of those examples, there is no such invariant measure, from which we must conclude that the factors constructed in those cases admit no normal, semifinite, tracial weight and are, accordingly, of type III.

§7. **Von Neumann algebras and their commutants.** In [27], Murray and von Neumann established that a factor of type I, II, or III has commutant a factor of the corresponding type, though a factor of type II_1 or II_∞ may have a factor of type II_∞ or II_1, respectively, as commutant. From the type I case, where it seems most natural to choose the representation of the factors as $\mathcal{B}(\mathcal{H})$, with $\mathcal{H}$ of the appropriate dimension, we might conclude that the "most natural" representation of a factor has as small a commutant as possible. With $\mathcal{M}$ of type II_1 or of type III, there is not much that can be sensibly effected in that direction. Some further reflection leads to the conclusion that the most "standard" representation to choose for a von Neumann algebra is one in which it and its commutant have the "same size." Just what that means is most simply illustrated in the case of a factor $\mathcal{M}$ of type I_n. In this case, $\mathcal{M}$ is isomorphic to $\mathcal{B}(\mathcal{H})$ with $\mathcal{H}$ of dimension n. Let $\mathcal{K}$ be the n-fold direct sum of $\mathcal{H}$ with itself. We can view $\mathcal{B}(\mathcal{K})$ as $n \times n$ matrices with entries from $\mathcal{B}(\mathcal{H})$. The algebra of matrices with the same element of $\mathcal{B}(\mathcal{H})$ at each diagonal entry and 0 at all off-diagonal positions is a factor isomorphic to $B(\mathcal{H})$ and to $\mathcal{M}$. Its commutant consists of those matrices all of whose entries are scalar multiples of I (in $\mathcal{B}(\mathcal{H})$); it, too, is a factor of type I_n. For $\mathcal{M}$ of type I_∞, $\mathcal{H}$ is infinite dimensional; we restrict ourselves to the separable case with the dimension of $\mathcal{H}$ the cardinality of the natural numbers. Following the same procedure as in the type I_n case, but paying attention to convergence, we arrive at a representation of $\mathcal{M}$ with commutant $\mathcal{M}'$ of type I_∞.

For the "standardization" we are after, it no longer suffices, in the case of a factor of type II_1, to work toward a representation in which the commutant is of type II_1. We notice, however, that in the type I case, "standardization" is reached (recall, we are in the *separable* case) precisely when there is a vector x in the representing Hilbert space $\mathcal{H}$ such that $\mathcal{M}x$ and $\mathcal{M}'x$ are both dense, linear subspaces of $\mathcal{H}$. When $\mathcal{M}$ is of type III, it is automatically represented in standard form (separable case): when $\mathcal{M}$ is of type II a standard representation can be arranged by a process of forming direct sums and "copying" on the diagonal (as we did in the type I_n case) and restricting $\mathcal{M}$ to the range of a projection in $\mathcal{M}'$.

The importance of studying $\mathcal{M}$ in its standard representation for understanding its possible action on a space can be seen from the following result (known as the Unitary Implementation Theorem).

THEOREM. *If $\mathcal{R}_1$ and $\mathcal{R}_2$ are von Neumann algebras acting on Hilbert spaces $\mathcal{H}_1$ and $\mathcal{H}_2$, respectively, in standard form and φ is an isomorphism of $\mathcal{R}_1$ onto $\mathcal{R}_2$, then there is a unitary transformation U of $\mathcal{H}_1$ onto $\mathcal{H}_2$ such that $\varphi(A) = UAU^{-1}$ for each A in $\mathcal{R}_1$.*

We say that U of this theorem *implements* the isomorphism φ. the uniqueness of the standard form of a von Neumann algebra (up to unitary equivalence) is assured by the Unitary Implementation Theorem.

When we discussed the GNS construction applied to the normalized trace τ on a factor $\mathcal{M}$ of type II_1, we arrived at a generating trace vector x_τ for $\mathcal{M}$. It follows (easily) from the fact that τ is a faithful state of $\mathcal{M}$ that x_τ is generating for $\mathcal{M}'$ as well. Thus, as represented, $\mathcal{M}$ is in standard form. Murray and von Neumann noted that, if $\mathcal{M}$ is a factor and x_τ is a generating trace vector for $\mathcal{M}$, then x_τ is a generating trace vector for $\mathcal{M}'$. Moreover, for each A in $\mathscr{M}$, there is a unique A' in $\mathscr{M}'$ such that $Ax_\tau = A'x_\tau$. Observing that

$$(AB)x_\tau = AB'x_\tau = B'Ax_\tau = (B'A')x_\tau,$$

they show that $A \to A'$ is an (adjoint-preserving) anti-isomorphism of $\mathscr{M}$ onto $\mathscr{M}'$. They were able to extend this result, using the techniques they had developed, to all factors of types I or II; in standard form, each is * anti-isomorphic to its commutant. Factors of type III were beyond their reach. They raised this question for all factors (indeed, for all von Neumann algebras in standard form).

This problem remained open until 1967 when M. Tomita, completing years of arduous research announced his affirmative solution [**47**]. This occurred at a general conference on the theory of operator algebras and its applications held at Louisiana State University in Baton Rouge, Louisiana (in honor of J. Dixmier who was visiting Tulane University in New Orleans that year). This same conference was attended by several of the physicists who had made great strides in applying operator algebras to the study of quantum mechanical systems with infinitely many degrees of freedom (cf. [**1, 4, 13, 19, 20, 49**]). In particular, R. Haag, N. Hugenholtz, and M. Winnink, who had just completed a penetrating analysis of the conditions for a state to be an equilibrium state of a quantum statistical mechanical system in the framework of the theory of operator algebras [**18**], were present.

A large contingent of the powerful Japanese Functional Analysis school was also present, among them M. Takesaki, an important young contributor to the subject of operator algebras, who was to become one of the great leaders of the subject. Takesaki listened carefully to Haag, Hugenholtz, and Winnink, and, later, during a yearlong visit to the University of Pennsylvania (1968–1969), refined the Tomita work and fused it with the equilibrium-state analysis of [**18**] to create the Tomita-Takesaki Modular Theory [**44**], the dominant theme in research on von Neumann algebras throughout the 1970s.

The modular theory of Tomita-Takesaki has as its basic ingredients a von Neumann algebra $\mathscr{R}$ and a normal, faithful, semifinite weight ρ. As discussed, the analogue for weights of the GNS construction produces a representation of $\mathscr{R}$ on a Hilbert space $\mathscr{H}_\rho$. Although no vector represents the weight (by contrast with the case of a state), the assumption about the weight yields a faithful representation of $\mathscr{R}$ as a von Neumann algebra in "standard form," in the sense of having a commutant of the "same size" as the algebra, although there may be no generating vector for both the representing algebra and its commutant. While each von Neumann algebra has a faithful, normal, semifinite weight, the existence of a faithful, normal state imposes a countability restriction on the von Neumann algebra (each orthogonal family of projections in the algebra is countable) that is automatically fulfilled for algebras acting on a separable Hilbert space. Even when the von Neumann algebra satisfies this countability restriction, it is advantageous, for certain purposes, to have the theory available for weights as opposed to states. Nevertheless, the main features of the theory can be described in terms of a faithful, normal *state*, and this simplifies the description considerably.

Suppose ω is a faithful, normal state of $\mathscr{R}$. The GNS construction applied to ω gives rise to a faithful representation φ of $\mathscr{R}$ on a Hilbert space $\mathscr{H}$ and a (unit) vector u generating for $\varphi(\mathscr{R})$ and $\varphi(\mathscr{R})'$ such that $\omega(A) = \langle\varphi(A)u, u\rangle$ for each A in $\mathscr{R}$. To simplify our notation, we write $\mathscr{R}$ for $\varphi(\mathscr{R})$ and ω for the restriction of the vector state ω_u to $\varphi(\mathscr{R})$. Tomita's key idea is to analyze the adjoint operations on $\mathscr{R}$ and $\mathscr{R}'$, reducing them to the Hilbert space level through the use of the vector u. Define S_0Au to be A^*u and $F_0A'u$ to be A'^*u for each A in $\mathscr{R}$ and A' in $\mathscr{R}'$. The mappings S_0 and F_0 are conjugate-linear ($S_0(ax + y) = \bar{a}S_0x + S_0y$) and defined only on the dense subspaces $\mathscr{R}u$ and $\mathscr{R}'u$, respectively. In general, they are not even bounded—a very unpromising start! Some simple calculation reveals that S_0 and F_0 behave, to a great extent, as adjoints of one another. Combined with this, the fact that they are densely defined implies that they have "closures" S and F, respectively. While not necessarily bounded, "closed operators" (those whose graphs are closed) that are densely defined are susceptible to some Hilbert space analysis. In particular, S has a polar decomposition $J\Delta^{\frac{1}{2}}$, as described in Section 4 for T in $\mathscr{B}(\mathscr{H})$, where $\Delta(= S^*S)$ is a (generally, unbounded) selfadjoint operator. Since S is conjugate-linear, densely defined, and $S^2Au = Au$ for each A in $\mathscr{R}$, J is a conjugate-linear, isometric mapping of $\mathscr{H}$ onto $\mathscr{H}$ such that $J^2 = I$. The main result of Tomita, stated in the notation we have developed, follows.

THEOREM. *For each A in $\mathscr{R}$ and each real t, $\Delta^{it}A\Delta^{-it} \in \mathscr{R}$ and $J\mathscr{R}J = \mathscr{R}'$.*

The notation $\Delta^{it}(= e^{it\log\Delta})$ indicates spectral theory and the function calculus applied to the (possibly unbounded) selfadjoint operator Δ, so that Δ^{it} is a unitary operator U_t. The mapping $t \to U_t$ is a homomorphism of the additive groups of $\mathbf{R}$ into the group of unitary operators on $\mathscr{H}$. Moreover, this mapping is continuous from $\mathbf{R}$ in its standard topology to the unitary group in the weak-operator topology. We refer to $t \to U_t$ as a *one-parameter unitary group* (on $\mathscr{H}$). The condition, $\Delta^{it}A\Delta^{-it} \in \mathscr{R}$, implies that $A \to U_tAU_{-t}$ is an (adjoint-preserving) automorphism σ_t^ω of $\mathscr{R}$. The mapping $t \to \sigma_t^\omega$ is, again, a homomorphism of $\mathbf{R}$, this time into the group of automorphisms of $\mathscr{R}$; it is a special one-parameter group of automorphisms of $\mathscr{R}$. We refer to it as the *modular group* of $\mathscr{R}$ corresponding to the (faithful, normal) state ω.

The mapping $A \to JA^*J$ is an (adjoint-preserving) anti-isomorphism of $\mathscr{R}$ onto $\mathscr{R}'$ by virtue of the equality, $J\mathscr{R}J = \mathscr{R}'$. This answers the question of Murray and von Neumann in the affirmative: If $\mathscr{R}$ is a von Neumann algebra in standard form, then $\mathscr{R}$ and $\mathscr{R}'$ are * anti-isomorphic. When $\mathscr{R}$ is a factor of type II_1 and ω is the normalized trace τ on $\mathscr{R}$, $\Delta = I$ and the mapping $A \to JA^*J$ is the mapping, $A \to A'$, where $Ax_\tau = A'x_\tau$ with A in $\mathscr{R}$, A' in $\mathscr{R}'$, and x_τ a trace vector for $\mathscr{R}$ and $\mathscr{R}'$ used to define S and F.

In the process of answering the Murray-von Neumann question, Tomita discovered the modular automorphism group; it has come to be recognized as an important tool in studying the structure of factors of type III. Takesaki's discussions with Haag, Hugenholtz, and Winnink (HHW) at the 1967 Baton Rouge conference led him to the crucial discovery that the modular automorphism group σ_t^ω and ω are related by the condition HHW described for a quantum statistical mechanical system whose (bounded) observables correspond to the selfadjoint elements in $\mathscr{R}$ and whose dynamical evolution has ω as its equilibrium state. (This comment makes only a fleeting allusion to the dynamical evolution-equilibrium state relation; temperature and other parameters must also be discussed in a more careful formulation.)

DEFINITION. A one-parameter group of (adjoint-preserving) automorphisms, $t \to \alpha_t$, of a von Neumann algebra $\mathscr{R}$ satisfies the *modular condition* relative to ω when, for each A and B in $\mathscr{R}$, there is a function F continuous and bounded on the (closed) strip $\{z \in \mathbf{C} : 0 \le \operatorname{Im} z \le 1\}$ holomorphic on the interior ($\{z \in \mathbf{C} : 0 < \operatorname{Im} z < 1\}$), and satisfying

$$F(t) = \omega(\alpha_t(A)B), \qquad F(t+i) = \omega(B\alpha_t(A))$$

for all real t.

THEOREM. *The modular automorphism group corresponding to ω is the unique one-parameter group of automorphisms of $\mathscr{R}$ that satisfies the modular condition relative to ω.*

The modular condition is called "the KMS boundary condition" by HHW, for Kubo, Martin, and Schwinger, who describe the corresponding condition in another context (in the more physics-oriented literature); ω is said to satisfy the KMS boundary condition relative to σ_t^ω. It should be remarked that neither Tomita's theorem nor the uniqueness theorem just stated are easy to prove; the ingredients of the arguments are spectral theory, von Neumann algebra theory, Fourier transforms, and complex variables.

In a major contribution to modular theory, A. Connes [7] analyzes the constructs of the theory in greater detail providing many new and powerful devices for classifying factors of type III. He notes that if ω_1 and ω_2 are faithful, normal states of $\mathscr{R}$ and $t \to \sigma_t^{(1)}$, $t \to \sigma_t^{(2)}$ are the corresponding modular automorphism groups, then for each t, $\sigma_t^{(1)}$ and $\sigma_t^{(2)}$ differ by an "inner automorphism" of $\mathscr{R}$, that is, there is a unitary operator U_t in $\mathscr{R}$ such that $\sigma_t^{(1)}(A) = U_t\sigma_t^{(2)}(A)U_t^*$ for each A in $\mathscr{R}$. In this case, $t \to U_t$, need not be a homomorphism of $\mathbf{R}$ into the unitary group of $\mathscr{R}$. It does satisfy the "cocycle" condition, $U_{s+t} = U_s\sigma_s^{(2)}(U_t)$ for each pair of real numbers s and t. (The mapping $t \to U_t$ is known as the *Connes cocycle*.)

Connes denotes by ε the quotient homomorphism of the group of (adjoint-preserving) automorphisms of $\mathscr{R}$ onto its quotient $\operatorname{Out}(\mathscr{R})$ by the (normal) subgroup of inner automorphisms of $\mathscr{R}$. From the preceding discussion, $\varepsilon(\sigma_t^{(1)}) = \varepsilon(\sigma_t^{(2)})$ for each real t. Thus, $t \to \varepsilon(\sigma_t^{(1)})$ defines a homomorphism, that Connes denotes by δ of $\mathbf{R}$ into $\mathrm{Out}(\mathscr{R})$. The mapping δ is called the *modular homomorphism* of $\mathscr{R}$; it is "invariantly" associated with $\mathscr{R}$. The kernel $T(\mathscr{R})$ of δ, a subgroup of $\mathbf{R}$, is an algebraic invariant of $\mathscr{R}$. When $\mathscr{R}$ is a factor of type I or II (that is, when $\mathscr{R}$ is *semifinite*), $T(\mathscr{R}) = \mathbf{R}$; each modular automorphism is inner. In the separable case, if each modular automorphism is inner the factor is semifinite.

Connes [7] defines another invariant for a factor $\mathscr{M}$. He denotes by $S(\mathscr{M})$ the intersection, over the faithful, normal weights ρ on $\mathscr{M}$, of the spectra of the operators Δ_ρ. The condition that $\mathscr{M}$ is not of type III is equivalent to the condition that $0 \notin S(\mathscr{M})$, in which case, $S(\mathscr{M}) = \{1\}$. In any event, $S(\mathscr{M})\backslash\{0\}$ is a closed subgroup **[11]** of the multiplicative group $\mathbf{R}_+^*$ of positive reals. Aside from $\{1\}$, the possibilities for this subgroup are $\mathbf{R}_+^*$ itself and $\{\lambda^n : n \in \mathbf{Z}\}$, where $\mathbf{Z}$ is the group of integers and $\lambda \in (0, 1)$. The factor $\mathscr{M}$ is said to be of type III_λ when the subgroup is the cyclic group with generator λ in $(0, 1)$; it is of type III_1 when the group is $\mathbf{R}_+^*$; it is of type III_0 when $S(\mathscr{M})$ consists of 0 and 1. In the final section, we shall see examples of these factors.

We conclude this section with a relation, established by Connes [7], between $S(\mathscr{M})$ and $T(\mathscr{M})$ for a factor $\mathscr{M}$ not of type III_0. In this case, $T(\mathscr{M})$ is the kernel of the homomorphism $t \to \chi_t$ of $\mathbf{R}$ into the dual of $S(\mathscr{M})\backslash\{0\}$, where $\chi_t(\lambda) = \lambda^{it}$. Thus $T(\mathscr{M}) = (-2\pi/\log\lambda)\mathbf{Z}$, when $\mathscr{M}$ is of type III_λ.

§8. **Crossed products.** In Section 5, we described a construction of factors arising from a homomorphism α of a discrete group G into the automorphism group of an abelian von Neumann algebra $\mathscr{A}$. That construction can be extended to the case where $\mathscr{A}$ is replaced by an arbitrary von Neumann algebra $\mathscr{R}$; everything remains the same, with $\mathscr{R}$ in place of $\mathscr{A}$. The resulting von Neumann algebra $\mathscr{R} \otimes \alpha$ is called the *crossed product* of $\mathscr{R}$ by G. In terms of this extended construction, Connes [7] presents an explicit "decomposition" of the factors of type III_λ, $\lambda \neq 1$.

THEOREM. *If $\mathscr{M}$ is a factor of type III_λ, ($\lambda \in (0, 1)$), there is a factor $\mathscr{N}$ of type II_∞ and an automorphism α of $\mathscr{N}$ for which $tr \circ \alpha = \lambda tr$, where tr is the (infinite) trace on $\mathscr{N}$, such that $\mathscr{M}$ is $\mathscr{N} \otimes_\tau \mathbf{Z} (= \mathscr{N} \otimes \alpha)$, where $\alpha(n) = \alpha^n$. Conversely, such a crossed product is a factor of type III_λ. Moreover, if $(\mathscr{N}_1, \alpha_1)$ and $(\mathscr{N}_2, \alpha_2)$ are two such pairs corresponding to λ in $(0, 1)$, then $\mathscr{N}_1 \otimes \alpha_1$ is isomorphic to $\mathscr{N}_2 \otimes \alpha_2$ if and only if there is an isomorphism of $\mathscr{N}_1$ onto $\mathscr{N}_2$ carrying α_1 onto $\alpha_2 \circ \beta$ where β is some inner automorphism of $\mathscr{N}_2$.*

The type III_0 factors are also crossed products—but this time of a II_∞ von Neumann algebra (not a factor) by α. The factors of type III_1 elude the (discrete) crossed product analysis of Connes. Takesaki **[45]** encompassed these factors in a crossed product description by extending the crossed product construction to apply to locally compact groups (especially **R**) and defining a "dual action" of these groups in the abelian case. This culminated in the celebrated Takesaki Duality Theorem. Suppose τ is a "continuous action" of the locally compact group G on a von Neumann algebra $\mathscr{R}$ acting on a Hilbert space $\mathscr{H}$; that is, τ is a homomorphism of G into the group of (adjoint-preserving) automorphisms of $\mathscr{R}$ such that $g \to \tau_g(A)x$ is a continuous mapping of G into $\mathscr{H}$ for each A in $\mathscr{R}$ and x in $\mathscr{H}$. An inessential change of representation for $\mathscr{R}$ allows us to assume that τ_g is implemented by a unitary operator U_g on $\mathscr{H}$. The crossed product $\mathscr{R} \otimes \tau$ acts on $\mathscr{H} \otimes L_2(G)$ (the space of square-integrable mappings of G, with Haar measure, into $\mathscr{H}$) and is generated by the operators $A \otimes I$ and $U_g \otimes L_g$, where L_g is the left-translation unitary operator on $L_2(G)((L_g\xi)(g') = \xi(g^{-1}g'))$.

Turning to the case of the action of **R** on a von Neumann algebra $\mathscr{R}$ by a modular automorphism group, we let ρ be a faithful normal weight on the von Neumann algebra $\mathscr{R}$ with $t \to \sigma_t$ its corresponding modular group. Then $t \to \sigma_t$ is a continuous action of **R** on $\mathscr{R}$. If $\mathscr{R}$ is of type III, $\mathscr{R} \otimes \sigma$ is of type II_∞. Define $(U_t\xi)(s)$ to be $e^{-its}\xi(s)$: $(\xi \in L_2(\mathbf{R}))$. Then $I \otimes U_t$ induces an automorphism $\hat{\sigma}_t$ of $\mathscr{R} \otimes \sigma$ and $t \to \hat{\sigma}_t$ is a continuous action (the action *dual* to σ) of $\hat{\mathbf{R}}$ on $\mathscr{R} \otimes \sigma$.

THEOREM (TAKESAKI DUALITY). *If $\mathscr{R}$ is of type III, $(\mathscr{R} \otimes \sigma) \otimes \hat{\sigma}$ is isomorphic to $\mathscr{R}$.*

Each type III_1 factor is a crossed product of a type II_∞ factor with **R**.

§**9. Tensor Products.** Murray and von Neumann [**27**] introduce a construction involving several Hilbert spaces that produces another Hilbert space, their *tensor product*. With the aid of this construction, it is possible to combine von Neumann algebras to form a *tensor product of von Neumann algebras*. The investigation of tensor products has had an important influence on the understanding we have of von Neumann algebras.

From one point of view, there is no difficulty in defining the algebraic tensor products (over **C**) of two von Neumann algebras $\mathscr{R}$ and $\mathscr{S}$; the problem arises in providing the algebra with an appropriate norm and completing. If $\mathscr{R}$ and $\mathscr{S}$ act on the Hilbert spaces $\mathscr{H}$ and $\mathscr{K}$, respectively, and we form the tensor product $\mathscr{H} \otimes \mathscr{K}$ in the style of Murray and von Neumann, there is a natural candidate for $\mathscr{R} \bar{\otimes} \mathscr{S}$, the von Neumann algebra tensor product of $\mathscr{R}$ and $\mathscr{S}$. To form that tensor product, observe that $\mathscr{R}$ and $\mathscr{S}$ have (natural) isomorphic images as von Neumann algebras $\mathscr{R}_0$ and $\mathscr{S}_0$ acting on $\mathscr{H} \otimes \mathscr{K}$. The von Neumann algebra generated by $\mathscr{R}_0$ and $\mathscr{S}_0$ is the one we choose for $\mathscr{R} \bar{\otimes} \mathscr{S}$. Other possibilities that present themselves as a tensor product of $\mathscr{R}$ and $\mathscr{S}$ (though not von Neumann algebras) are $\mathscr{R} \otimes_a \mathscr{S}$, the algebra generated by $\mathscr{R}_0$ and $\mathscr{S}_0$, and $\mathscr{R} \otimes \mathscr{S}$, the C^*-algebra generated by $\mathscr{R}_0$ and $\mathscr{S}_0$.

The tensor product $\mathscr{H} \otimes \mathscr{K}$ is best described, for accurate mathematical purposes, in terms of functionals on the spaces $\mathscr{H}$ and $\mathscr{K}$. For our purposes, and most easily described, we can think of the basic elements of $\mathscr{H} \otimes \mathscr{K}$ as the "simple tensors" $x \otimes y$, with x in $\mathscr{H}$ and y in $\mathscr{K}$, and the algebraic tensor product $\mathscr{H} \otimes_a \mathscr{K}$ as being the linear span of these. Certain identifications must be made in accordance with such rules as $(ax+x')\otimes y = a(x\otimes y)+x'\otimes y$. Providing $\mathscr{H} \otimes_a \mathscr{K}$ with an inner product defined by

$$\langle x \otimes y, z \otimes w\rangle = \langle x, z\rangle\langle y, w\rangle$$

and the associated norm, we can complete $\mathscr{H} \otimes_a \mathscr{K}$ relative to the norm to produce $\mathscr{H} \otimes \mathscr{K}$, the desired Hilbert-space tensor product.

A major purpose to constructing the tensor product $\mathscr{H} \otimes \mathscr{K}$ is to develop a tool for analyzing multilinear mappings from $\mathscr{H} \times \mathscr{K}$, the Cartesian product of $\mathscr{H}$ and $\mathscr{K}$, into some Hilbert space $\mathscr{L}$. Suppose φ is such a mapping (so that $x \to \varphi(x, y_0)$ and $y \to \varphi(x_0, y)$ are linear mappings of $\mathscr{H}$ and $\mathscr{K}$ into $\mathscr{L}$, respectively, for each fixed y_0 in $\mathscr{K}$ and x_0 in $\mathscr{H}$). If no topological assumptions are made about φ, the appropriate tensor product is $\mathscr{H} \otimes_a \mathscr{K}$; there is a unique *linear* mapping η of $\mathscr{H} \otimes_a \mathscr{K}$ into $\mathscr{L}$ with the property that $\varphi(x,y) = \eta(x \otimes y)$ for all x in $\mathscr{H}$ and y in $\mathscr{K}$. Moreover, $\mathscr{H} \otimes_a \mathscr{K}$ is the unique linear space with this "factorization" property. If φ is required to satisfy certain continuity restrictions (a boundedness condition among them), there is a unique *bounded* linear mapping η of $\mathscr{H} \otimes_a \mathscr{K}$ into $\mathscr{L}$ with the property that $\varphi(x,y) = \eta(x \otimes y)$ for all x in $\mathscr{H}$ and y in $\mathscr{K}$.

If A is a bounded linear operator on $\mathscr{H}$, there is a unique linear operator $A \otimes I$ on $\mathscr{H} \otimes \mathscr{K}$, with the same bound as A, determined by $(A \otimes I)(x \otimes y) = (Ax) \otimes y$ for each x in $\mathscr{H}$ and y in $\mathscr{K}$. Similarly, with B a bounded linear operator on $\mathscr{K}$, we can define $I \otimes B$ on $\mathscr{H} \otimes \mathscr{K}$. The algebras $\mathscr{R}_0$ and $\mathscr{S}_0$, alluded to before, are the von Neumann algebras $\{A \otimes I : A \in \mathscr{R}\}$ and $\{I \otimes B : B \in \mathscr{S}\}$. Each element of $\mathscr{R}_0$ commutes with all elements of $\mathscr{S}_0$ (that is, $\mathscr{R}_0 \subseteq \mathscr{S}_0'$). Despite the apparent dependence of $\mathscr{R}\overline{\otimes}\mathscr{S}$ on the underlying Hilbert spaces $\mathscr{H}$ and $\mathscr{K}$, if $\mathscr{R}$ is isomorphic to $\mathscr{R}_1$ and $\mathscr{S}$ is isomorphic to $\mathscr{S}_1$, then $\mathscr{R}\overline{\otimes}\mathscr{S}$ is isomorphic to $\mathscr{R}_1\overline{\otimes}\mathscr{S}_1$. Of course, these constructions can be iterated to yield tensor products of any finite number of Hilbert spaces or von Neumann algebras; this iterated process is "associative."

In **[27]**, Murray and von Neumann observe that a factor $\mathscr{M}$ acting on a Hilbert space $\mathscr{H}$ and its commutant $\mathscr{M}'$ generate an algebra $\mathscr{B}$ isomorphic to $\mathscr{M} \otimes_a \mathscr{M}'$. By virtue of the Double Commutant Theorem, the von Neumann algebra generated by $\mathscr{M}$ and $\mathscr{M}'$ (the weak-operator closure of $\mathscr{B}$) is $\mathscr{B}(\mathscr{H})$. Murray and von Neumann note that $\mathscr{M}$ and $\mathscr{M}'$ generate a von Neumann algebra isomorphic to $\mathscr{M}\overline{\otimes}\mathscr{M}'$ just in the case where $\mathscr{M}$ is of type I.

In **[D]**, Dixmier poses the problem of determining the commutant $(\mathscr{R}\overline{\otimes}\mathscr{S})'$ of the tensor product $\mathscr{R}\overline{\otimes}\mathscr{S}$. In the finite-dimensional case, it is easy to establish the formula

$$(\mathscr{R}\overline{\otimes}\mathscr{S})' = \mathscr{R}'\overline{\otimes}\mathscr{S}'.$$

It is a simple matter to establish the inclusion

$$\mathscr{R}'\overline{\otimes}\mathscr{S}' \subseteq (\mathscr{R}\overline{\otimes}\mathscr{S})'.$$

The validity of the reverse inclusion was an open question for a number of years. The difficult case occurs when $\mathscr{R}$ and $\mathscr{S}$ are factors of type III (the other cases were settled in several articles). The existence of a * anti-isomorphism between a von Neumann algebra and its commutant in standard form is basic to the problem. The tensor-product, commutant formula was established by Tomita's theorem [**44, 47**].

If $\mathscr{R}$ and $\mathscr{S}$ are factors, $\mathscr{R}\bar{\otimes}\mathscr{S}$ is a factor. The type of $\mathscr{R}\bar{\otimes}\mathscr{S}$ is certainly determined by the types of $\mathscr{R}$ and $\mathscr{S}$; if either is of infinite type, so is the product, and if both are of type I, so is the product. When $\mathscr{R}$ and $\mathscr{S}$ are of type II, the same is true of $\mathscr{R}\bar{\otimes}\mathscr{S}$. When both $\mathscr{R}$ and $\mathscr{S}$ are of type III, the type of $\mathscr{R}\bar{\otimes}\mathscr{S}$ remained unclarified until Sakai [**37**] proved that $\mathscr{R}\bar{\otimes}\mathscr{S}$ is a factor of type III.

It is also of interest to define a tensor product for C^*- algebras. The right choices are more involved in this case. Once again, we can consider the C^*-algebras $\mathscr{A}$ and $\mathscr{B}$ as acting on the Hilbert spaces $\mathscr{H}$ and $\mathscr{K}$, respectively, form the C^*- algebras $\{A \otimes I : A \in \mathscr{A}\}$ and $\{I \otimes B : B \in \mathscr{B}\}$ on $\mathscr{H} \otimes \mathscr{K}$ and the C^*-algebra $\mathscr{A} \otimes \mathscr{B}$ generated by these C^*-algebras. If $\mathscr{A}_1$ and $\mathscr{B}_1$ are C^*-algebras acting on Hilbert spaces $\mathscr{H}_1$ and $\mathscr{K}_1$, respectively, and we form $\mathscr{A}_1 \otimes \mathscr{B}_1$, then $\mathscr{A} \otimes \mathscr{B}$ is * isomorphic to $\mathscr{A}_1 \otimes \mathscr{B}_1$ when $\mathscr{A}$ and $\mathscr{B}$ are * isomorphic to $\mathscr{A}_1$ and $\mathscr{B}_1$, respectively. The algebraic tensor product $\mathscr{A} \otimes_a \mathscr{B}$ of the C^*-algebras $\mathscr{A}$ and $\mathscr{B}$ is * isomorphic to the algebra generated by $\mathscr{A}_1$ and $\mathscr{B}_1$ (as a consequence of the results in [**27**]). Thus the norm on $\mathscr{A} \otimes \mathscr{B}$ induces on $\mathscr{A} \otimes_a \mathscr{B}$ a norm with the properties that $\mathscr{A} \otimes_a \mathscr{B}$ becomes a normed algebra and, for each T in $\mathscr{A} \otimes_a \mathscr{B}$, $\|T^*T\| = \|T\|^2$. A norm on $\mathscr{A} \otimes_a \mathscr{B}$ with these properties is called a *C^*-norm*. The completion of $\mathscr{A} \otimes_a \mathscr{B}$ relative to a C^*-norm is a reasonable choice as the C^*-algebra tensor product of $\mathscr{A}$ and $\mathscr{B}$. The question arises, of course, as to whether or not all these choices are the same; equivalently, is there just one C^*-norm on $\mathscr{A} \otimes_a \mathscr{B}$. In general, there are many C^*-norms on $\mathscr{A} \otimes_a \mathscr{B}$, among them the *spatial norm* σ, described before, obtained from faithful representations of $\mathscr{A}$ and $\mathscr{B}$ on Hilbert spaces. It is a nontrivial fact that σ is the smallest of all the C^*-norms on $\mathscr{A} \otimes_a \mathscr{B}$. Each C^*-norm α is a *cross norm*; that is, $\alpha(A \otimes B) = \alpha(A)\alpha(B)$ for all A in $\mathscr{A}$ and B in $\mathscr{B}$. These properties of tensor products of C^*-algebras and C^*-norms require some clever techniques. They were developed in an article by Takesaki [**46**]. The tensor product and its basic properties were introduced by Turumaru in [**48**].

It is the case, though not trivial, that for $\mathscr{A}$ finite-dimensional or abelian, $\mathscr{A} \otimes_a \mathscr{B}$ admits just on C^*-norm no matter which C^*-algebra $\mathscr{B}$ we choose. In case $\mathscr{A}$ is finite-dimensional, $\mathscr{A} \otimes_a \mathscr{B} = \mathscr{A} \otimes \mathscr{B}$; that is, $\mathscr{A} \otimes_a \mathscr{B}$ is complete relative to its (unique) C^*-norm, the spatial norm. When $\mathscr{A}$ has the property that $\mathscr{A} \otimes_a \mathscr{B}$ has a unique C^*-norm for all C^*-algebras $\mathscr{B}$, we say that $\mathscr{A}$ is a *nuclear* C^*-algebra. It is easy to show that if $\mathscr{A}$ is the norm closure of a family of nuclear C^*-subalgebras directed by inclusion, then $\mathscr{A}$ is nuclear. Thus, if $\mathscr{A}$ has as a norm-dense subalgebra the union of an ascending sequence of C^*-algebras $\mathscr{A}_k$, each containing I and each isomorphic to some $\mathscr{M}_{n(k)}(\mathbf{C})$, then $\mathscr{A}$ is nuclear.

This class of C^*-algebras, the *matricial* C^*-algebras (also called *uhf algebras*), developed and classified by Glimm [**17**], is a basic class for the study of operator algebras. We shall return to it in the next section. In [**46**], Takesaki shows that the C^*-algebra generated by $\{L_g : g \in \mathscr{F}_2\}$, in the notation of the group-algebra examples described at the end of **Examples**, is not nuclear.

In [**32**], von Neumann introduces infinite tensor products of finite type I factors. The infinite tensor product of a family of C^*- algebras can be defined as the (C^*-algebra) inductive limit of the family of (spatial) tensor products of finite subfamilies. With $\{\mathscr{A}_a : a \in \mathbb{A}\}$ a family of C^*-algebras, we denote this tensor product as $\otimes_a \mathscr{A}_a$. If ρ_a is a state of $\mathscr{A}_a$ for each a in $\mathbb{A}$, there is a unique state $\otimes_{a\in\mathbb{A}}\rho_a (= \rho)$ of the tensor product such that

$$\rho(A_1 \otimes \cdots \otimes A_n) = \rho_{a(1)}(A_1) \cdots \rho_{a(n)}(A_n)$$

whenever $A_j \in \mathscr{A}_{a(j)}$, $\{a(1), \dots, a(n)\} \subseteq \mathbb{A}$. Such states ρ are called *product states* of the tensor product. They occupy a central position in the analysis of infinite tensor products.

§**10. Matricial algebras.** In [**29**], Murray and von Neumann introduce and study a family of factors of type II_1 very closely related to matrix algebras. In the terminology of Section 9, each of these factors is the weak-operator closure of a matricial C^*-algebra. Following an impressive sequence of technical lemmas, Murray and von Neumann prove that all matricial factors of type II_1 are * isomorphic. They also show that none of the factors $\mathscr{L}_{\mathscr{F}_n}$ is matricial and that $\mathscr{L}_\Pi$ is matricial (cf. Section 5).

If $\mathscr{A}$ is a matricial C^*-algebra and $\{\mathscr{A}_k\}$ is a family of C^*-subalgebras with $\mathscr{A}_k$ isomorphic to $\mathscr{M}_{n(k)}(\mathbf{C})$ such that $\mathscr{A}_k \subseteq \mathscr{A}_{k+1}$, then $n(k)$ divides $n(k+1)$ (recall that $I \in \mathscr{A}_k$). If $\mathscr{M}_{k+1}$ is the subalgebra of $\mathscr{A}_{k+1}$ consisting of those elements that commute with all elements of $\mathscr{A}_k$, then $\mathscr{M}_{k+1}$ is itself isomorphic to the full matrix algebra of order $n(k+1)/n(k)$ and $\mathscr{A}$ is isomorphic to the (infinite) tensor product of the family $\{\mathscr{M}_k\}$ (where $\mathscr{M}_1 = \mathscr{A}_1$). If $n(k')$ can be factored as $m \cdot n$, then $\mathscr{M}_{k'}$ has a subfactor $\mathscr{M}$ of type I_m and the set of elements in $\mathscr{M}_{k'}$ commuting with $\mathscr{M}$ is a subfactor $\mathscr{N}$ of type I_n. Replacing $\mathscr{M}_{k'}$ by $\mathscr{M}$ and $\mathscr{N}$ in the family $\{\mathscr{M}_k\}$ and forming the tensor product, we arrive, again, at the (matricial) C^*-algebra $\mathscr{A}$. Thus the orders $n(k+1)/n(k)$ of the algebras $\mathscr{M}_{k+1}$ are not, themselves, invariants for the algebra, though the "total" power to which a given prime may appear in these orders may be. The remaining problem in analyzing matricial C^*-algebras can be illustrated by the question of whether the tensor product of a countable family of $\mathscr{M}_2(\mathbf{C})$ algebras is isomorphic to the tensor-product of that algebra with $\mathscr{M}_3(\mathbf{C})$—or even whether that algebra contains a C^*-subalgebra (containing I) isomorphic to $\mathscr{M}_3(\mathbf{C})$. Using approximation techniques combined with the trace and projections in the algebra, it can be shown that both questions have a negative answer.

With $\mathscr{A}$, $\mathscr{A}_k$, and $n(k)$ as before, let m_p be $\sup\{r : (\exists k)(p^r | n(k))\}$, where p is a prime. Thus m_p may be 0 or ∞, as well as any positive integer. Let $n(\mathscr{A})$ be $2^{m_2}3^{m_3}5^{m_5}\cdots$. Glimm **[17]** proves the following result.

THEOREM. *Two matricial C^*-algebras $\mathscr{A}$ and $\mathscr{B}$ are * isomorphic if and only if $n(\mathscr{A}) = n(\mathscr{B})$.*

By forming the (infinite) tensor product of matrix algebras of suitable orders, a matricial C^*-algebra $\mathscr{A}$ can be constructed with any given invariant $n(\mathscr{A})$; there are many nonisomorphic matricial C^*-algebras. The result of Murray and von Neumann that all matricial II_1 factors are * isomorphic should be contrasted with this situation for matricial C^*-algebras; a substantial "coalescing" takes place under the "umbrella" of the trace.

If τ_k is the normalized trace on $\mathscr{M}_k$, then $\otimes_k \tau_k$ is the unique (normalized) trace τ on $\mathscr{A}$. The GNS construction applied to τ and $\mathscr{A}$ yields a representation of $\mathscr{A}$ on a Hilbert space $\mathscr{H}$. It is not immediate, though not difficult, to show that each matricial C^*-algebra is simple. Thus each representation of $\mathscr{A}$ is faithful. For simpler notation, we may think of $\mathscr{A}$ as acting on $\mathscr{H}$ and τ as $\omega_u|\mathscr{A}$ for some unit generating vector u for $\mathscr{A}$. It follows that the weak-operator closure $\mathscr{M}$ of $\mathscr{A}$ is the matricial factor of type II_1 with u as a generating trace vector. Thus $\mathscr{M}'$ is of type II_1 with u as a trace vector.

Viewed in this way, it is natural to study the weak-operator closure of the image of a matricial algebra $\mathscr{A}$ under the representation obtained from the GNS construction applied to a product state of $\mathscr{A}$. As a byproduct of his thesis work on the (infinite) canonical anticommutation relations (CAR), Powers **[34]** did just this, at the same time, bringing into focus the pivotal role of product states. (See also **[35]**.) Powers works with the matricial C^*-algebra $\mathscr{A}$ for which $n(\mathscr{A}) = 2^\infty$. This algebra is the (infinite) tensor product of algebras $\mathscr{M}_k$ each of which is isomorphic to $\mathscr{M}_2(\mathbf{C})$. It is referred to as the *CAR algebra*; its representations are naturally associated with the representations of the CAR. With an isomorphism between $\mathscr{M}_k$ and $\mathscr{M}_2(\mathbf{C})$ specified, Powers considers the state ρ_k of $\mathscr{M}_k$ that assigns to A the value $ta + (1-t)b$, where $t \in [0, \frac{1}{2}]$ and a and b are the upper and lower diagonal entries of the matrix in $\mathscr{M}_k$ corresponding to A. The product state he studies is $\otimes_k \rho_k (= \rho_t)$. He shows that the (faithful) image of $\mathscr{A}$ under the GNS representation corresponding to ρ_t has weak-operator closure a (matricial) factor $\mathscr{M}_t$ of type III. He proves the following result.

THEOREM. *If $t, t' \in (0, \frac{1}{2})$ and $t \neq t'$, then $\mathscr{M}_t$ and $\mathscr{M}_{t'}$ are not isomorphic.*

In the notation of Section 7, $\mathscr{M}_t$ is a factor of type III_λ, where $\lambda = t/1 - t$. Although Pukanszky **[36]** had found two nonisomorphic factors of type III (along the lines of the Murray-von Neumann factors of type II_1 that are not isomorphic) at an early stage, the factors $\mathscr{M}_t$, the *Powers factors*, exhibited a continuum of nonisomorphic matricial factors of type III.

Inspired by the results of Powers, Araki and Woods [**2**, **3**] undertook a detailed analysis of the factors arising from the product states of matricial algebras. It is a fact, requiring an involved argument, that each product state of a matricial algebra $\mathscr{A}$ gives rise to a representation of $\mathscr{A}$ whose image has weak-operator closure a factor. From Araki and Woods, these factors have come to be known as *ITPFI factors* (for 'infinite tensor product of finite type I factors'). Araki and Woods develop the technique of Powers involving the states ρ_k on $\mathscr{M}_k$. As in the situation of quantum statistical mechanics, each state ρ on $\mathscr{M}_n(\mathbf{C})$ has a "density matrix" H (such that $\rho(A) = \tau(HA)$ for each A in $\mathscr{M}_n(\mathbf{C})$). Powers considers the eigenvalues of H. Araki and Woods note that the ratios of the eigenvalues of the density matrices for the 'tail' of the tensor product is what really counts in the arguments of Powers, and they define an "asymptotic ratio set" in terms of which they classify many of the ITPFI factors.

A vexing problem that emerged from these investigations concerned the relation between ITPFI factors of type III and matricial factors of type III. Certainly each ITPFI factor of type III is matricial; but is each matricial factor of type III an ITPFI factor? Krieger [**23**] constructed certain matricial factors and established results indicating that they might not be ITPFI factors. In [**7**], Connes completes the process showing that the Krieger factors are, indeed, not ITPFI.

In connection with the Powers article [**34**], another vexing problem arose that was to puzzle most of the research workers in the area for the next eight years. The von Neumann algebra tensor product of the matricial factor of type II_1 and $\mathscr{B}(\mathscr{H})$, where $\mathscr{H}$ is an infinite-dimensional, separable Hilbert space, is a factor of type II_∞. There is no difficulty in seeing that it is matricial. Are all matricial factors of type II_∞ isomorphic to that tensor product? Connes, along with a host of other experts, set their sights on that question. Connes had established that a number of other fascinating questions about matricial factors are equivalent to this question. Among them are the following problems: Is each subfactor of the matricial II_1 factor either matricial or of type I_n (such a subfactor must be of finite type)? Is the von Neumann group algebra of an amenable *i.c.c.* group matricial? Is a factor $\mathscr{M}$, acting on (separable) $\mathscr{H}$, with the property that there is a conditional expectation mapping $\mathscr{B}(\mathscr{H})$ onto $\mathscr{M}$ (such factors are said to be *injective*) matricial? In [**8**], Connes answers all these questions in the affirmative, and does much more. His argument is a brilliant *tour de force* relying heavily on his penetrating analysis of the automorphism group of certain factors. The intervening years have seen simpler proofs of these results [**21**, **33**] (none of which is very simple!), but the work of Connes [**8**] is so deep and rich in ideas and techniques that it remains a basic resource in the subject.

As noted, Connes had shown that some matricial factors of type III escape the ITPFI net and, consequently, escape the results on classification by Araki-Woods [**3**]. In [**8**], Connes shows that all matricial factors of type III_λ,

$\lambda \in (0, 1)$, are isomorphic. He had noted, in **[7]**, that there are many matricial type III_0 factors depending on the action of an ergodic automorphism of an abelian von Neumann algebra. In **[8]**, he completes the reduction to the (measure-theoretic) classification of such actions by showing that each type III_0 matricial factor is the crossed product of an abelian algebra and ($\mathbf{Z}$ through) an ergodic automorphism of the algebra.

The question of whether all matricial factors of type III_1 are isomorphic remained the most tantalizing open question of the subject until Haagerup **[22]** settled it in the affirmative.

§**11. Conclusion.** If we reckon the birth of the theory of operator algebras as occurring in **[31]**, the subject is sixty years old. That is surely old enough to permit us to "look back" and assess the legacy of von Neumann in this area of mathematical thought. A substantial part of that legacy must be a collection of mathematical results of surpassing depth and beauty, many of them contributed by von Neumann himself and jointly with F. J. Murray, but a great many of them due to generations of mathematicians that followed von Neumann. Another significant part of the legacy is the host of applications and spin-offs of the theory of operator algebras; this theory touches virtually all areas of mathematics and many areas of physics. Legions of scientists were inspired by the ideas basic to the theory and fascinated by the power and elegance of the subject. A great many highly talented mathematicians and physicists have devoted much time and effort to bringing the subject to its present state. Several of them I do not (nor, I imagine, would von Neumann) hesitate to describe as brilliant. Perhaps that, the people and the enduring inspiration, is the most important part of the legacy; through them the ideas and work of von Neumann remain a living and vital force in present day science.

General References

[D] J. Dixmier, *Les algèbres d'opérateurs dans l'espace Hilbertien*, Gauthier-Villars, Paris, 1957, 2nd ed. 1969.

[D*] ———, *Les C^*-algèbres et leurs réprésentations*, Gauthier-Villars, Paris, 1964; English transl. C^*-Algebras, North-Holland Mathematical Library, vol. 15, North Holland Pub., Amsterdam, 1977.

[K-R] R. V. Kadison and J. R. Ringrose, *Fundamentals of the theory of operator algebras*, **I**, Academic Press, New York, 1983, vol. II, 1986.

[P] G. K. Pedersen, *C^*-algebras and their automorphism groups*, London Mathematical Society Monographs, vol. 14, Academic Press, London, 1979.

[S] S. Sakai,, *C^*-algebras and W^*-algebras*, Ergebnisse der Mathematik und ihrer Grenzgebiete, **60**, Springer-Verlag, Heidelberg, 1971.

[T] M. Takesaki, *Theory of operator algebras* I, Springer-Verlag, Heidelberg, 1979.

REFERENCES

1. H. Araki, *On the algebra of all local observables*, Publ. Res. Inst. Math Sci. Kyoto **5** (1964), 1–16.
2. H. Araki and E. J. Woods, *Complete Boolean algebras of type* I *factors*, Publ. Res. Inst. Math. Sci. Kyoto **2** (1966), 157–242.
3. ———, *A classifiction of factors*, Publ. Res. Inst. Math. Sci., Kyoto **4** (1968), 51–130.
4. H. J. Borchers, *On the structure of the algebra of field operators.* II, Comm. Math. Phys. **1** (1965), 49–56.
5. W-M. Ching, *Nonisomorphic nonhyperfinite factors*, Canad. J. Math. **21** (1969), 1293–1308.
6. E. Christensen, *Measures on projections and physical states*, Comm. Math. Phys. **86** (1982), 529–538.
7. A. Connes, *Une classification des facteurs de type* III, Ann. Sci. École Norm. Sup. Paris **6** (1973), 133–252.
8. ———, *Classification of injective factors, Cases* II_1, II_∞, III_λ, $\lambda \neq 1$, Ann. of Math. **104** (1976), 73–115.
9. ———, *Sur la classification des facteurs de type* II, C. R. Acad. Sci. Paris **281** (1975), 13–15.
10. ———, *On hyperfinite factors of type* III_0 *and Krieger's factors*, J. Funct. Anal. **18** (1975), 318–327.
11. A. Connes and A. Van Daele, *The group property of the invariant S of von Neumann algebras*, Math. Scand. **32** (1973), 187–192.
12. A. Connes and E. J. Woods, *A construction of approximately finite-dimensional non-ITPFI factors*, Canad. Math. Bull. **23** (1980), 227–230.
13. G-F. Dell'Antonio, S. Doplicher, and D. Ruelle, *A theorem on canonical commutation and anticommutation relations*, Comm. Math. Phys. **2** (1966), 223–230.
14. J. Dixmier and E. C. Lance, *Deux nouveaux facteurs de type* II_1, Invent. Math. **7** (1969), 226–234.
15. I. M. Gelfand and M. A. Neumark, *On the imbedding of normed rings into the ring of operators in Hilbert space*, Mat. Sb. **12** (1943), 197-213.
16. A. M. Gleason, *Measures on the closed subspaces of a Hilbert space*, J. Rational Mechanics and Anal. **6** (1957), 885–894.
17. J. G. Glimm, *On a certain class of operator algebras*, Trans. Amer. Math. Soc. **95** (1960), 318–340.
18. R. Haag, N. M. Hugenholtz and M. Winnink, *On the equilibrium states in quantum statistical mechanics*, Comm. Math. Phys. **5** (1967), 215–236.
19. R. Haag and D. Kastler, *An algebraic approach to quantum field theory*, J. Math. Phys. **5** (1964), 848–861.
20. R. Haag and B. Schroer, *Postulates of quantum field theory*, J. Math. Phys. **3** (1962), 248–256.
21. U. Haagerup, *A new proof of the equivalence of injectivity and hyperfiniteness for factors on a separable Hilbert space*, J. Funct. Anal. **62** (1985), 160–201.
22. ———, *Connes' bicentralizer problem and uniqueness of the injective factor of type* III_1, Acta Math. **158** (1987), 95–148.
23. W. Krieger, *On the infinite product construction of nonsingular transformations of a measure space*, Invent. Math. **15** (1972), 144–163.
24. G. W. Mackey, *Mathematical foundations of quantum mechanics*, W. A. Benjamin, Inc., New York, 1963.
25. D. McDuff, *A countable infinity of* II_1 *factors*, Ann. of Math. **90** (1969), 361–371.
26. ———, *Uncountably many* II_1 *factors*, Ann. of Math. **90** (1969), 372–377.
27. F. J. Murray and J. von Neumann, *On rings of operators*, Ann. of Math. **37** (1936), 116–229.
28. ———, *On rings of operators*, II, Trans. Amer. Math. Soc. **41** (1937), 208–248.
29. ———, *On rings of operators*, IV, Ann. of Math. **44** (1943), 716–808.
30. J. von Neumann, *On rings of operators*, III, Ann. of Math. **41** (1940), 94–161.

31. ———, *Zur Algebra der Funktionaloperationen und Theorie der normalen Operatoren*, Math. Ann. **102** (1929–30), 370–427.
32. ———, *On infinite direct products*, Compositio Math. **6** (1938), 1–77.
33. S. Popa, *A short proof of injectivity implies hyperfiniteness for finite von Neumann algebras*, J. Operator Theory **16** (1986), 261–272.
34. R. T. Powers, *Representations of uniformly hyperfinite algebras and their associated von Neumann rings*, Ann. of Math. **86** (1967), 138–171.
35. R. T. Powers and E. Størmer, *Free states of the canonical anticommutation relations*, Comm. Math. Phys. **16** (1970), 1–33.
36. L. Pukanszky, *Some examples of factors*, Publ. Math., Debrecen **4** (1956), 135–156.
37. S. Sakai, *On topological properties of W^*-algebras*, Proc. Japan Acad. **33** (1957), 439–444.
38. ———, *Asymptotically abelian* II_1*-factors*, Publ. Res. Inst. Math., Kyoto Univ. **4** (1968–1969), 299–307.
39. ———, *An uncountable number of* II_1 *and* II_∞ *factors*, J. Funct. Anal. **5** (1970), 236–246.
40. J. T. Schwartz, *Two finite, non-hyperfinite, non-isomorphic factors*, Comm. Pure Appl. Math **16** (1963), 19–26.
41. I. E. Segal, *Irreducible representations of operator algebras*, Bull. Amer. Math. Soc. **53** (1947), 73–88.
42. M. H. Stone, *Applications of the theory of Boolean rings to general topology*, Trans. Amer. Math. Soc. **41** (1937), 375–481.
43. ———, *A general theory of spectra*, I, Proc. Nat. Acad. Sci. U.S.A. **26** (1940), 280–283.
44. M. Takesaki, *Tomita's theory of modular Hilbert algebras and its applications*, Lecture Notes in Mathematics, vol. 128, Springer-Verlag, Heidelberg, 1970.
45. ———, *Duality for corssed products and the structure of von Neumann algebras of type* III, Acta. Math. **131** (1973), 249–310.
46. ———, *On the cross-norm of the direct product of C^*-algebras*, Tohoku Math. J. **16** (1964), 111–122.
47. M. Tomita, *Standard forms of von Neumann algebras,*, Fifth Functional Analysis Symposium of the Math. Soc. of Japan, Sendai, 1967.
48. T. Turumaru, *On the direct product of operator algebras*. I, Tohoku Math. J. **2** (1952), 242–251.
49. A. S. Wightman, *La théorie quantique locale et la théorie quantique des champs*, Ann. Inst. Henri Poincaré **1** (1964), 403–420.
50. F. J. Yeadon, *Measures on projections in W^*-algebras of type* II_1, Bull. London Math. Soc. **15** (1983), 139-145.
51. G. Zeller-Meier, *Deux autres facteurs de type* II_1, Invent. Math. **7** (1969), 235–242.

UNIVERSITY OF PENNSYLVANIA

Proceedings of Symposia in Pure Mathematics
Volume **50** (1990)

Introduction à la géométrie non-commutative

ALAIN CONNES

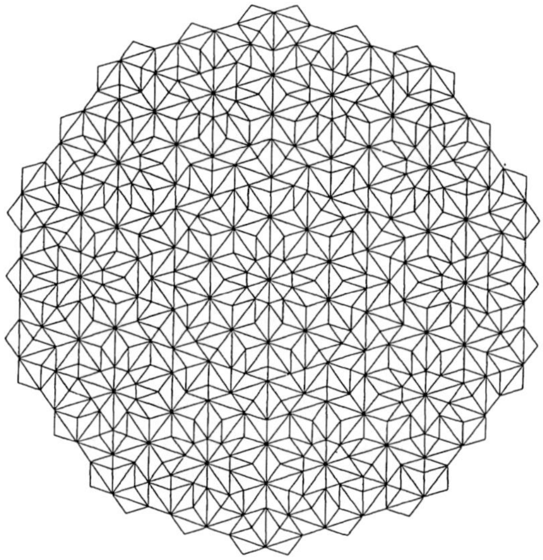

1. Un exemple d'espace quantique: l'ensemble des pavages de Penrose. Le but de la géométrie "non-commutative" est de comprendre la structure d'espaces géométriques plus complexes que les espaces ordinaires, dont la description ne peut plus se faire à partir d'un ensemble X et de classes de fonctions $f\colon X \to \mathbf{R}$ ayant certains types de régularité: mesurabilité, continuité, différentiabilité.

Pour fixer les idées nous commencerons par décrire un exemple simple d'espace non-commutatif X en montrant la différence entre le point de vue classique ou commutatif et notre point de vue. L'espace X est l'ensemble des univers de Penrose ou pavages quasi périodiques du plan.

1980 *Mathematics Subject Classification* (1985 *Revision*). Primary 46L80, 46L55, 46L40, 51Nxx.

A detailed version of this paper will appear in A. Connes, *Géométrie non-commutative*, Inter Editions, Paris, 1990, to appear.

J'ai eu la chance d'assister, au cours d'une conférence, à un exposé de R. Penrose sur les pavages quasi périodiques du plan (Figure 5). Penrose a construit de tels pavages à l'aide de deux motifs simples A et B. La construction n'est pas unique et le pavage obtenu T dépend de choix successifs effectués au cours de sa construction. De manière plus précise il faut pour construire un tel pavage disposer d'une suite $(z_n)_{n=0,1,2,\ldots}$ de 0 et 1 qui vérifie la règle de cohérence:

$$(1) \qquad z_n = 1 \Rightarrow z_{n+1} = 0.$$

On utilise la valeur de z_n comme indiqué dans l'appendice dans la construction de T. Cependant il se peut que deux suites $z = (z_n)$ et $z' = (z'_n)$ distinctes conduisent au même pavage et on a en fait:

$$(2) \qquad T \text{ identique à } T' \quad \text{ssi} \quad \exists n \quad \text{avec} \quad z_j = z'_j \quad \forall j \geq n.$$

Ainsi l'espace X des pavages obtenus est l'ensemble $K/\mathfrak{R}$ quotient d'un compact K par une relation d'équivalence. Le compact K, homéomorphe à l'ensemble de Cantor, est l'ensemble des suites $z = (z_n)_{n\in\mathbf{N}}$ de 0 et 1 qui vérifient la règle (1). C'est par construction un fermé du produit d'une infinité d'ensembles à deux éléments. La relation d'équivalence $\mathfrak{R}$ est donnée par:

$$(3) \qquad z \sim z' \quad \text{ssi} \quad \exists n, z_j = z'_j \quad \forall j \geq n.$$

On a donc $X = K/\mathfrak{R}$ et c'est exactement sous cette forme que R. Penrose présentait l'ensemble de ses pavages dans son exposé. Si l'on cherche à comprendre l'espace X comme un espace ordinaire, on s'aperçoit vite que les outils classiques ne fonctionnent pas et ne distinguent pas X de l'espace réduit à un point. Par exemple, on peut, (Cf. **[28]**) étant donnés deux pavages T_1 et T_2 et une portion finie quelconque P de T_1, retrouver la même configuration P dans T_2, de sorte qu'aucune portion finie ne distingue T_1 de T_2. Cette propriété géométrique se traduit par la trivialité de la topologie de X. La topologie naturelle de X a pour sous ensembles fermés $F \subset X$ les sous ensembles fermés de K et saturés pour $\mathfrak{R}$. Or toute classe d'équivalence pour $\mathfrak{R}$ est dense dans K et il en résulte que les seuls sous ensembles fermés $F \subset X$ sont $F = \emptyset$, $F = X$. Ainsi la topologie de X est triviale et ne distingue pas X d'un point, elle n'est bien entendu pas séparée, et ne contient aucune information intéressante.

Une attitude possible devant un tel exemple serait de dire qu'à des fluctuations près il y a *un seul* pavage T du plan et de ne pas s'inquiéter de la distinction entre X et un point. En fait nous verrons que X est un espace "non-commutatif" ou espace "quantique" très intéressant et que l'un de ses invariants topologiques, le groupe des dimensions, est le sous groupe de $\mathbf{R}$ engendré par $\mathbf{Z}$ et le nombre d'or $(1 + \sqrt{5})/2$. Ceci montrera, en particulier, pourquoi uniquement à cause de la topologie de X, la densité ou fréquence d'apparition d'un motif dans le pavage doit être un élément du groupe $\mathbf{Z} + ((1 + \sqrt{5})/2)\mathbf{Z}$. Expliquons d'abord pourquoi la description de X

comme espace topologique se fait en remplaçant l'algèbre $C(X)$ des fonctions continues sur X par une C^*-algèbre non-commutative, ce qui est à l'origine de la terminologie "géométrie non-commutative". Etant donné Y un espace (localement) compact ordinaire, le lemme d'Urysohn montre qu'il existe suffisamment de fonctions continues $f \in C(Y)$ sur Y pour déterminer uniquement la topologie de Y. Ceci est vrai pour les fonctions à valeurs réelles et à fortiori celles à valeurs complexes. Ainsi la topologie de Y détermine et est déterminée par l'algèbre $A = C(Y)$ des fonctions continue à valeurs complexes sur Y, munie de l'involution $f \to f^*$ où $F^*(y) = \overline{f(y)}\ \forall y \in Y$. (Ici, un énoncé analogue serait faux si l'on remplaçait **R** ou **C** par un corps local totalement discontinu). De plus la classe de **C**-algèbres involutives A obtenue se caractérise très simplement. Il s'agit des C^*-algèbres (avec unité) *commutatives*. Les C^*-algèbres ont une caracterérisation axiomatique très simple qui exprime exactement ce que l'on attend des fonctions de classe C^0. Une C^*-algèbre A est une algèbre involutive qui est une algèbre de Banach pour la norme $x \to \|x\|$ uniquement déterminée par la structure d'algèbre comme le rayon spectral de x^*x: $\|x\|^2 =$ rayon spectral $(x^*x) = \sup\{|\lambda|, x^*x - \lambda$ non inversible$\}$. Pour qu'une algèbre involutive A soit une C^*-algèbre il faut et il suffit qu'elle admette une représentation involutive π dans un espace de Hilbert telle que:

(4)
$$(1)\ \pi(x) = 0 \Rightarrow x = 0;$$
$$(2)\ \pi(A) \text{ est fermé en norme.}$$

Le théorème de Gelfand, qui s'appuie sur l'analyse complexe et la structure des idéaux maximaux de A, montre que si A est une C^*-algèbre *commutative* il existe un espace topologique compact $Y = \mathrm{SP}(A)$ tel que $A = C(Y)$. La raison d'être de la géométrie non-commutative est la possibilité d'adapter la plupart des outils classiques tels mesures de Radon, K-théorie, cohomologie... nécessaires à l'étude d'un espace compact Y au cas où la C^*-algèbre $A = C(Y)$ est remplacée par une C^*-algèbre non-commutative.

Bien entendu cette adaptation au cas non-commutatif n'est jamais immédiate et s'accompagne souvent de phénomènes totalement nouveaux. Même dans le paysage de la théorie de la mesure, si simple dans le cas commutatif, l'on voit apparaître l'implication suivante:

$$\text{Non-commutativité} \Rightarrow \text{Groupe à 1 paramètre d'évolution}$$

dont nous aurons l'occasion de reparler.

Quand la C^*-algèbre A n'est pas commutative le spectre $Y = \mathrm{Sp}(A)$ de Gelfand est remplacé par l'ensemble Y des représentations irréductibles π de A dans l'espace de Hilbert. Mais, bien que l'on puisse toujours démontrer l'existence de suffisamment de représentations irréductibles de A, dès que A est *simple*, i.e., n'a pas d'idéal bilatère non-trivial, la topologie naturelle de l'espace Y a la même propriété de trivialité que celle de l'ensemble X des pavages de Penrose rencontré ci-dessus. Deux représentations irréductibles

π_1 et π_2 de A peuvent par une transformation unitaire être rendues aussi ressemblantes que l'on veut. C'est le théorème de transitivité de R. V. Kadison [**33**]. Il est donc naturel de chercher a associer à l'espace X des pavages de Penrose une C^*-algèbre *non-commutative* qui remplace $C(X)$.

Comme la dichotomie entre espaces classiques et espaces "quantiques" n'est pas exactement celle entre algèbre commutative et algèbre non-commutative, il est d'abord nécessaire d'expliquer en quel sens certaines algèbres non-commutatives équivalents à des algèbres commutatives. Toutes les représentations irréductibles de la C^*-algèbre $M_n(\mathbf{C})$ des matrices $n \times n$ sur $\mathbf{C}$ sont deux à deux équivalentes, et il en résulte que l'on ne change pas l'espace $\mathrm{Sp}(A)$ des représentations irréductibles de A quand on remplace la C^*-algèbre A par $M_n(A) = M_n(\mathbf{C}) \otimes A$. Plus généralement on ne change aucune des propriétés importantes de A relativement à la théorie de la mesure, K-théorie, cohomologie ... quand on remplace A par $M_n(A)$ ou par une C^*-algèbre *équivalente au sens de Morita.* Sans donner la définition technique de cette équivalence (Cf. [**45**]) notons que comme elle ne change pas l'espace $\mathrm{Sp}(A)$ des représentations irréductibles de A, il est facile de voir que:

(5) Il existe au plus une C^*-algèbre commutative $C(Y)$ équivalente au sens de Morita, à une C^*-algèbre A.

Ainsi le cas classique ne contient pas que des C^*-algèbres commutatives mais il est limité aux C^*-algèbres qui sont équivalentes à une C^*-algèbre commutative. Or il est clair que cette classe de C^*-algèbres n'a rien de naturel et n'est même pas fermée pour l'opération élémentaire de réunion croissante. Pour voir à quelle opération géométrique cela correspond sur les espaces Y, il faut d'abord comprendre à quelle situation géométrique correspond la description d'un compact Y non par $C(Y)$ mais par une C^*-algébre non-commutative équivalence à $C(Y)$. Le point crucial est que l'opération ensembliste qui consiste à *identifier deux points* x et y se traduit algébriquement par le remplacement de l'algèbre commutative des fonctions sur $\{x\} \cup \{y\}$ par l'algèbre des matrices 2×2, $\begin{bmatrix} a_{xx} & a_{xy} \\ a_{yx} & a_{yy} \end{bmatrix}$ à coefficients complexes. Dans le spectre de cette algèbre les états purs ω_x, ω_y associés à x et y par les formules $\omega_x(a) = a_{xx}$; $\omega_y(a) = a_{yy}$ donnent des représentations irréductibles équivalentes d'où l'identification ensembliste $x \sim y$. Prenons par exemple une variété Y compacte obtenue en recollant des morceaux ouverts U_i de l'espace Euclidien, $U_i \subset Y$ selon les intersections $U_i \cap U_j$. Il faut pour obtenir Y à partir de l'ouvert Euclidien $V = \coprod U_i$ réunion disjointe des U_i, identifier les couples (z, z') d'éléments de V où le collage a lieu. On vérifie facilement que la C^*-algèbre $C(Y)$ est Morita équivalente à la C^*-algèbre A des matrices $a = a_{(z,z')}$ qui dépendent continuement des couples (z, z') ci-dessus et où le produit est défini par la règle:

$$(ab)_{(z,z')} = \sum_{z''} a_{(z,z'')} b_{(z'',z')}.$$

En résumé dès qu'un espace ordinaire Y est *surparamétré*, i.e., est présenté comme un quotient ou par recollement, il est naturel de lui associer une algèbre non-commutative A qui code cette information, mais est néanmoins Morita équivalente à $C(Y)$. Mais il peut très bien arriver qu'à vouloir recoller des morceaux d'espace Euclidien pour construire une variété on obtienne un espace analogue à l'espace X des pavages de Penrose. La construction de la C^*-algèbre A des matrices $a = a_{(z,z')}$ restera cependant valable et nous allons maintenant la décrire dans le cas de cet espace $X = K/\mathfrak{R}$. Un élément général $a \in A$ est donné par une matrice $(a_{z,z'})$ de nombres complexes, indexée par les couples $(z, z') \in \mathfrak{R}$. Le produit de deux éléments de A des donné par le produit des matrices

$$(ab)_{z,z''} = \sum_{Z'} a_{z,z'} b_{z',z''}.$$

A chaque élément $x \in X$ de X correspond une classe d'équivalence pour $\mathfrak{R}$ qui est un sous-ensemble dénombrable de K, on peut donc associer à x l'espace de Hilbert ℓ^2_x ayant pour base orthonormale cet ensemble dénombrable. Tout élément a de A définit un opérateur dans ℓ^2_x par l'égalité.

$$(a(x)\zeta)_z = \sum_{Z'} a_{z,z'} \zeta_{z'}.$$

Pour $a \in A$ la norme $\|a(x)\|$ de l'opérateur $a(x)$, est finie et ne dépend pas de $x \in X$, c'est la norme de C^*-algèbre. Bien entendu, il y a un point technique qui consiste à donner la définition rigoureuse de la classe de matrices a que nous considérons. Mais la relation $\mathfrak{R} \subset K \times K$ est réunion croissante des relations $\mathfrak{R}_n = \{(z, z'); z_j = z'_j \quad \forall j \geq n\}$, et on définit A comme la fermeture normique de l'algèbre obtenue à partir des matrices a où a est continue sur $\mathfrak{R}_n$ et nulle sur $\mathfrak{R}_n^c$ pour n assez grand. Comme il n'y a pas de formule explicite simple pour la norme d'une matrice en fonction de ses coefficients il est difficile de décrire exactement la régularité de la matrice générale $a \in A$ mais en tous les cas toute matrice continue sur $\mathfrak{R}_n$ et nulle sur $\mathfrak{R}_n^c$ appartient à A et tout élément de A correspond à une matrice $a_{(z,z')} = a$ continue sur chaque $\mathfrak{R}_n$. On peut résumer la construction ci-dessus en disant qu'alors que l'espace X ne peut être non-trivialement décrit en utilisant des fonctions à valeurs dans $\mathbf{C}$, il existe sur X une classe très riche de fonctions à valeurs opérateurs:

$$a(x) \in \ell(\ell^2_x) \quad \forall x \in X.$$

La structure algébrique de A est dictée par ce point de vue car on a:

$$(\lambda a + \mu b)(x) = \lambda a(x) + \mu b(x) \qquad \forall a, \quad b \in A, \quad x \in X$$
$$(ab)(x) = a(x)b(x) \qquad \forall x \in X.$$

Pour bien montrer la richesse de la C^*-algèbre A nous en donnons maintenant une autre description [**22**] comme limite inductive d'algèbres de dimension finie. Le Cantor K est par construction la limite projective des ensembles finis K_n, où K_n est l'ensemble des suites de $n+1$ éléments: $(z_j)_{j=0,1,\ldots,n}$ de 0 et 1 qui vérifient la règle $z_j = 1 \Rightarrow z_{j+1} = 0$. On a une projection évidente $K_{n+1} \to K_n$ qui consiste à oublier le dernier z_{n+1}. Considérons sur l'ensemble fini K_n la relation $\mathfrak{R}^n = \{(z, z') \in K_n \times K_n; z_n = z'_n\}$. Toute fonction $a = a_{(z,z')}$ sur $\mathfrak{R}^n$ définit un élément $\tilde{a}$ de la C^*-algèbre A par l'égalité:

$$\tilde{a}_{(z,z')} = a_{((z_0,\ldots,z_n)(z'_0,\ldots,z'_n))} \qquad \forall (z, z') \in \mathfrak{R}_n$$
$$\tilde{a}_{(z,z')} = 0 \qquad \forall (z, z') \notin \mathfrak{R}_n.$$

De plus la sous-algèbre de A obtenue à partir des fonctions sur $\mathfrak{R}^n$ se calcule facilement, c'est une somme directe, $A_n = M_{k_n}(\mathbf{C}) \oplus M_{k'_n}(\mathbf{C})$ de deux algèbres de matrices, où k_n (resp. k'_n) est le nombre d'éléments de K_n qui se terminent par 0 (resp. 1). Enfin l'inclusion $A_n \to A_{n+1}$ est uniquement déterminée par le diagramme (de Bratteli [7])

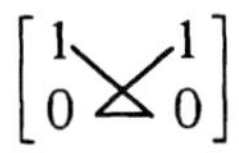

(i.e., l'égalité $k_{n+1} = k_n + k'_n$, $k'_{n+1} = k_n$ permet de plonger $M_{k_n} \oplus M_{k'_n}$ comme matrices par blocs dans M_{k_n} et par l'homomorphisme $(a, a') \to a$ dans $M_{k'_{n+1}}$).

La C^*-algèbre A est alors la limite inductive des algèbres de dimension finie A_n et on peut calculer les invariants, pour la classification de Bratteli, Elliott, Effros, Chen, Handelman ([7], [24], [23], [30]) de ces C^*-algèbres particulières. L'invariant à calculer est un groupe, muni d'une relation d'ordre. Comme il garde un sens pour toute C^*-algèbre A nous allons donner sa définition en général. Le groupe $K_0(A)$ est le groupe engendré par les classes d'équivalence de projecteurs, comme dans les travaux de Murray et von Neumann sur les facteurs. Un projecteur $e \in A$ est un élément de A tel que $e^2 = e$ et $e = e^*$. Deux projecteurs e, f sont équivalents s'il existe $u \in A$ tel que $u^*u = e$, et $uu^* = f$. Pour pouvoir additionner les classes d'équivalence de projecteurs on utilise les matrices $M_n(A)$, n arbitraire ce qui permet de donner un sens à $e \oplus f = \left[\begin{smallmatrix} e & 0 \\ 0 & f \end{smallmatrix}\right]$ pour deux projecteurs e, $f \in M_k(A)$. Le groupe ordonné $K_0(A)$, $K_0(A)^+$ est obtenu canoniquement à partir du semi-groupe des classes d'équivalence de projecteur $e \in M_n(A)$ par l'opération usuelle de symétrisation qui fait passer du semi-groupe $\mathbf{N}$ des entiers positifs au groupe ordonné $\mathbf{Z}$, $\mathbf{Z}^+$ des entiers relatifs.

Ce groupe ordonné est très facile à calculer pour des algèbres de dimension finie comme les algèbres A_n, $UA_n = A$ rencontrées ci-dessus. Comme A_n est la somme directe de deux algèbres de matrices, on obtient:

$$K_0(A_n) = \mathbf{Z}^2, \quad K_0(A_n)^+ = \mathbf{Z}^+ \oplus \mathbf{Z}^+ \subset \mathbf{Z} \oplus \mathbf{Z}.$$

Le groupe ordonné $(K_0(A), K_0(A)^+)$ est alors la limite inductive des groupes ordonnés $(\mathbf{Z} \oplus \mathbf{Z}, \mathbf{Z}^+ \oplus \mathbf{Z}^+)$, où l'inclusion du $n$ième dans le $n$ième est donné par la matrice $\left[\begin{smallmatrix} 1 & 1 \\ 0 & 1 \end{smallmatrix}\right]$ qui correspond à l'inclusion $A_n \subset A_{n+1}$ décrite ci-dessus.

Comme cette matrice définit une bijection $(a, b) \to (a + b, a)$ de $\mathbf{Z}^2$ sur $\mathbf{Z}^2$ la limite inductive cherchée est le groupe $K_0(A) = \mathbf{Z}^2$. Mais cette bijection n'est pas une bijection de $\mathbf{Z}^+ \oplus \mathbf{Z}^+$ sur $\mathbf{Z}^+ \oplus \mathbf{Z}^+$, et à la limite le semi groupe

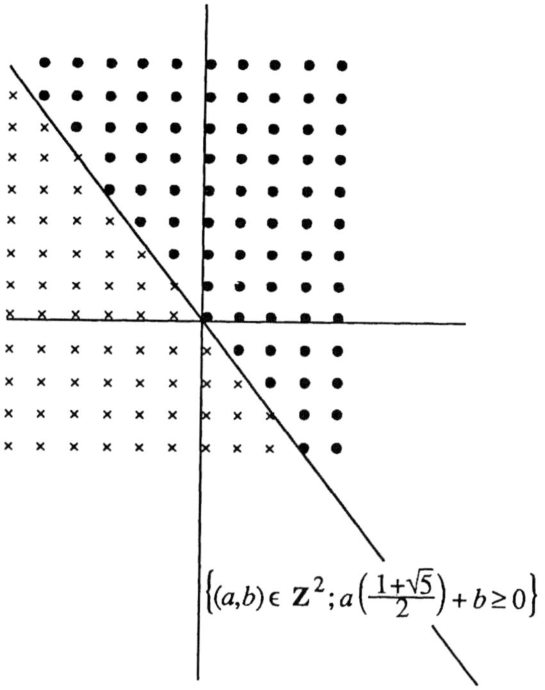

FIGURE 1. Le groupe $K_0(A)$; Chaque point de réseau entier représente un élement de $K_0(A)$, les élements de $K_0(A)$ situés à droite de Δ correspondent à des idempotents.

$K_0(A)^+$ devient: $K_0(A)^+ = \{(a,b) \in \mathbf{Z}^2; ((1/\sqrt{5})/2)a + b \geq 0\}$ comme on le voit en diagonalisant la matrice $\left[\begin{smallmatrix} 0 & 1 \\ 1 & 1 \end{smallmatrix}\right]$ (Cf. Figure 1.)

Il en résulte que modulo le choix d'une base dans $\mathbf{Z}^2$, i.e., modulo $\mathrm{PSL}(2, \mathbf{Z})$ on trouve le nombre d'or $((1 + \sqrt{5})/2)$ comme un invariant topologique de l'espace X à travers la C^*-algèbre A. Sous une autre formulation, on montre que cette C^*-algèbre a une unique trace τ, $\tau(1) = 1$, i.e., il existe sur A une unique forme linéaire τ telle que $\tau(xy) = \tau(yx)$ pour tous $x, y \in A$, et les valeurs prises par τ sur les projecteurs donne le sous-groupe $\mathbf{Z} + ((1 + \sqrt{5})/2)\mathbf{Z} \subset \mathbf{R}$. Cette trace τ sur A est *positive*, i.e., elle vérifie

$$\tau(a^*a) \geq 0 \quad \forall a \in A$$

et elle se calcule directement pour $a = a_{(z,z')}$ comme intégrale des coefficients diagonaux de la matrice a:

$$\tau(a) = \int_K a_{(z,z)}\, d\mu(z)$$

où la mesure de probabilité μ sur K est uniquement déterminée par la condition $\tau(ab) = \tau(ba)$ $\forall a, b \in A$. En théorie classique de la mesure, étant donnée une mesure de Radon ρ sur un espace compact Y, on obtient l'espace de Hilbert $L^2(Y, \rho)$ comme complétion de $C(Y)$ pour le produit scalaire $\langle f, g\rangle = \int f\bar{g}\, d\rho$, et la fermeture faible de l'algèbre $C(Y)$ agissant dans

$L^2(Y, \rho)$. Dans le cas qui nous intéresse, l'algèbre A remplace $C(Y)$, la trace positive τ remplace la mesure de Radon ρ et la fermeture faible de l'action de A par multiplication à gauche dans $L^2(A, \tau)$ (complétion de A pour le produit scalaire $\langle a, b\rangle = \tau(a^*b)$) est le facteur hyperfini de type II_1 de Murray et von Neumann (Cf. [**38**]), que nous noterons R. Il nous reste à expliquer, toujours sur notre exemple des pavages du plan comment la notation de dimension continue de Murray et von Neumann est reliée à la notion de densité d'un sous-ensemble discret d'un pavage. Une propriété extraordinaire d'un facteur de type II_1 comme le facteur R est la classification complète des classes d'équivalence de projecteurs $e \in R$ par un nombre réel $\dim(e) \in [0, 1]$ qui peut prendre *toute valeur continue* entre 0 et 1. La Grassmannienne des projecteurs $e \in \mathbf{R}$ décrit ainsi, non plus les droites, plans, ... de la géométrie usuelle, mais des "espaces de dimension $\alpha \in [0, 1]$" i.e., une géométrie continue. La force de cette découverte transparaît très clairement quand on lit le texte original de Murray et von Neumann ([**38**], [**39**], [**40**], [**41**]). On peut comme dans le cas usuel parler de l'intersection de "sous-espaces" et le projecteur correspondant $e \vee f$. est le plus grand projecteur majoré par e et f. On peut de même parler du sous-espace engendré par e et f et le projecteur correspondant est noté $e \wedge f$ L'égalité fondamentale est alors

$$\dim(e \wedge f) + \dim(e \vee f) = \dim(e) + \dim(f) \quad \forall e, f.$$

Ces propriétés permettent de prolonger la notion de dimension continue aux représentations de R dans l'espace de Hilbert, disons aux R-modules, de telle sorte que ces modules sont classifiés exactement par leur dimension: $\dim_R(\mathfrak{h})$ qui peut être n'importe quel réel $\in [0, +\infty]$.

Supposons maintenant avoir pour chaque pavage $T \in X$ un ensemble dénombrable Z_T comme l'ensemble Δ^0 des sommets du pavage T. On peut alors, modulo des problèmes de mesurabilité, définir un module sur R de la manière suivante. Pour tout $z \in K$ considérons l'espace de Hilbert

$$\mathfrak{h}_z = \ell^2(Z_{T_z}) \quad \text{où} \quad T_z \text{ est le pavage associé à } z$$

toujours modulo les problèmes de mesurabilité, on peut alors considérer l'espace de Hilbert $\mathfrak{h}$ des sections de carré intégrable pour la mesure μ de la famille $\mathfrak{h}_z$, i.e., un élément de $\mathfrak{h}$ est une section $\xi_z \in \mathfrak{h}_z \quad \forall z \in K$ et sa norme est donnée par $\int \|\xi_z\|^2 \, d\mu(z) = \|\xi\|^2$. L'espace $\mathfrak{h}$ est un R-module car lorsque $z, z' \in K$ sont équivalents pour la relation d'équivalence $\mathfrak{R}$ les pavages T_z et $T_{z'}$ sont les mêmes, ce qui montre que l'on a une identification canonique des espaces $\mathfrak{h}_z$ et $\mathfrak{h}_{z'}$. On peut donc donner un sens à la formule:

$$(a\xi)_z = \sum a_{(z,z')}\xi_{z'}$$

qui définit l'action de R dans $\mathfrak{h}$. Bien sûr nous avons allègrement négligé le problème de mesurabilité, et pour le traiter il faut que l'ensemble dénombrable Z_T associé au pavage T dépende en quelque sorte mesurablement de T. Au lieu de donner une définition il vaut mieux donner un exemple, T

est construit à partir de deux motifs et on peut par exemple associer à T l'ensemble des motifs A qui apparaissent dans T ou l'ensemble des motifs B, des motifs AB, des arêtes $\Delta^1(T)$ du pavage etc.... La théorie de la dimension continue de Murray et von Neumann montre alors que étant donné deux familles $(Z_T)_{T\in X}$ et $(Z'_T)_{T\in X}$ d'ensembles dénombrables, pour que l'on puisse construire pour μ presque tout $T \in X$ une bijection mesurable de Z_T avec Z'_T il faut et il suffit que les *dimensions continues* $\dim_R(\mathfrak{h})$ et $\dim_R(\mathfrak{h}')$ soient égales. On peut montrer en utilisant le théorème ergodique que par exemple pour l'ensemble Z_T des motifs d'un certain type t dans T, la dimension continue $\dim_R(\mathfrak{h})$ s'obtient comme la *densité ou fréquence d'apparition* de t dans T, i.e., on prend un très grand disque D dans T et on divise le nombre de motifs de type t qui apparaissent dans D par la surface du disque, quand le rayon de D tend vers ∞, la limite est la densité cherchée, elle existe presque surement modulo μ. L'on voit donc dans cet exemple que les *dimensions continues* de Murray et von Neumann ont la même relation par rapport aux dimensions entières classiques, que la notion de densité d'un ensemble infini a par rapport à la notion classique de cardinalité d'un ensemble fini. Ainsi il est facile d'imaginer un sous ensemble Z de $\mathbf{N}$ de densité un nombre $\alpha \in [0,1]$, il suffit que

$$\lim_{N\to\infty} \frac{1}{N}\text{Card}(Z \cap \{0,1,\dots,N\}) = \alpha.$$

La notion de dimension continue est exactement analogue et il est clair que dans l'exemple ci-dessus α est la dimension "relative" de $\ell^2(Z)$ par rapport à $\ell^2(\mathbf{N})$. Mais revenons à la C^*-algèbre A qui remplace $C(X)$ de notre point de vue. Alors que la dimension continue de Murray et von Neumann peut prendre toutes les valeurs réelles positives pour les projecteurs de R (ou des matrices $M_n(R)$), comme $\dim(e) = \tau(e)$, cette dimension ne peut pour les projecteurs qui appartiennent à A prendre ses valeurs que dans le sous groupe $\mathbf{Z} + ((1+\sqrt{5})/2)\mathbf{Z}$. En résumé, nous avons dans cet exemple montré que la "topologie" de X est loin d'être triviale, qu'elle donne naissance à la C^*-algèbre A qui est une C^*-algèbre *simple*, uniquement caractérisée comme C^*-algèbre (à équivalence de Morita près) par:

(1) A est limite inductive d'algèbres de dimension finie, on dit que A est approximativement finie (ou AF).

(2) $$(K_0(A), K_0(A)^+ (= \left(\mathbf{Z}^2, \left\{(a,b); \left(\frac{1\sqrt{5}}{2}\right) a + b \geq 0\right\}\right).$$

Notons enfin que la C^*-algèbre A est exactement celle qui intervient dans la construction des sous-facteurs d'indice inférieur à 4 de von Jones **[32]**, pour l'indice égal au nombre d'or. On peut exploiter cela pour décrire explicitement le sours-facteur dans cette situation géométrique et cela suggère une explication générale pour l'apparence des nombres: $sin\frac{\pi}{n}$ conjecturée dans **[28]** pour les pavages quasi périodiques plus généraux.

La géométrie non-commutative a pour but de comprendre et d'utiliser des espaces semblables à l'espace X ci-dessus, il faut pour cela:

(1) Disposer d'un nombre suffisant d'exemples non-triviaux.
(2) Développer l'analogue non-commutatif des outils classiques y compris le calcul différentiel et la géométrie Riemannienne.

Pour les applications il faut distinguer celles qui portent

(a) sur la géométrie classique de *familles* Y_x de variétés ordinaires en général non-compactes, paramétrées par $x \in X$
(b) sur la géométrie moins classique de l'espace X des paramètres.

Nous allons maintenant, en faisant référence à des articles développés esquisser quel est l'état de la situation en ce qui concerne (1), (2), (a), et (b).

2. Feuilletages et théorème de l'indice avec mesure transverse. Une construction assez générale d'espaces X semblable à l'espace des pavage de Penrose est celle des feuilletages qui est traitée dans **[12]**, **[13]**, **[16]**, et **[18]**. Si M est une variété (compacte) et F un sous-fibré intégrable de son fibré tangent TM, l'espace X des feuilles (du feuilletage de M (où $\dim L = \dim F$ et $T_xL = F_x$ pour tout $x \in L$) est en général un espace dont la topologie à la même type de pauvreté que celle de l'espace des pavages de Penrose. On peut récupérer une structure non-triviale grâce aux topos de Grothendieck mais la difficulté consiste à garder le contact avec les outils puissants de l'analyse fonctionnelle (positivité, mesure, espaces de Hilbert...). C'est exactement ce que réalise la construction de la C^*-algèbre associée à un feuilletage, et le point de vue des topos n'apparaît que plus tard pour faire des calculs de cohomologie cyclique, analogue (au sens de (2)) de l'homologie de de Rham pour les espaces ordinaires. J'ai expliqué dans mon cours au Collège de France en 1986 comment calculer en général la cohomologie cyclique de l'algèbre des éléments suffisamment réguliers dans la C^*-algèbre de (M, F) en relation avec le topos étale naturel du feuilletage. Pour illustrer deux applications dans cet exemple des feuilletages nous allons expliquer brièvement le théorème de l'indice longitudinal et la relation des facteurs de type III avec l'invariant de Godbillon-Vey.

La notion de trace (positive) sur la C^*-algèbre A du feuilletage (M, F) correspond à la notion géométrique de mesure transverse pour le feuilletage. Une telle mesure est donnée par un courant de de Rham C sur M de dimension égale à celle de F et qui a pour propriétés

(α) $\langle C, \omega \rangle \geq 0$ pour tout forme différentielle ω, dont la restriction aux feuilles (orientées) est positive
(β) $bC = 0$ i.e., C est un courant fermé.

Grâce à C on définit une trace positive sur la C^*-algèbre A et grâce à la théorie de Murray et von Neumann on définit une dimension continue, $\dim_C$ pour une famille d'espaces de Hilbert $(\mathscr{H}_L)_{L \in X}$ indexée mesurablement par l'espace X des feuilles du feuilletage, comme nous l'avions expliqué plus haut. On peut alors énoncer une formule qui permet pour tout opérateur

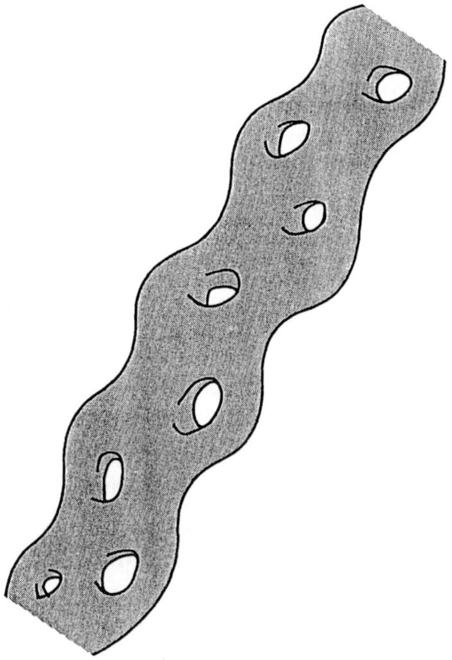

FIGURE 2. Feuille générique.

différentiel $D\colon E^+ \to E^-$ sur M qui est *longitudinalement elliptique*, i.e., se restreint aux feuilles en une famille $(D_L)_{L\in X}$ d'opérateurs elliptiques, de calculer son indice:

$$\mathrm{Ind}_C(D_L)_{L\in X} = \dim_C(\mathrm{Ker}(D_L)_{L\in X} - \dim_C(\mathrm{Ker}(D_L^*)_{L\in X}.$$

La formule est l'exact analogue de la formule de l'indice d'Atiyah-Singer: $(\mathrm{Ind}(D) = \langle \mathrm{ch}(\sigma_D) T\,d(T^*M), [T^*M]\rangle)$ à la différence près qu'il faut utiliser la classe d'homologie du courant C au lieu de la classe fondamentale de M:

$$\mathrm{Ind}_C(D_L)_{L\in X} = \langle \mathrm{ch}(\sigma_D) T\,d(T^*M), C\rangle.$$

En général la classe d'homologie du courant fermé C n'est pas rationnelle et les dimensions continues de Murray et von Neumann sont bien nécessaires dans l'énoncé. Ce théorème permet d'étendre au cadre des variétés non-compactes un énoncé comme le théorème de Riemann-Roch: $\ell(D) - \ell(\Delta - D) = \deg(D) + 1 - g$; M courbe complexe de genre g. On y remplace le degré $\deg D = \sum n_i$ d'un diviseur $D = \sum n_i P_i$, $P_i \in M$, $n_i \in \mathbf{Z}$ par la densité i.e., une limite convenable $\frac{1}{|S|} \sum_{P_i \in S} n_i$ (où S est un disque de surface $|S|$ dans M). On y remplace de même la dimension $\ell(D)$ de l'espace des sections holomorphes de carré intégrable du fibré L_D par la dimension de Murray et von Neumann de l'espace des sections holomorphes de carré intégrable du fibré L_D. Enfin le genre s'obtient à partir de la dimension de Murray et von Neumann des 1-formes harmoniques de carré intégrable et n'est pas toujours (dans la situation non-moyennable) obtenu à partir de la densité des (Cf. Figure 2) défauts topologiques.

Le théorème de l'indice "longitudinal" mentionné ci-dessus, n'utilise que très peu la structure transverse du feuilletage car il n'utilise que la *théorie de la mesure* dans le sens transverse, i.e., l'algèbre de von Neumann de type II associée à une trace sur la C^*-algèbre du feuilletage. On peut alors en déduire des résultats (John Roe [**47**]) portant sur les variétés non-compactes en utilisant le même procédé de construction de mesures que celui qui permet par exemple la démonstration du théorème combinatoire de Szemeredi "Tout sous ensemble $Z \subset \mathbf{N}$ de densité positive contient une progression arithmétique de longueur arbitraire" en utilisant la théorie ergodique [**26**].

3. Facteurs de type III, cohomologie cyclique et invariant de Godbillon-Vey. Passons maintenant à la relation entre facteurs de type III et l'invariant de Godbillon-Vey. Sous sa première forme elle est due à S. Hurder et A. Katok [**31**], mais j'expliquerai mon point de vue. La théorie modulaire des facteurs de type III a comme outil principal l'*existence* pour toute algèbre de von Neumann M (l'analogue non-commutatif de l'algèbre des fonctions mesurables essentiellement bornées) et tout poids φ sur M (L'analogue non-commutatif d'une mesure positive) d'un groupe à un paramètre $(\sigma_t^\varphi)_{t\in\mathbf{R}}$ d'automorphismes de M, le groupe d'automorphismes modulaires de φ, qui est trivial si et seulement si φ est une trace, i.e., $\varphi(xy) = \varphi(yx)$ pour tous $x, y \in M$. Ce résultat est dû aux mathematiciens Japonais M. Tomita et M. Takesaki. J'ai montré dans ma thèse, que si l'on change le choix du poids φ sur M en le remplaçant par un autre poids ψ, on ne change le groupe à un paramètre (σ_t^φ) que par un *automorphisme intérieur* de M. En d'autres termes si l'on divise le groupe $\mathrm{Aut}(M)$ des automorphismes de M par le sous-groupe normal $\mathrm{Int}(M)$ des automorphismes intérieurs: $\alpha(x) = uxu^* \forall x \in M$, on obtient un homomorphisme complètement canonique:

$$\delta\colon \mathbf{R} \to \mathrm{Out}M = \mathrm{Aut}\, M\; \mathrm{Int}\, M$$

qui ne dépend d'aucun choix auxiliaire. Ainsi la seule donnée de l' algèbre (de von Neumann) M détermine uniquement une *dynamique*, un groupe à un paramètre d'automorphismes de M, modulo les automorphismes automatiquement présents dans le cas non-commutatif que sont les automorphismes intérieurs. Ce résultat est le point de départ de la classification des facteurs de type III, i.e., des algèbres de von Neumann dont le centre $Z(M)$ est réduit à $\mathbf{C}$ et pour lesquelles δ n'est pas égal à 1. Il est évident par exemple que le noyau $T(M) = \mathrm{Ker}\,\delta$ est un sous-groupe de $\mathbf{R}$ qui est un invariant de M et les résultats de Powers, Araki et Woods sur les facteurs obtenus comme produits tensoriels infinis d'algèbres de matrices montraient que $T(M)$ distingue de très nombreux facteurs. (Pour $M = \otimes_{\nu=1}^{\infty}(M_{n_\nu}(\mathbf{C}), \varphi_\nu)$ où φ_ν est l'état sur M_{n_ν} associé à la liste de valeurs propres $\lambda_{\nu,1}, \ldots, \lambda_{\nu,n_\nu}$), réels positifs de somme 1, on a $t \in T(M)$ ssi $\sum_{\nu=1}^{\infty}(1 - |\sum \lambda_{\nu,j}^{1+it}|) < \infty$.)

Cette dynamique canonique δ des facteurs de type III permet de définir un invariant du facteur M qui est un *flot ergodique* $(W(M), W_\lambda)$ où $W(M)$ est un espace mesurable ordinaire muni d'une classe de mesure et $(W_\lambda)_{\lambda \in \mathbf{R}_+^*}$ est un groupe à un paramètre de transformations, ou flot, sur $W(M)$, dont le paramètre naturel est le groupe dual $\mathbf{R}_+^*$ de $\mathbf{R}$. Comme ce flot a une description intrinsèque en termes de classes de poids sur M, l'action de $\mathbf{R}_+^*$ étant simplement la multiplication $\varphi \in \lambda\varphi$, il porte le nom de *flot des poids* de M. [**21**], [**35**].

Mais revenons aux feuilletages (V, F) de variétés compactes et au lien entre type III et invariant de Godbillon-Vey. Soit X l'espace des feuilles d'un feuilletage (V, F). A la classe de la mesure de Lebesgue sur V correspond une "classe de mesure" sur X et donc une algèbre de von Neumann, en général non-commutative, obtenue en complétant la C^*-algèbre du feuilletage.

La dynamique canonique δ de l'algèbre de von Neumann $M =$ "$L^\infty(X)$" est un phénomène purement non-commutatif et nous allons obtenir la classe de Godbillon-Vey comme classe d'homologie de X à partir de cette dynamique et de la classe fondamentale de X. En supposant pour simplifier que F est de codimension 1 et en parlant abusivement de courants (de de Rham) pour repésenter l'homologie de X on aboutit à la formule suivante:

$$(*) \qquad \text{Classe de Godbillon-Vey} = i_H[\mathring{X}]$$

où:

(a) $[X]$ désigne la classe fondamentale de X.

(b) $[\mathring{X}]$ désigne la derivée en $t = 0$ de la transformée de cette classe par la dynamique δ de M.

(c) i_H désigne la contraction du courant $[\mathring{X}]$ par le "champ de vecteur" engendrant la dynamique δ.

On démontre alors que l'on peut localiser GV par rapport à n'importe quel sous-ensemble mesurable $P \subset W(M)$ de l'espace du flot des poids de M et que la classe obtenue est une classe d'homologie sur X, invariant quand on remplace P par $W_\lambda(P)$ où W_λ est le flot des poids. Il en résulte que dès que la classe GV de Godbillon-Vey est non nulle on peut construire sur le flot des poids une *mesure finie invariante* et en particulier que M est de type II car le flot des poids d'un facteur de type II est celui des translations de $\mathbf{R}$ et n'a pas de mesure invariante.

Bien sûr l'égalité (*) ci-dessus surréaliste tant que l'on n'a pas défini proprement ce qu'est l'homologie de l'espace X, et en fait ce qu'est la théorie des courants de de Rham sur X. C'est l'explication de ce point fort important qui nous conduit à aborder la question (2) ci-dessus et à parler de la cohomologie cyclique.

Pour une variété différentiable ordinaire Y on peut s'amuser à donner diverses présentations du calcul des formes différentielles ne s'appuyant que sur l'algèbre $\mathfrak{a} = C^\infty(Y)$ des fonctions de classe C^∞ sur Y. La plus ordinaire aurait pour "mots clefs" dérivations et algèbre extérieure. Je vais essayer d'expliquer pourquoi ce point de vue est, à mon avis, inadapté au cas non-commutatif, puis pourquoi l'on est conduit à la cohomologie cyclique. D'abord, la notion de dérivation d'une algèbre $\mathfrak{a}$, i.e., d'application linéaire $D: \mathfrak{a} \to \mathfrak{a}$ vérifiant $D(xy) = D(x)y + xD(y) \forall x, y \in \mathfrak{a}$ correspond à la notion d'homomorphisme ρ d'un groupe (discret Γ à valeurs dans $\mathbf{C}$: $\rho: \Gamma \to \mathbf{C}$. (Si par exemple on associe à Γ l'algèbre $\mathfrak{a} = \mathbf{C}\Gamma$ du groupe, à tout homomorphisme ρ correspond la dérivation D_ρ, $(D_\rho a)(g) = \rho(g)a(g)$ pour tout $a \in \mathbf{C}\Gamma$). Or comme $\mathbf{C}$ est commutatif il n'existe en général, pour Γ non-commutatif, que peu d'homomorphismes de Γ dans $\mathbf{C}$. Leur noyau contient toujours le sous-groupe des commutateurs et si Γ est parfait, i.e., égal au sous-groupe engendré par les commutateurs $g_1 g_2 g_1^{-1} g_2^{-1}$, il n'y aura pas d'homomorphisme non-trivial de Γ dans $\mathbf{C}$. Par contre, si l'on interprète les homomorphismes de Γ dans $\mathbf{C}$ comme la 1-cohomologie $H^1(\Gamma, \mathbf{C})$ de Γ à coefficients dans $\mathbf{C}$ on voit que dans le cas général, les groupes de cohomologie $H^n(\Gamma, \mathbf{C})$ sont parfaitement adaptés aux groupes non-commutatifs. La seule erreur à éviter est de croire que H^1 et une opération convenable de produit extérieur engendrent H^n en général.

Montrons, toujours dans le cas des algèbres de groupes, comment associer à un cocycle de groupe $c \in Z^n(\Gamma, \mathbf{C})$, l'analogue d'un courant de de Rham fermé sur le dual $\widehat{\Gamma}$ de Γ. Rappelons d'abord qu'un cocycle de groupe est donné de manière équivalente par:

(a) Une fonction de $n+1$ éléments de Γ, $\gamma(g_0, g_1, \dots, g_n) \in \mathbf{C}$ invariante par translations à gauche $\gamma(gg_0, gg_1, \dots, gg_n) = \gamma(g_0, \dots, g_n) \forall g \in \Gamma$, totalement antisymétrique, et fermée:

$$\sum (-1)^j \gamma(g_0, \dots, \check{g}_j, \dots, g_{n+1}) = 0 \quad \forall g_0, \dots, g_{n+1} \in \Gamma.$$

(b) Une fonction de n éléments de Γ,

$$c(g_1, \dots, g_n) = \gamma(1, g_1, g_1, g_2, \dots, g_1 g_2, \dots, g_n)$$

qui vérifie $bc = 0$, où:

$$(*) \quad (bc)(g_1, \dots, g_{n+1}) = c(g_2, \dots, g_{n+1}) - c(g_1 g_2, g_3, \dots, g_{n+1}) + \cdots \\ + (-1)^j c(g_1, \dots, g_j g_{j+1}, \dots, g_{n+1}) + \cdots \\ + (-1)^n c(g_1, \dots, g_n g_{n+1}) + (-1)^{n+1} c(g_1, \dots, g_n).$$

Un courant de de Rham C sur une variété V est entièrement caractérisé par la fonctionnelle multilinéaire suivante sur l'algèbre des fonctions de classe C^∞ sur V:

$$\tau(f^0, \dots, f^n) = \langle C, f^0 \, df^1 \wedge \cdots \wedge df^n \rangle.$$

L'égalité $d(fg) = (df)g + fdg$ montre que cette fonctionnelle vérifie $b\tau = 0$ où l'opérateur de cobord b est donné par

$(*)'$

$$\begin{aligned}(b\tau)(f^0,\dots,f^{n+1}) = \tau(f^0f^1,\dots,f^{n+1}) - \tau(f^0,f^1f^2,\dots,f^{n+1}) + \cdots \\ + (-1)^j\tau(f^0,\dots,f^jf^{j+1},\dots,f^{n+1}) + \cdots \\ + (-1)^n\tau(f^0,\dots,f^nf^{n+1}) + (-1)^{n+1}\tau(f^{n+1}f^0,\dots,f^n).\end{aligned}$$

De plus le courant C est un courant fermé au sens de de Rham si et seulement si la fonctionelle τ est cyclique, i.e., vérifie

$$\tau(f^1,\dots,f^n,f^0) = (-1)^n\tau(f^0,\dots,f^n).$$

Utilisons la ressemblance évidente entre $(*)$ et $(*)'$ pour associer à un n-cocycle $c \in Z^n(\Gamma,\mathbf{C})$ une fonctionnelle τ vérifiant $(*)'$ sur l'algèbre $\mathfrak{a} = \mathbf{C}\Gamma$. Comme τ est *multilinéaire* et $\mathbf{C}\Gamma$ est linéairement engendrée par Γ il suffit de définir $\tau(a^0,\dots,a^n)$ lorsque les a^i sont des éléments g_i de Γ. On pose alors:

$$\begin{aligned}&\tau(g^0,\dots,g^n) = c(g^1,\dots,g^n) \quad &&\text{si } g^0\dots g^n = 1\\ &\tau(g^0,\dots,g^n) = 0 &&\text{si } g^0\dots g^n \neq 1.\end{aligned}$$

On vérifie alors en utilisant $(*)$ que τ vérifie $(*)'$, et de plus que comme c provient de γ qui est antisymétrique, τ est *cyclique*, $\tau(a^1,\dots,a^n,a^0) = (-1)^n\tau(a^0,a^1,\dots,a^n)\forall a^i \in \mathfrak{a}$. Ainsi tout cocycle de groupe $c \in Z^n(\Gamma,\mathbf{C})$ définit sur l'algèbre du groupe $\mathfrak{a} = \mathbf{C}\Gamma$ un *n-cocycle cyclique* τ, et cette construction n'est plus dépendante de la notion troup unidimensionnelle de dérivation.

Mais revenons à l'espace X des feuilles d'un feuilletage (V,F) de codimension 1, transversalement orienté, et montrons comment construire un 1-cocycle, représentant la "classe fondamentale" $[X]$, sur l'algèbre du feuilletage. Si τ est un n-cocycle cyclique sur une algèbre $\mathfrak{a}$ et l'on remplace $\mathfrak{a}$ par l'algèbre des matrices $M_k(\mathfrak{a}) = B$, on obtient un cocycle cyclique τ' sur B en posant:

$$\tau'(\mu_0\otimes a_0,\dots,\mu_n\otimes a_n) = \operatorname{Trace}(\mu_0\dots\mu_n)\tau(a_0,\dots,a_n)$$

où les $\mu_i \in M_k(\mathbf{C})$ et $a_i \in \mathfrak{a}$. Plus généralement, si $\mathfrak{a}$ et B sont des algèbres Morita équivalentes les cocycles cycliques sur $\mathfrak{a}$ et B se correspondent de manière naturelle. Nous utiliserons ce principe pour remplacer l'algèbre du feuilletage par l'algèbre équivalente $\mathfrak{a}$ obtenue en prenant une sous-variété T de V transverse au feuilletage (Cf. Figure 3) telle que toute feuille rencontre T (on ne suppose T ni fermée ni connexe). L'espace X des feuilles du feuilletage est alors le quotient de T par la relation d'équivalence: $\mathfrak{R}x \sim y$ si et seulement si x et y sont sur la même feuille L de (V,F). En fait, dès que la feuille L a une holonomie non-triviale il faut être plus précis et donner une classe d'holonomie de chemin γ dans L, allant de x à y. On obtient ainsi un paramètre discret supplémentaire qui permet de remplacer la relation $\mathfrak{R}$ par une variété $\widetilde{\mathfrak{R}}$ qui est un groupoïde différentiable. Toute fonction de classe

C^∞ à support compact $(a(\gamma), \gamma \in \widetilde{\mathfrak{R}})$ définit un élément de l'algèbre $\mathfrak{a}$ et le produit est donné par:

$$(ab)(\gamma) = \sum_{\gamma_1\gamma_2=\gamma} a(\gamma_1)b(\gamma_2).$$

Comme F est de codimension 1, la variété T est de dimension 1, de même que la variété $\widetilde{\mathfrak{R}}$ qui est orientée.

La classe fondamentale de l'espace $X = V/F$ des feuilles est donnée par le 1-*cocycle cyclique* τ sur $\mathfrak{a}$:

$$\tau(a,b) = \int_{\widetilde{\mathfrak{R}}} a(\gamma^{-1})\, db(\gamma).$$

Dans le terme de droite, db est la différentielle ordinaire de la fonction b sur la variété $\widetilde{\mathfrak{R}}$, c'est donc une 1-forme sur $\widetilde{\mathfrak{R}}$, de même que le produit $a(\gamma^{-1})\, db(\gamma)$ que l'on intègre sur la variété orientée $\widetilde{\mathfrak{R}}$.

On vérifie directement que τ est un 1-cocycle cyclique sur $\mathfrak{a}$ ce qui permet de donner un sens à la classe fondamentale de X et de démontrer la formule $GV = i_H\dot{\tau}$, dont on déduit que si la classe de Godbillon-Vey est non-nulle l'algèbre de von Neumann du feuilletage est de type III. Ainsi pour un espace "quantique" comme l'espace X on dispose de l'analogue non-commutatif de la théorie de la mesure et de la théorie des courants de de Rham. Mais comme une classe de mesure (celle de Lebesgue) engendre une dynamique non-trivale δ, les relations entre les deux théories sont beaucoup plus intéressantes que dans le cas classique. Par exemple la non-nullité d'une classe telle celle de Godbillon-Vey impose une propriété qualitative (type III) de la classe de mesure.

4. Cohomologie cyclique et K-théorie des espaces quantiques. Pour utiliser la cohomologie cyclique pour un espace "quantique" X, il est crucial de disposer de l'accouplement entre cocycles cycliques et "fibrés vectoriels" sur X. Dans le cas d'un espace compact X ordinaire il y a équivalence entre la donnée d'un fibré vectoriel (complexe) E sur X et celle d'un module projectif de type fini $\mathscr{E}$ sur l'algèbre $A = C(X)$. La correspondance est obtenue en prenant pour $\mathscr{E}$ le module $C(X, E)$ sur $C(X)$ des sections continues de E. Or la notion de module projectif de type fini sur une algèbre A ne demande en rien que cette algèbre soit commutative. De plus, à isomorphisme près tout module projectif de type fini sur A est donné par un idempotent $e \in M_n(A)$ de l'algèbre des matrices sur A, dont la classe d'équivalence, au sens de Murray et von Neumann est uniquement déterminée par la classe d'isomorphisme du module. Ainsi la K-théorie de l'espace X s'obtient comme le groupe déduit du semi-groupe des classes d'équivalence d'idempotents $e \in M_n(A)$. La construction classique du caractère de Chern d'un fibré vectoriel s'adapte très simplement pour donner l'accouplement suivant entre cocycles cycliques (de dimension paire) et K-théorie:

$$\langle \tau, e\rangle = \tau(e, \dots, e)$$

On démontre en effet que le nombre obtenu ne dépend que de la classe de l'idempotent e dans le groupe $K_0(A)$ et de plus que si $e(t)$ est une famille de classe C^1 d'idempotent de a, la valeur de $\langle \tau, e(t) \rangle$ ne dépend pas de t.

Dans le cas où X est une variété ordinaire et $\tau = \tau_c$ le cocycle cyclique sur l'algèbre $\mathfrak{a} = C^\infty(X)$ associé à un courant de de Rham C tel que $bC = 0$, le caractère de Chern d'un fibré vectoriel E sur X défini par un idempotent $e \in M_n(\mathfrak{a})$ est déterminé comme forme différentielle sur X par la connexion Grassmannienne, et l'accouplement ci-dessus est l'évaluation $\langle C, \operatorname{ch} E \rangle$ de C sur cette forme différentielle.

La construction de Chern Weil du caractère de Chern d'un fibré vectoriel E par connexion et courbure apparait ainsi comme un cas particulier de l'accouplement entre K-théorie et cohomologie cyclique, lequel se décrit également en termes de connexion et courbure (Cf. [**10**]). Comme la valeur $\langle \tau, e \rangle$ de cet accouplement ne dépend que de la classe de cohomologie de τ et de la classe de K-théorie, $[e] \in K_0(\mathfrak{a})$, il permet en particulier de détecter la non-trivialité d'une telle classe. Dans le cas d'une variété X on a $\mathfrak{a} = C^\infty(X)$ mais la K-théorie de l'algèbre $\mathfrak{a}$ ne dépend que de la topologie de X. En effet l'inclusion naturelle de $\mathfrak{a}$ dans la C^*-algèbre $A = C(X)$ définit un isomorphisme en K-théorie. L'égalité $K_0(\mathfrak{a}) \cong K_0(A)$ et l'accouplement ci-dessus permettent donc de définir des invariants *topologiques* de fibrés vectoriels sur X. L'utilisation de la structure différentiable de X pour obtenir des renseignements purement topologiques est un exemple typique des méthodes de la *topologie différentielle.* En général un espace quantique X tel que l'espace des feuilles d'un feuilletage ou le dual d'un groupe discret possède une structure suffisamment riche pour que l'on dispose non seulement de la C^*-algèbre $A =$ "$C(X)$" mais aussi d'une sous-algèbre dense $\mathfrak{a}$ de A qui joue le rôle de l'algèbre des fonctions de classe C^∞ sur X, ou tout au moins est contenue dans l'algèbre des fonctions de classe C^∞. Par exemple, si Γ est un groupe discret, la C^*-algèbre réduite $C_r^*(\Gamma)$ de Γ, (engendrée par la représentation régulière gauche de Γ dans $\ell^2(\Gamma)$) contient somme sous- algèbre dense $\mathfrak{a} = \mathbf{C}\Gamma$. Si Γ est commutatif de type fini les éléments de $\mathfrak{a}$ sont des fonctions analytiques réelles sur le tore dual de Γ.

La stratégie générale pour analyser la K-théorie et l'homologie d'un espace quantique X comprend les étapes suivantes:

(α) Calcul de la cohomologie cyclique $HC(\mathfrak{a})$
(β) Construction géométrique d'éléments de la K-théorie $K_0(A)$
(γ) Prolongement de l'accouplement $\langle \tau, [e] \rangle$ de $K_0(\mathfrak{a})$ à $K_0(A)$.

Une fois ces trois étapes réalisées on peut alors l'espace X comme un espace ordinaire, en particulier on peut utiliser le théorème de l'indice à la "Atiyah-Singer" pour les familles d'opérateurs elliptiques $(D_x)_{x \in X}$ paramétrées par X. Nous allons voir en effet que l'indice d'une telle famille est un élément de $K_0(A) =$ "$K^0(X)$" et ce n'est que dans ce groupe de K-théorie (qui est un groupe abélien dénombrable) et non dans le groupe $K_0(\mathfrak{a})$, que se manifestent les propriétés d'invariance topologique ou d'annulation telles que:

(a) L'invariance par homotopie de l'indice de l'opérateur de signature.
(b) L'annulation de l'indice de l'opérateur de Dirac quand la courbure scalaire est > 0.

Il nous reste à décrire trois étapes (α), (β), (γ).

α. **Cohomologie cyclique** $HC(\mathfrak{a})$. Le cas des groupes discrets: $\mathfrak{a} = \mathbf{C}\Gamma$ était une motivation importante pour la cohomologie cyclique. Comme nous l'avons montré plus haut (Cf. [**10**]) tout cocycle de groupe donne naissance à un cocycle cyclique τ_c sur $\mathfrak{a}$.

Réciproquement D. Burghelea a calculé en général [**18**] la cohomologie cyclique de $\mathfrak{a} = \mathbf{C}\Gamma$. En particulier il en résulte que $H^*(B\Gamma, \mathbf{C})$, i.e., la cohomologie de Γ en est un facteur direct.

Passons au cas des feuilletages, le rôle de Γ est alors tenu par le graphe G du feuilletage [**48**] qui est une variété et un groupoïde différentiable. En particulier l'algèbre $\mathfrak{a}$ est l'algèbre de convolution des fonctions C^∞ à support compact sur G. La cohomologie cyclique de $\mathfrak{a}$ est alors reliée à la cohomologie $H^*_\tau(BG, \mathbf{C})$ de l'espace classifiant BG de la catégorie topologique ou groupoïde G. Plutôt que de décrire la théorie générale (Cf. [**4**]) nous allons expliquer la signification de H^*_τ et la flèche de "localisation" de $HC(\mathfrak{a})$ vers H^*_τ avec des hypothèses simplificatrices.

En général de nombreux feuilletages différents correspondent au même espace quantique X comme espace de feuilles. On peut par exemple partir d'une variété feuilletée (V, F) et d'une transversale fermée T et effectuer de la chirurgie le long des feuilles L de (V, F) en chaque point de $L \cap T$. Cela modifie la variété V, les feuilles L mais laisse intact l'espace X des feuilles.

Le rôle de l'espace BG est le suivant: c'est l'espace total V d'un feuilletage (V, F) ayant X comme espace de feuilles mais dont chaque feuille $L^{(*)}$ est *contractile*. En général on ne peut pas toujours trouver un tel couple (V, F) de dimension finie mais nous ferons cette hypothèse pour simplifier la discussion. On a donc $BG = V$. Dans la notation $H^*_\tau(BG, \mathbf{C})$, τ désigne le fibré transverse au feuilletage, i.e., $\tau_x = T_x(V)/F_x \forall x \in V$. Ce fibré τ n'est pas nécessairement orienté ou orientable et $H^*_\tau(V, \mathbf{C})$ désigne la cohomologie tordue correspondante. Supposons pour simplifier que τ est orienté. On désigne alors par Ω^k le faisceau sur V des sections généralisées (i.e., de classe $C^{-\infty}$) du fibré $\Lambda^k \tau^*$ qui sont *constantes le long des feuilles*, i.e., dont la différentielle longitudinale est nulle. Le faisceau Ω^k joue (Cf. [**29**]) un rôle analogue au faisceau des formes différentielles holomorphes sur une variété complexe. En utilisant la différentiation transverse et la trivialité locale du feuilletage on obtient une résolution du faisceau constant de la forme $\mathbf{C} \to \Omega^0 \xrightarrow{d} \Omega^1 \longrightarrow \cdots \xrightarrow{d} \Omega^t \longrightarrow 0$[1] où $t = \dim \tau$ est la dimension transverse.

[1] En fait le revêtement d'holonomie de chaque feuille.

La cohomologie de chacun des faisceaux Ω^k est en général non-triviale et se calcule (Cf. **[29]**) en résolvant chaque Ω^k par des formes différentielles longitudinales à valeurs dans $\Lambda^k \tau^*$. La suite spectrale correspondant à la résolution (*) est exactement adaptée à la description de la flèche de localisation:

$$HC(\mathfrak{a}) \xrightarrow{\ell} H_\tau^*(V, \mathbf{C}) .$$

Pour localiser un n-cocycle cyclique φ sur $\mathfrak{a}$ on commence par le restreindre aux sous-algèbres $\mathfrak{a}(U)$ de $\mathfrak{a}$ qui correspondent à la restriction du feuilletage F à un petit ouvert U. Comme le feuilletage est localement trivial la situation locale est facile à analyser et on obtient ainsi une section globale du faisceau Ω^{t-n} sur V. (Localement on obtient un courant transverse de dimension n mais en utilisant l'orientation de τ on le considère comme section généralisée de $\Lambda^{t-n}(\tau^*)$).

Si par exemple φ est la trace sur $\mathfrak{a}$ associée à une mesure tranverse invariante par holonomie, φ est un o-cocycle cyclique et li lui correspond une section globale du faisceau Ω^t sur V qui n'est autre que la mesure transverse. Si φ est le cycle fondamental de l'espace V/F, dont nous avons donné un exemple plus haut dans le cas de codimension un, φ est un cocycle cyclique de dimension t sur $\mathfrak{a}$ et la section globale correspondante de Ω^0 sur V est la fonction constante égale à 1.

En affinant la construction ci-dessus on en déduit une surjection ℓ de la cohomologie cyclique $HC(\mathfrak{a})$ vers la cohomologie $H_\tau^*(V, \mathbf{C})$ et l'on démontre que c'est un isomorphisme pour la cohomologie cyclique périodique $H(\mathfrak{a})$. L'originalité d'espaces quantiques comme $X = V/F$ se manifeste en particulier par l'existence de cycles non-triviaux sur X dont la dimension excède la dimension apparente t de X. Par cycle sur X nous entendons un cocycle cyclique $\varphi \in HC^n(\mathfrak{a})$, l'homologie entre deux cycles étant donnée par l'image de l'opérateur B (Cf. **[10]**) et la notion de dimension correspondant à la filtration de la cohomologie cyclique périodique par l'image des $HC^n(\mathfrak{a})$.

Nous renvoyons à **[4]**, **[20]**, et **[18]** pour plus de détails.

β. **K-théorie $K(A)$.** Dans le cas des groupes discrets, $A = C_r^*(\Gamma)$, la construction d'éléments non-triviaux de la K-théorie de la C^*-algèbre A a pour origine les travaux de Mishchenko et Kasparov sur la conjecture des hautes signatures de Novikov **[34]**, **[36]**. Cela permet de définir un élément $\mathrm{Ind}_\Gamma(D) \in K_0(A)$ chaque fois que l'on part d'une variété compacte M, d'un opérateur elliptique D sur M et d'un revêtement Galoisien de M de groupe Γ. Cette construction se retrouve sous diverses formes dans **[4]**, **[20]**, et **[19]**, mais en l'affinant suffisamment **[19]** on montre que cet *indice* Γ *équivariant* du relevé $\widetilde{D}$ de D sur le revêtement $\widetilde{M}$ définit en fait un élément du groupe: $K_0(\mathfrak{a} \otimes \mathfrak{R})$ où $\mathfrak{a} = \mathbf{C}\Gamma$ et $\mathfrak{R}$ est une algèbre universelle, celle des matrices infinies $(a_{ij})_{i,j \in \mathbf{N}}$ de nombres complexes, qui sont à décroissance rapide. On obtient ainsi en général une flèche μ de la K-homologie $K_*(B\Gamma)$ vers la K-théorie de la C^*-algèbre $C_r^*(\Gamma)$. La conjecture de Novikov forte énonce l'injectivité rationnelle de μ et a été démontrée par Mischchenko et Kasparov dans de nombreux cas par les méthodes de la K-théorie bivariante et l'accouplement avec la K-homologie de $C_r^*(\Gamma)$ **[37]**, **[34]**.

Pour appliquer à ce problème les méthodes de topologie différentielle il est nécessaire de résoudre le problème (3) ci-dessus, mais le théorème de l'indice suivant montre bien que la cohomologie cyclique de $\mathfrak{a}$ est suffisamment riche pour séparer les éléments de $K_*(B\Gamma)$. Ce théorème **[19]** calcule l'accouplement $\langle \tau_c, \operatorname{Ind}_\Gamma(D)\rangle$ entre le cocycle cyclique sur $\mathfrak{a} \otimes \mathfrak{R}$ associé à un cocycle de groupe $c \in Z^{2q}(B\Gamma, \mathbf{C})$ et l'indice équivariant d'un opérateur elliptique D sur M. Au recouvrement Γ-équivariant $\widetilde{M}$ de M correspond une application classifiante: $\Psi\colon M \to B\Gamma$ et la formule de l'indice est la suivante:

$$\langle \tau_c, \operatorname{Ind}_\Gamma(D)\rangle = \frac{1}{(2\pi)^q}\frac{q!}{(2q)!}\langle \operatorname{ch} D\, T\, dTM\Psi^*(c), [M]\rangle .$$

Les seules nouveautés dans le terme de droite par rapport au théorème d'Atiyah-Singer sont la constante numérique $\frac{1}{(2\pi)^q}q!(2q)!$ et la présence de la classe de cohomologie $\Psi^*(c)$. En affinant quelque peu (Cf. **[4]**) la construction de μ pour le cas où le groupe Γ a de la torsion on peut raisonnablement penser que cette flèche est un isomorphisme (Cf. **[4]**). Ce résultat aurait de nombreuses implications et la recherche dans ce sens est très active.

Passons au cas des feuilletages, i.e., des espaces quantiques $X = V/F$. On construit à nouveau une flèche μ de la K-homologie tordue $K_{*,\tau}(BG)$ vers la K-théorie de la C^*-algèbre $A = C^*(V, F)$ du feuilletage.

Cela permet d'associer à toute famille d'opérateurs elliptiques $(D_x)_{x\in X}$ un indice, $\operatorname{Ind}(D) \in K_0(A)$. L'exemple typique d'une telle famille est obtenu en restreignant à chaque feuille L du feuilletage $(V, F)^2$ un opérateur différentiel sur V qui est *longitudinalement elliptique* (Cf. **[20]**). Si les feuilles de (V, F) sont contractiles on a $BG = V$ et en supposant que le fibré F est K-orienté on a alors:

$$K_{*,\tau}(BG) = K_{*,\tau}(V) = K^*(V), \quad \text{la } K\text{-théorie de } V.$$

La flèche μ associe alors à tout fibré E sur V, $[E] \in K^*(V)$ l'indice $\operatorname{Ind}(D) \in K_0(A)$ de la famille $(\mathscr{D}_E)_{L\in V/F}$ des opérateurs de Dirac longitudinaux à coefficient dans le fibré E. (La K-orientation de F est utilisée pour définir l'opérateur de Dirac).

Le premier exemple non-trivial de feuilletage est le feuilletage de Kronecker:$dy = \theta\, dx$ du tore $\mathbf{T}^2$. La C^*-algèbre correspondante A est Morita équivalente à la C^*-algèbre A_θ **[46]** engendrée par deux unitaires u, v tels que $uv = \exp(2\pi i\theta)vu$ dès que θ est irrationnel. (Si θ est rationnel A est Morita équivalente à $C(S^1)$). La K-théorie de la C^*-algèbre A_θ se calcule grâce au résultat remarquable de Pimsner et Voiculescu **[43]** sur les produits croisés. En particulier le groupe $K_0(A_\theta)$ est égal à $\mathbf{Z}^2$ ordonné par le demi

[2] Plus précisément au revêtement d'holonomie $\widetilde{L}$ de L.

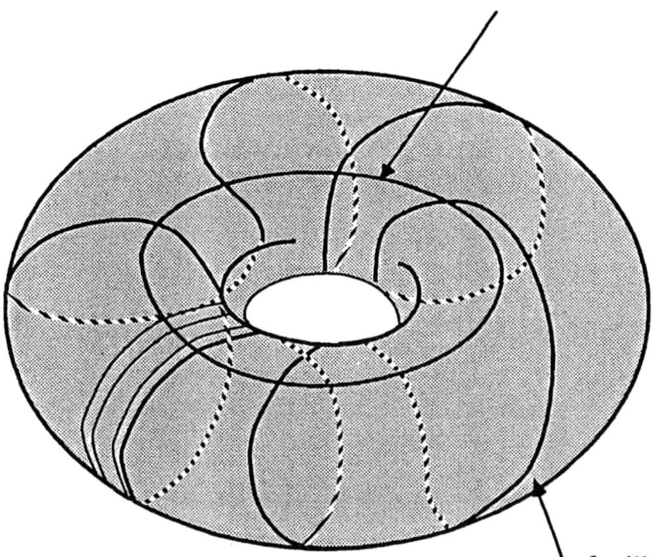

FIGURE 3. Feuilletage du tore.

plan $\{(n,m); n+m\theta \geqq 0\}$. Ceci permet de vérifier que la flèche μ est un isomorphisme, ce qui est un résultat général pour les feuilletages de dimension 1 (Cf. [**15**]). En fait il est possible dans cet exemple de décrire de manière simple et géométrique les éléments de $K_0(A)$ à partir des *transversales fermées* au feuilletage (V,F).

On démontre qu'à toute transversale T au feuilletage (V,F) correspond un homomorphisme injectif d'une C^*-algèbre Morita équivalente à $C_0(T)$ dans $C^*(V,F)$. Il en résulte que si T est fermée, l'élément $1 \in C(T)$ définit un idempotent, unique à équivalence près, $e(T) \in C^*(V,F)$. Ainsi, dans le cas du feuilletage de Kronecker avec $\theta \notin \mathcal{Q}$, chaque géodésique fermée du tore plat $\mathbf{T}^2$, donnée par $T_{n,m} = \{(nt,mt), t \in S^1\}$ (n et m premiers entre eux) définit une transversale et donc un idempotent $e_{n,m} = e(T_{n,m})$. Ce idempotents engendrent la partie positive de $K_0(A)$.

Cette construction, à partir de transversales, d'éléments, de $K_0(A)$ où $A = C^*(V,F)$ permet de formuler le théorème de l'indice directement au niveau de la K-théorie [**20**]. Pour formuler ce théorème utilisons la notation $X = V/F$ pour l'espace des feuilles et $K^0(X) = K_0(A)$, $A = C^*(V,F)$ pour la K-théorie de cet espace. Etant donnée une famille d'opérateurs elliptiques paramétrée par X, $(D_x)_{x\in X}$ telle que la famille des restrictions $(D_L)_{L\in X}$ aux feuilles d'un opérateur longitudinalement elliptique, on a défini son *indice analytique* $\mathrm{Ind}(D) \in K^0(X)$. Il reste à définir *l'indice topologique* d'une telle famille en utilisant le *symbole principal* (longitudinal) des opérateurs D_L. Or presque par définition ce symbole définit un élément de la K-théorie à support compact $K^0(F)$ où F désigne l'espace total sur V du fibré tangent aux feuilles. On utilise alors un plongement de la variété V dans $\mathbf{R}^{2N}$ pour faire de la variété ouverte F une transversale au feuilletage de $V \times \mathbf{R}^{2N}$ par $F \times \{0\}$ dont l'espace des feuilles est $Y = X \times \mathbf{R}^{2N}$. La construction ci-dessus fournit une application i de $K^0(F)$ dans $K^0(Y)$, l'indice topologique de D est alors:

$$\mathrm{Ind}_t(D) = \beta^{-1}(i\sigma(D)) \in K^0(X)$$

où $\beta\colon K^0(X) \simeq K^0(Y)$ est l'isomorphisme de périodicité de Bott qui reste vrai pour les C^*-algèbres et établit un isomorphisme entre $K_0(A)$ et $K_0(A \otimes C_0(\mathbb{R}^{2N}))$.

On peut déduire de ce théorème l'accouplement entre l'image de la flèche μ et la cohomologie cyclique $HC(\mathfrak{a})$. Dans le cas le plus simple, i.e., quand $\varphi \in HC^0(\mathfrak{a})$ est la trace sur $\mathfrak{a}$ associée à une mesure transverse on retrouve le théorème de l'indice cité plus haut:

$$\mathrm{Ind}_C(D_L)_{L\in X} = \langle \mathrm{ch}\, \sigma_D T\, d(T^*M), C\rangle\,.$$

Mais le cas général est beaucoup plus riche. Par exemple, si le feuilletage (V, F) est de codimension 1, orienté transversalement et l'on prend pour φ le 2-cocycle $\varphi = i_H \dot{\tau}$ que nous avons discuté plus haut à propos de l'invariant de Godbillon-Vey, on a:

$$\langle \varphi, \mu(\alpha)\rangle = \langle GV, \mathrm{ch}(\alpha)\rangle$$

pour toute classe de K-homologie $\alpha \in K_*(BG)$, où GV désigne la classe de Godbillon-Vey qui est une classe de cohomologie de dimension 3 sur BG.

γ. **Prolongement de l'accouplement de $K_0(\mathfrak{a})$ à $K_0(A)$.** Pour pouvoir appliquer la méthode de la topologie différentielle encore faut-il démontrer que l'accouplement entre un cocycle cyclique $\varphi \in HC(\mathfrak{a})$ et le groupe de K-théorie $K_0(\mathfrak{a})$ se prolonge au groupe $K_0(A)$ qui lui ne dépend que de la *topologie* de l'espace quantique X.

La stratégie générale consiste à prolonger le cocycle φ à une algèbre intermédiaire entre $\mathfrak{a}$ et A: $\mathfrak{a} \subset \mathfrak{a}' \subset A$ pour laquelle l'injection $i\colon \mathfrak{a}' \to A$ est un isomorphisme en K-théorie. Ceci est le cas dès que $\mathfrak{a}'$ est *stable par calcul fonctionnel holomorphe*, i.e.,:

Pour toute fonction holomorphe sur le spectre d'un élément $a \in \mathfrak{a}'$, $\mathrm{Sp}_A(a)$, on a $f(a) \in \mathfrak{a}'$.

Il n'est pas nécessaire que le cocycle φ soit continue par rapport à la norme de C^*-algèbre (cela n'arrive jamais pour les cocycles intéressants) pour qu'il se prolonge à fermeture $\mathfrak{a}'$ de $\mathfrak{a}$ par calcul fonctionnel holomorphe. Il suffit pour cela que φ vérifie certaines inégalités **[18]** qui sont par exemple automatiques pour tout 1-cocycle cyclique $\varphi(a^0, a^1)$ qui est continu en a^0 (pour la norme de A) pour tout a^1 fixé, $a^1 \in \mathfrak{a}$. Avec pas mal de travail supplémentaire (Cf. **[18]**) on en déduit que la propriété de prolongement a lieu, dans le cas des feuilletages, pour tous les cocycles cycliques qui sont *purement transverses*, i.e., qui correspondent à des sections globales fermées des faisceaux Ω^{t-n} définis plus haut. Cela inclus bien sur le cas des mesures transverses mais aussi celui de la classe fondamentale transverse $[V/F] = [X]$. Un autre cas particulier important de ce résultat est celui des classes de Gelfand-Fuchs [**27**]. En le combinant avec l'invariance par homotopie de l'indice de l'opérateur de signature on en déduit que ces classes de cohomologie $\omega \in H^*(B\,\mathrm{Diff})$ verifient la conjecture de Novikov, i.e., que l'égalité

$$\langle L_M \Psi^*(\omega), [M]\rangle$$

définit un invariant d'homotopie de tout fibré plat sur M de groupe structural le groupe Diff muni de la topologie discrète. (Ici M est compacte orientée et $\Psi\colon M \to B\,\mathrm{Diff}$ est l'application classifiante du fibré).

Pour les cocycles cycliques φ, sur l'algèbre d'un feuilletage, qui ne sont pas purement transverses la difficulté est essentiellement la même que pour les groupes discrets, cas que nous discutons maintenant.

Soient donc Γ un groupe discret et $\mathfrak{a} = \mathbf{C}\Gamma$ l'algèbre du groupe, $A = C_r^*(\Gamma)$ son adhérence normique dans l'algèbre des opérateurs dans l'espace de Hilbert $\ell^2(\Gamma)$, où Γ agit par la représentation régulière gauche. Supposons Γ de type fini et soit $F \subset \Gamma$ un système fini de générateurs, $g \to |g|$ la longueur des mots correspondante. Considérons l'opérateur D dans $\mathfrak{l}^2(\Gamma)$, de multiplication par $\ell(g) = |g|$, i.e., dans la base orthonormale $(\varepsilon_g)_{g\in\Gamma}$ canonique de $\ell^2(\Gamma)$ on a:

$$D\varepsilon_g = \ell(g)\varepsilon_g\,.$$

L'application $a \in \mathfrak{a} \xrightarrow{\delta} [D,a] \in \mathscr{L}(\ell^2)$ définit une dérivation de $\mathfrak{a}$ dans $\mathfrak{L}(\ell^2)$, l'algèbre des opérateurs bornés dans l'espace de Hilbert ℓ^2 et on définit l'espace de Schwartz $\mathscr{S}(\Gamma)$ comme la sous-algèbre de A qui est l'intersection $\bigcap_{n\in\mathbf{N}} \mathrm{Dom}\,\delta^n$ des domaines des puissances de la dérivation δ. (Il est possible d'itérer δ en itérant dans $\mathscr{L}(\ell^2)$ l'opération $x \to [D,x]$).

Il n'est pas difficile de montrer que l'algèbre $\mathscr{S}(\Gamma)$ est stable par calcul fonctionnel holomorphe dans $A = C_r^*(\Gamma)$ de sorte que:

$$K_0(\mathscr{S}(\Gamma)) = K_0(A)\,.$$

Enfin, grâce aux inégalités de U. Haagerup, P. Jolissaint, et P. de la Harpe, il est possible de prolonger le cocycle cyclique τ_c associé à un cocycle de groupe *borné* c, à toute l'algèbre $\mathscr{S}(\Gamma)$, dès que le groupe Γ est un *groupe hyperbolique*. Cela permet en particulier de démontrer la conjecture de Novikov pour ces groupes très généraux [**19**] dont la définition est implicite, et qui sont génériques parmi les groupes discrets de présentation finie.

Il existe un autre résultat remarquable (dû à J. B. Bost [**6**]) qui permet de démontrer l'égalité $K_0(\mathfrak{a}') = K_0(A)$ pour des sous-algèbres $\mathfrak{a}'$ qui ne sont plus stables par calcul fonctionnel holomorphe dans A. Il s'agit d'une adaptation au cas non-commutatif (aux espaces quantiques) du principe d'Oka

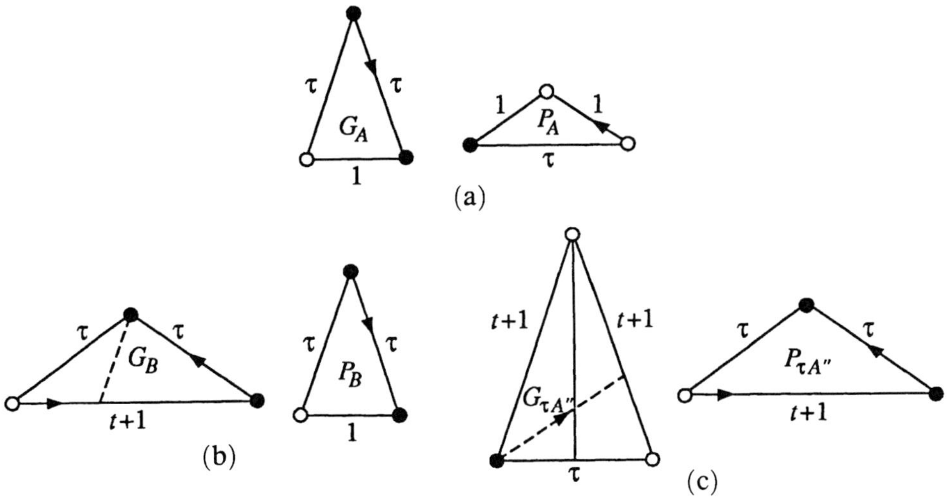

FIGURE 4

en géométrie analytique complexe. Ce résultat permet de résoudre complètement le problème de prolongement (3) pour les feuilletages de dimension 1 et pour les groupes discrets résolubles.

Pour terminer, signalons qu'au prix du passage à des cocycles de dimension infinie (où une série $\sum_n \varphi_{2n}$ remplace un cocycle φ_{2m}) on peut appliquer directement la cohomologie cyclique aux C^*-algèbres (Cf. [**17**]).

Appendice: Les pavages de Penrose. Dans cette section nous rappelons pour la commodité du lecteur des résultats classiques (Cf. [**28**]) sur les pavages quasi périodiques du plan de R. Penrose.

On considère d'abord deux types G_A et P_A de tuiles triangulaires représentées dans la Figure 4. Les sommets sont colorés: blanc ou noir, et l'arête entre deux sommets d'une même couleur est orientée. La tuile G_A a deux sommets noirs et un blanc, la tuile P_A a deux sommets blancs et un noir.

On appelle pavage du plan de type A, toute triangulation du plan par des triangles isométriques à G_A ou P_A, de telle sorte que les couleurs des sommets communs soient les mêmes ainsi que les orientations des arêtes communes.

Nous renvoyons à [**28**] pour la démonstration de l'existence de tels pavages du plan (Cf. Figure 2). Nous dirons que deux pavages T et T' sont identiques s'ils se déduisent l'un de l'autre par une isométrie du plan Euclidien.

Soit X l'ensemble des pavages du plan de type A (à isométrie près), le résultat essentiel que nous utiliserons est la possibilité de paramétrer X par l'ensemble K des suites infinies $(a_n)_{n\in\mathbf{N}}$, $a_n \in \{0,1\}$ de 0 et de 1 telles que $a_n = 1 \Rightarrow a_{n+1} = 0$, et ce de telle sorte que tout pavage T de type A provienne d'une suite: $T = T(a)$ et que deux suites a et b donnent le même pavage si et seulement s'il existe N tel que $a_n = b_n$ pour tout $n \geqq N$. Comme il s'agit d'un point crucial pour la suite nous allons expliciter la correspondance $a \to T(a)$

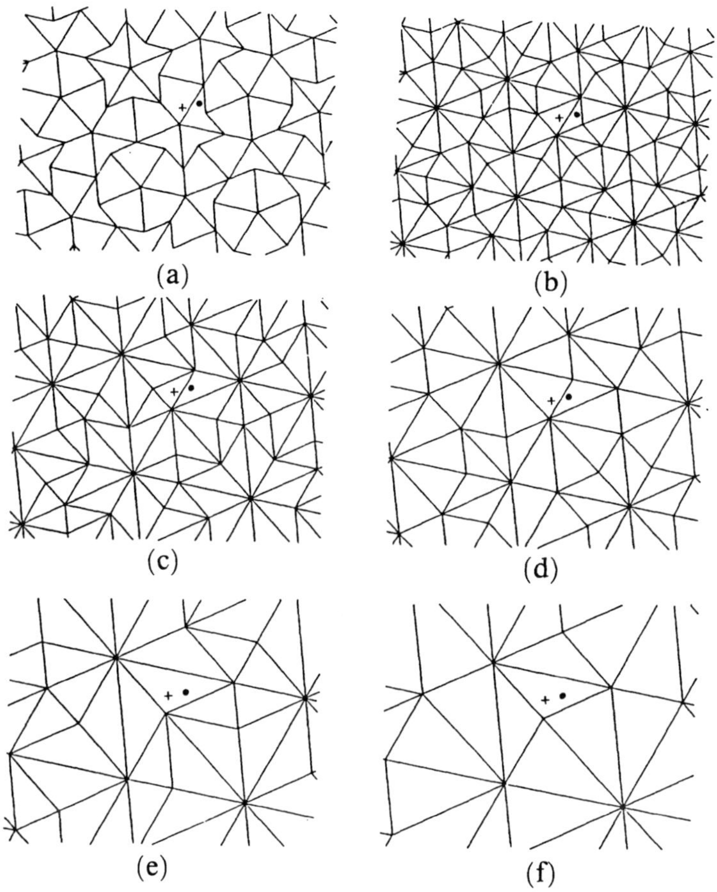

FIGURE 5

(Cf. [**28**, p. 586]). Si T est un pavage de type A et si l'on efface dans la triangulation T toutes arêtes courtes qui joignent deux sommets de couleurs différentes (Cf. Figure 5), on obtient une nouvelle triangulation T_1 du plan dont tous les triangles sont isométriques à l'un des deux triangles G_B, P_B de la figure. On obtient ainsi un pavage du plan de type B, c'est-à-dire une triangulation du plan par les triangles isométriques à G_B ou P_B de telle sorte que les couleurs des sommets communs soient les mêmes ainsi que les orientations des arêtes communes. Si l'on efface dans cette triangulation T_2 toutes les arêtes courtes qui joignent deux sommets de mêmes couleurs on obtient une nouvelle triangulation T_3 dont tous les triangles sont isométriques à l'un des triangles $G_{\tau A'}$, $P_{\tau A'}$ de la Figure 4. En itérant ce procédé on obtient donc une suite T_n de triangulations du plan. A chacune d'entre elle correspondent des triangles G_n et P_n.

Etant donné un pavage T de type A et un triangle α de la triangulation T (Cf. Figure 5) on lui associe la suite $(a_n)_{n\in\mathbf{N}}$, où a_n vaut 0 ou 1 selon que le triangle de T_n qui contient α est grand (i.e., isométrique à G_n) ou petit. Notons $i(T,\alpha)$ la suite ainsi obtenue.

Désignons alors par K l'ensemble des suites $(a_n)_{n\in\mathbf{N}} a_n \in \{0,1\}$, telles que $a_n = 1 \Rightarrow a_{n+1} = 0$. Il est clair que toute suite de la forme $i(T,\alpha)$ appartient à K car le triangle P_n n'est pas utilisé pour construire $P_{n+1} = G_n$. Réciproquement (Cf. [**28**, p. 568]) on démontre que tout élément a de K est de la forme $i(T,\alpha)$ pour un pavage de type A et un α convenables.

Deux triangles α, β d'un même pavage T se retrouvent pour n assez grand dans le même triangle de T_n. Il en résulte que les suites $a = i(T,\alpha)$ et $b = i(T,\beta)$ vérifient:

$$(*) \qquad a_m = b_m \quad \forall m \geqq n\,.$$

Réciproquement (Cf. [**28**, p. 568]), on d'émontre que si $a = i(T,\alpha)$ et $b \in K$ vérifient $(*)$ il existe un triangle β de T tel que $b = i(T,\beta)$.

On obtient ainsi une bijection entre l'ensemble X des pavages de Penrose et le quotient $K/\mathfrak{R}$ de l'ensemble K par la relation d'equivalence $\mathfrak{R}$ définie par la relation $(*)$ (Cf. [**28**, p. 568]).

Bibliographie

1. M. F. Atiyah, *K-Theory*. W. A. Benjamin, Inc., New York, Amsterdam, 1967.
2. M. F. Atiyah and I. M. Singer, *The index of elliptic operators* I, Ann. of Math. **87** (1968), 484–530.
3. ——, *The index of elliptic operators* IV, Ann. of Math. **93** (1971), 119–138.
4. P. Baum and A. Connes, *Geometric K-theory for Lie groups and foliations*, preprint Institute des Hautes Études Scientifiques, 1982.
5. P. Baum and R. Douglas, "*K*-homology and index theory," in *Operator algebras and applications* (R. Kadison, Ed.), Proc. Sympos. Pure Math., vol. 38, part I, Amer. Math. Soc., Providence, R.I., 1982, pp. 117–173.
6. J. B. Bost, *Principe d'Oka et K-théorie des algèbres de Banach non-commutatives*, A paraître dans Inventiones Math.
7. O. Bratteli, *Inductive limits of finite dimensional C*-algebras*, Trans. Amer. Math. Soc. **171** (1972), 195–234.
8. D. Burghelea, *The cyclic cohomology of the group rings*, Comment. Math. Helv. **60** (1985), 354–365.
9. P. Cartier, *Homologie cyclique*, Exposé 621, Séminaire Bourbaki, Février 1984.
10. A. Connes, *Noncommutative differential geometry*, part I, *The Chern character in K-homology*, part II, *de Rham homology and noncommutative algebra*, preprint Institute des hautes études scientifiques (1983), Publ. Math. **62** (1986), 44–144.
11. ——, *C*-algèbres et géométrie différentielle*, C. R. Acad. Sci. Paris Sér. I Math. **290** (1980), 599–604.
12. ——, "Sur la théorie non-commutative de l'intégration," in *Algèbres d'opérateurs* (P. de la Harpe, Ed.), Lecture Notes in Math., vol. 725, Springer-Verlag, Berlin and New York,
13. ——, *The von Neumann algebra of a foliation*, Lecture Notes in Physics, vol. 80, Springer-Verlag, Berlin and New York, 1978, pp. 145–151.
14. ——, *Feuilletages et algèbres d'opérateurs*, Séminaire Bourbaki, Février 1980, 551–01.
15. ——, *An analogue of the Thom isomorphism for crossed products of a C*-algebra by an action of* **R** Adv. in Math. **39** (1981), 31–55.

16. ——, "A survey of foliations and operator algebras," in *Operator algebras and applications* (R. Kadison, Ed.), Proc. Symp. Pure Math., vol. 38, part I, Amer. Math. Soc., Providence, R.I., 1982, pp. 521–628.
17. ——, *Entire cyclic cohomology of Banach algebras and characters of θ-summable Fredholm modules*, *K* theory **1** (1988), 519–548.
18. ——, "Cyclic cohomology and the transverse fundamental class of a foliation," in *geometric methods in operator algebras* (H. Araki and E. G. Effros, Eds), Pitman Research Notes in Mathematics Series, **123**, Longman Scientific & Technical and John Wiley & Sons, Inc., New York, 1986, pp. 52–144.
19. A. Connes and M. H. Moscovici, *Cyclic cohomology, the Novikov conjecture and hyperbolic groups*, To appear in Topology.
20. A. Connes and G. Skandalis, *The longitudinal index theorem for foliations*, Publ. Res. Inst. Math. Sci. **20** (1984), 1139–1183.
21. A. Connes and M. Takesaki, *The flow of weights on factors of type* III, Tohoku Math. J., vol. 29, no. 4 (1977), pp. 473–575.
22. J. Cuntz and W. Krieger, *A class of C^*-algebras and topological Markov chains*, Invent. Math. **56** (1980), 251–268.
23. E. G. Effros, D. E. Handelman and C. -L. Shen, *Dimension groups and their affine representations*, Amer. J. Math. **102** (1980), 385–407.
24. G. A. Elliott, *On totally ordered groups and K_0*, Ring Theory (Proc. Waterloo, 1987), Lecture Notes in Math., vol. 734, Springer-Verlag, Berlin and New York, 1979, pp. 1–34.
25. ——, *On the classification of inductive limits of sequences of semisimple finite-dimensional algebras*, J. Algebra **38** (1976), 29–44.
26. H. Furstenberg, *Recurrence in ergodic theory and combinatorial number theory*, Princeton Univ. Press, Princeton, N.J. 1981.
27. I. Gelfand and D. Fuchs, *The cohomology of the Lie algebra of formal vector fields*, Izv. Ann. SSSR **34** (1970), pp. 327–342.
28. B. Grünbaum and G. C. Shepard, *Tilings and patterns*, Freeman and Company, New York, 1987.
29. A. Haefliger, *Differentiable cohomology*. Cours donné au C.I.M.E. (1976).
30. D. Handelman, "Positive matrices and dimensions groups affiliated to topological Markov chains," in *Operator algebras and applications* (R. Kadison, Ed.), Proc. Pure Sympos. Math., vol. 38, part I, Amer. Math. Soc., Providence, R.I. 1982, pp. 191–194.
31. S. Hurder, *Secondary classes and the von Neumann algebra of a foliation*, preprint MSRI Berkeley (1983).
32. V. Jones, *Index of subfactors*, Invent. Math. **72** (1983), 1–25.
33. R. V. Kadison, *Irreducible operator algebras*, Proc. Nat. Acad. Sci. U.S.A. **43** (1957), 273–276.
34. G. G. Kasparov, *Equivariant KK-theory and the Novikov conjecture*, Invent. Math. **91** (1988), 147–201.
35. W. Krieger, *On ergodic flows and the isomorphism of factors*, Math. Ann. **223** (1976), 19–70.
36. A. S. Mishchenko, *Homotopy invariants of non simply connected manifolds, Rational invariants* I. Izv. Akad. Nauk. SSSR **34** (1970) no. 3, 501–514.
37. ——, *C^*-algebras and K-theory*, Algebraic Topology (Proc. Aarhus, 1978), Lecture Notes in Math., vol. 763, Springer-Verlag, Berlin and New York, 1979.
38. F. Murray and J. von Neumann, *On rings of operators*, Ann. Math. **37** (1936), 116–229.
39. ——, *On rings of operators* II, Trans. Amer. Math. Soc. **41** (1937), 208–248.
40. ——, *On rings of operators* IV, Ann. Math. **44** (1943), 716–808.
41. J. von Neumann, *On rings of operators* III, Ann. Math. **41** (1940), 94–161.
42. G. K. Pedersen, *C^*-algebras and their automorphism groups*, Academic Press, London, 1979.
43. M. Pimsner and D. Voiculescu, *Exact sequences for K-groups and Ext-groups of certain cross-product C^*-algebras*, J. Operator Theory **4** (1980), 93–118.
44. J. Renault, *A groupoid approach to C^*-algebras*, Lecture Notes in Math., vol. 793, Springer-Verlag, Berlin and New York, 1980.

45. M. A. Rieffel, *Morita equivalence for C^*-algebras and W^*-algebras,* J. Pure Appl. Algebra **5** (1974), 51–96.
46. ——, *C^*-algebras associated with irrational rotations,* Pacific J. Math. **93** (1981), pp. 415–429.
47. J. Roe, *An index theorem for open manifolds* I, J. Diff. Geometry **27** (1988), 87–113; II, J. Diff. Geometry **27** (1988), 115–136.
48. E. Winkelnkemper, *The graph of a foliation* (preprint).

INSTITUTE DES HAUTES ETUDES SCIENTIFIQUES, FRANCE

Proceedings of Symposia in Pure Mathematics
Volume **50** (1990)

Some of the Legacy of John von Neumann in Physics: Theory of Measurement, Quantum Logic, and von Neumann Algebras in Physics

HUZIHIRO ARAKI

1. Introduction. As in many other areas, the legacy of John von Neumann in physics is rich and widespread. I will divide my exposition into three parts, indicated by the subtitles.

In his book on quantum mechanics [**1**], von Neumann developed a mathematical theory of measurement in quantum mechanics which culminates in a touchy subject, symbolized by the catch phrase "reduction of wave packets". Recently this subject has been revived due to advances in precision measurement techniques which enable one to measure macroscopic quantum phenomena. A brief account of some of the typical issues will be discussed in §2.

Quantum logic is another product of the interest and concern of von Neumann about the foundation of quantum mechanics. The continuous geometry in another book of von Neumann [**2**] is a mathematically natural alternative for quantum logic and a specific example provided by type II von Neumann algebras seemed to be a (mathematical) Utopia for quantum calculus, i.e., calculus of unbounded operators. However this line of research has not so far been taken up by physicists. Instead, more recent (mathematical) interest and developments are centered around states on the quantum logic. This will be briefly discussed in §3.

The von Neumann proof of the essential uniqueness of representations of the canonical commutation relations [**3**] has given the mathematical foundation for the equivalence between the algebraic formulation of quantum mechanics by Heisenberg based on canonical commutation relations of canonical variables and the analytic formulation of quantum mechanics by Schrödinger

---

1980 *Mathematics Subject Classification* (1985 *Revision*). Primary 46L60, 81E05, 82A15, 81B10, 81B05.

This paper is in final form and no version of it will be submitted for publication elsewhere.

based on Schrödinger equations in terms of the Schrödinger representation of canonical variables.

This uniqueness result holds only for a system of a finite degree of freedom (i.e., finite number of canonical variables). For a system of an infinite degree of freedom, one can start with the algebra of observables in the Heisenberg picture and discuss a large number of mutually inequivalent representations associated with states of physical interests. This is the C^*-algebra approach which will be discussed in the remaining sections. For the analysis of each state of physical interest and associated representations, the von Neumann algebra developed by Murray and von Neumann [**4–7**] is an essential tool.

Quantum field theory is a typical example of a system of an infinite degree of freedom. The formulation of a basic framework in terms of a net of the (von Neumann) algebras of local observables will be described in §5.

As typical examples of application of algebraic techniques, the Doplicher-Haag-Roberts theory of (superselection) sectors and the Doplicher-Longo theory of local current algebra via the use of the standard split inclusion will be briefly explained in §§6 and 7.

Quantum statistical mechanics in an infinitely extended space is another typical example of a system of an infinite degree of freedom and will be described as a C^*-dynamical system. The KMS condition which links the equilibrium states with the Tomita-Takesaki theory of the modular structure of von Neumann algebras will be briefly explained in §8.

Use of an algebraic approach to some exactly soluble model of phase transition in statistical mechanics will be briefly described also in §8.

Use of the cyclic cocycle and the noncommutative Chern character for a theory of quantum Hall effort will be mentioned in §9.

2. Theory of measurement. The measurement process in the orthodox quantum mechanics as described by von Neumann [**1**] consists of the following three stages.

Let k be the index distinguishing the possible measured value of the observable being measured. Let $\Phi = \sum c_k \Phi_k$ be an (unknown) vector in a Hilbert space H_1 describing the measured system, expressed as a superposition of eigenvectors Φ_k of the observables. Let χ_0 be a vector in another Hilbert space H_2 describing the measurement apparatus.

The measured system and measurement apparatus will be noninteracting at the beginning and so the initial state of the combined system is assumed to be given by a tensor product as follows:

$$\psi(A) = (A\Psi, \Psi), \qquad A \in \mathfrak{A} = B(H_1 \otimes H_2), \tag{2.1}$$

$$\Psi = \Phi \otimes \chi_0. \tag{2.2}$$

The measured system and the apparatus will then interact with each other during the measurement and are supposed to result in the vector

$$\Psi' = \sum_k c_k u_k \otimes \Phi_k \tag{2.3}$$

(if the apparatus are chosen correctly) so that different eigenvalues of the measured observable are now distinguishable in terms of mutually orthogonal vectors Φ_k, representing distinguishable states of the apparatus. If the combined system can be described by quantum mechanics, then Ψ' should be reached by a unitary time development from the initial (state) vector Ψ. This is supposed to be the first stage of the measuring process.

The standard interpretation of quantum mechanical measurement requires the second stage, which is the development of the pure state $\psi'(A) = (\Psi', A\Psi')$ into the following mixture:

$$\rho(A) = \mathrm{Tr}(\rho A), \qquad A \in \mathfrak{A}, \tag{2.4}$$

$$\rho = \sum_k |c_k|^2 P(u_k) \otimes P(\Phi_k), \tag{2.5}$$

where $P(\xi)$ denotes the orthogonal projection operator onto the one-dimensional space spanned by a vector ξ, and the positive operator ρ given by (2.5) (called the density matrix for the state ρ given by (2.4)) is in the trace class with $\mathrm{Tr}\,\rho = \sum |c_k|^2 = \|\Psi'\|^2 = 1$.

Actually, any reasonable explanation of the change from the initial state ψ to the mixture ρ would be fine without going through the intermediate state ψ' but breaking the process into the two stages through an intermediate pure state ψ' makes the argument more well defined.

The third and final stage will be the change (shrinkage) of ρ to the pure state given by $u_k \otimes \Phi_k$ (or $P(u_k) \otimes P(\Phi_k)$) as a result of the "reading" or "recording" of a measured result k.

The cat paradox of Schrödinger **[8]** shows that common sense (based on classical experiences) demands the explanation of the entire process starting from the initial state ψ and ending with the final state given by $u_k \otimes \Phi_k$ corresponding to the measured value labeled by k. In Schrödinger's example, the initial state ψ would be a superposition of states of a radioactive nuclei emitting or not yet emitting γ rays and the final Φ_k would be a living or dead cat depending on k.

The third stage described above can be dealt with if we take the standard, completely probabilistic interpretation of quantum mechanics and declare that we simply take one sample out of an ensemble described by a mixture ρ, although some people do not like it. A further discussion of this point will invariably get into a philosophical issue.

The first stage appears innocent at first look but there is an impossibility theorem in cases where the measured observable (such as the component of magnetic moment along a fixed direction) does not commute with a universal symmetry (such as the rotation symmetry). This is not necessarily a catastrophe for the measurement theory because the first stage can be (theoretically) achieved in an approximate sense if sufficiently large (in terms of superposition of quantum numbers for the conserved symmetry) measurement apparatus is used. Such an impossibility theorem was first proved by Wigner **[9]** and generalized by Araki and Yanase **[10]**. (See **[11]** for a review on this subject.)

The largest amount of literature and heated discussions are centered around the second stage. A mixture cannot be achieved starting from a pure state by a unitary time development. Even if one changes formalism in one way or another, the impossibility within the standard quantum mechanics persists, as is shown by Wigner **[12]**. (See also **[13, 14]**.)

The main issue here can be described in terms of the "phase" and the interference pattern. In taking a superposition (the vector sum) of two pure states given by vectors Ψ_1 and Ψ_2 (the correspondence of the vectors to states being modulo the "phase factor" $e^{i\theta}$ due to the insensitivity of the state (2.1) to its multiplication on the vector), we can explicitly exhibit the interference phenomena as

$$\|r_1\Psi_1 + e^{i\theta} r_2\Psi_2\|^2 = r_1^2\|\Psi_1\|^2 + r_2^2\|\Psi_2\|^2 + 2r_1r_2\,\mathrm{Re}(e^{i\theta}(\Psi_1, \Psi_2)), \tag{2.6}$$

where r_1 and r_2 are positive numbers. The last term (the cross term) depends on values of θ and will vanish if averaged over θ (with the uniform measure). The difference between the pure state given by Ψ' of (2.3) and the mixture ρ given by (2.4) and (2.5) is exactly this cross term. The issue is how this cross term effectively vanishes or remains alive in actual measurement procedure. Various theories about it have been proposed and some heated discussions have occurred in conferences. But this will bring us too far away.

However, we should mention a recent surge of various precision measurements aimed at clarification of basic issues in quantum measurement theory, which are always concerned with the detection of the above cross term and hence the phase factor. Roughly speaking, an incident beam of particles such as neutrons are split into two channels. After changing the phase of wave function in one channel relative to that in the other channel by some method, the beams in the two channels are joined together and the interference (i.e., the effect of the cross term) is measured. If the intensity of the beam is made low so that there is only one particle in one wave packet going through the apparatus at a time, then the classical particle picture in which a particle goes through either one of the channels does not look compatible with the wave picture of the material (quantum) wave going through both channels and producing an interference. The detection measurement of the particle in one of the channels is incompatible with the measurement of the interference but the choice among these two possible measurements (at the end point of the channels) can be made after the particle wave already passed the initial branching point of the channels so that the particle should have already decided which of the channels or both of the channels it will go through according to the classical picture. This is called the delayed choice experiment. The shift of the phase in one channel can be made by magnetic field which never touches the paths of the particle. This is essentially the Bohm-Aharonov effects. The shift of the phase in one channel can also be made by a 360° rotation in the case of spin 1/2 particles such as neutrons. Thus the factor (-1) due to 360° rotation of the (spin 1/2) projective representation of the rotation group is detected.

During recent years, there have been quite a number of conferences on foundation of quantum mechanics. Reports and discussions of experiments of the type described above occupied a major portion in some of these conferences [**15–19**], while new precision experiments gave great impact to theoretical and philosophical discussions in other conferences [**20–26**].

3. Quantum logic. Classical logic may be formulated in terms of propositional calculus of a Boolean lattice. Von Neumann and Birkoff [**27**] recognized the lattice of (orthogonal) projection operators as the corresponding object for quantum mechanics and called it quantum logic. In the finite dimensional case (for example, the spin of a particle without taking its position into account), the lattice of orthogonal projections can be characterized as an orthocomplemented modular lattice. The extension of the modular lattice to the infinite dimensional case led von Neumann to continuous geometry [**2**] and concrete examples are provided by type II_1 von Neumann algebras.

Canonical variables in quantum mechanics such as position and momentum are represented by unbounded selfadjoint operators. The calculus of unbounded operators is tricky due to the domain problems and a simple polynomial of noncommuting unbounded operators may not be (or possess) selfadjoint (extensions) even if it is formally so. However, the calculus of unbounded selfadjoint operators affiliated with a type II_1 von Neumann algebra was discovered to be much easier by Murray and von Neumann in §16 of [**4**]. We quote the following sentences from that paper. ($U(M)$ denotes the set of all densely defined linear closed operators affiliated with M, and $B = B(H)$.)

"The operators in $U(B)$ (H a Hilbert space) behave very pathologically from the point of view of operator algebra: Thus if $A, B \in U(B)$, the $A + B$ and AB cannot be formed in general, and the entire mechanism of replacing operators by their extension leads to very paradoxic results." This applies, of course, to every $U(M)$, M in case (I_∞). In cases (I_n) and (II_1) again it is not so: The algebra of $U(M)$ works without any difficulties, and if A is an extension of $B, A,\ B \in U(M)$, then $A = B$."

While von Neumann seems to have liked the type II_1 case, physics had not developed in that direction so far. The lattice of all (orthogonal) projection operators in a Hilbert space (the quantum logic for a type I_∞ factor) is not modular. Piron [**28**] has characterized it as a complete atomic orthomodular orthocomplemented lattice with a covering property, providing a characterization of quantum logic for ordinary quantum mechanics.

Projections are the simplest among physical observables. (Projections represent "questions" in the terminology of Mackey [**29**].) Since linear combinations of projections in a von Neumann algebra are dense in it, axiomatization of quantum mechanics in terms of quantum logic may be considered as an axiomatization of observables in quantum mechanics. As the discussion of states has to go hand in hand with that of observables, it is important to characterize states from the viewpoint of quantum logic.

A normal state φ on a von Neumann algebra M (states in the sense of a positive linear functional with norm 1) gives a positive number $\varphi(p)$ for each projection p in M, which is a "probability" measure on the lattice $P(M)$ of projections in M in the sense that (1) it is totally additive for mutually orthogonal projections p_α,

$$\varphi\left(\sum p_\alpha\right) = \sum \varphi(p_\alpha) \quad \text{if } p_\alpha \perp p_\beta \text{ for } \alpha \neq \beta, \tag{3.1}$$

and (2) it is normalized,

$$\varphi(1) = 1. \tag{3.2}$$

Gleason [**30**] proved the converse for $M = B(H)$ (the set of all bounded linear operators on a Hilbert space H), $\dim H \geq 3$. (Easy counterexamples exist in abundance for $\dim H = 2$. Trivially true for $\dim H = 1$). Recently Gleason's Theorem has been generalized to all von Neumann algebras M which have no type I_2 direct summand. (By Christensen [**31**] for properly infinite type and by Yeadon [**32**, **33**] for finite type.) Therefore we can say that a normal state of von Neumann algebra is a probability measure on the corresponding quantum logic.

The affine structure of normal states has a direct physical interpretation in terms of the mixture of states and so is the linear structure of observables which is the (real) Banach dual of the (linear span of) normal states. A function of an observable can be interpreted as a kind of relabeling of the measured results and hence has a direct physical relevance. In particular, the function x^2 together with the linear structure gives rise to the Jordan product

$$x \circ y = \frac{1}{2}((x+y)^2 - x^2 - y^2).$$

From this point of view, Jordan algebra structure seems to be physically relevant. For von Neumann algebras, Kadison [**34**] has shown that its Jordan algebra structure essentially determines its algebraic structure up to the order of product, namely a Jordan isomorphism of von Neumann algebras is a direct sum of a *-isomorphism and *-anti-isomorphism ("anti" refers to the reversed order of the product.)

A symmetry in physics is a mapping of the physical system onto itself preserving the relevant physical structure. For the projection lattice, the relevant structure may be taken to be the structure of an orthocomplemented lattice (i.e., the order and the complement). Then a symmetry can be characterized by the Feldman-Dye Theorem [**35**, **36**]: an orthoisomorphism of projection lattices is induced by and is in one-to-one correspondence with a Jordan isomorphism of the corresponding von Neumann algebras.

Earlier, Wigner formulated a physical symmetry in the case of $M = B(H)$ as a bijective mapping of unit rays (or equivalently one-dimensional subspaces of H) onto themselves preserving the "transition probability" $|(\psi, \varphi)|^2$ for unit vectors ψ and φ (taken from unit rays or one-dimensional subspaces). Wigner [**37**] shows that it is induced by either a unitary or an antiunitary operator on H. It is later realized that preservation of orthogonality is sufficient if $\dim H \geq 3$ [**38**].

In the case $M = B(H)$, the unit rays are in one-to-one correspondence with the normal pure states of M and hence with minimal elements in the projection lattice $P(M)$. By defining the image of $p \in P(M)$ as the largest element orthogonal to images of all minimal elements orthogonal to p, the given bijection of unit rays can be lifted to an orthoisomorphism of $P(M)$ and hence Wigner's Theorem (in the strengthened form) for $\dim H \geq 3$ follows from the Feldman-Dye Theorem in view of the known result that any $*$-automorphism of $B(H)$ is inner. (The anti-isomorphism can be reduced to the automorphism by multiplying a known anti-isomorphism $x \to Jx^*J$, where J is the complex conjugation on H.)

A physical symmetry can also be formulated as a bijection of the normal states onto themselves preserving mixture. Kadison [**39**] shows that any bijective affine map between normal states of von Neumann algebras is the transpose of a Jordan isomorphism.[1] Slightly generalizing the problem, we can consider a bijective, linear, norm continuous, order isomorphism between preduals of von Neumann algebras. If the orthogonality (in terms of the support projection) of states is preserved, then it is shown in the case of type I von Neumann algebras to be the transpose of a Jordan isomorphism times the multiplication by a positive invertible central element [**40**]. Whether this holds for general von Neumann algebras with or without orthogonality preservation assumption seems to be open.

The surjectivity of the map is crucial for the results in [**38** and **39**]. However, Wigner's Theorem essentially holds without bijectivity if surjectivity is also dropped in the conclusion [**41**].

4. C^*-algebra approach in physics. Heisenberg formulated quantum mechanics on the basis of the canonical commutation relations (CCR)

[1]In [**41**], we quoted this result incorrectly with an additional w^*-continuity assumption.

$$[q_j, p_k] = \begin{cases} 0 & (j \neq k), \\ i\hbar\mathbf{1} & (j = k) \end{cases} \tag{4.1}$$

($j, k = 1, 2, \ldots, n$, $\hbar = h/2\pi$, h is the Planck constant) between the canonical variables q_j (such as position coordinates of particles) and their canonical conjugate variables p_k, both of which are mathematically families of mutually commuting selfadjoint operators. In the exponentiated form, the continuous one-parameter groups of unitaries

$$U(\alpha) = \exp i \sum \alpha_j q_j, \qquad V(\beta) = \exp i \sum \beta_k p_k \tag{4.2}$$

are to satisfy the following Weyl form of CCR:

$$\begin{aligned} &U(\alpha)V(\beta)U(-\alpha)V(-\beta) = \exp -i\hbar(\alpha, \beta), \\ &U(\alpha)U(\alpha') = U(\alpha + \alpha'), \qquad V(\alpha)V(\alpha') = V(\alpha + \alpha'). \end{aligned} \tag{4.3}$$

Here $\alpha \in \mathbf{R}^n$, $\beta \in \mathbf{R}^n$, and $(\alpha, \beta) = \sum \alpha_j \beta_j$.

Von Neumann [3] proved that any such U and V (called a representation of CCR in the Weyl form) are unitarily equivalent to a direct sum of a unique standard representation (called the Schrödinger representation) given on the Hilbert space $H = L^2(\mathbf{R}^n)$ of L^2 functions on $\mathbf{R}^n$ by

$$\begin{aligned} &[U(\alpha)f\,](x) = e^{i(\alpha,x)} f(x), \\ &[V(\beta)f\,](x) = f(x + \hbar\beta) \end{aligned} \qquad (x \in \mathbf{R}^n) \tag{4.4}$$

for $f \in H$, or, in terms of the canonical variables,

$$q_j = \text{ the multiplication of } x_j, \qquad p_k = -i\hbar(\partial/\partial x_k). \tag{4.5}$$

This is the quantization used by Schrödinger for the wave equation in his wave mechanics. The above uniqueness theorem of von Neumann (the uniqueness for irreducible representations) gave the mathematical foundation for the equivalence of this wave mechanics and the more algebraic matrix mechanics of Heisenberg which starts from CCR (4.1).

The uniqueness result is limited to the case of a finite n. For an infinite degree of freedom ($n = \infty$), there are a large number of mutually nonequivalent irreducible representations. We then take an algebraic point of view and start from a C^*-algebra $\mathfrak{A}$ of observables. Physical states will be described by mathematical states of the algebra, i.e., positive linear functionals with norm 1. The cyclic (GNS-)representation of the algebra associated with different states can be mutually nonequivalent. For example, equilibrium states with different temperatures yield mutually disjoint representations [**43**]. This kind of formulation of observables and states of a physical system is called the C^*-algebra approach.

To study each state of interest in detail, the von Neumann algebra obtained as the weak closure M of the cyclic (GNS-)representation of the C^*-algebra of observables associated with the state becomes relevant. In the case of ground states, the extremal ground states are pure states and M is the type

I factor $B(H)$. For quantum field theory, the ground state is the vacuum state, for which M is of type I [**42**] and the extremal vacuum state is pure. In these cases the whole $B(H)$ becomes physically relevant as in the ordinary quantum mechanics.

For the algebra of observables localized in some space-time regions, the von Neumann algebra generated by them in the irreducible vacuum states are of type III in examples [**44**] and in a specific region with some symmetry [**45–49**]. For an equilibrium state of a finite temperature, M is of type III_1 if it is invariant under the spatial translation [**50, 51**]. The abundance of type III algebras in physics actually turned people's attention to them in the 1960s. The type II_1 algebra on the other hand appears for the ∞- temperature equilibrium states (almost by definition).

5. Quantum field theory. Quantum field theory can be formulated on the basis of the notion of local observables. For each (four-dimensional) space-time bounded region $\mathscr{O}$ in the Minkowski space M, let $\mathfrak{A}(\mathscr{O})$ be the set of all observables which can be measured by some instrument in $\mathscr{O}$. The union $\mathfrak{A}$ of $\mathfrak{A}(\mathscr{O})$ for all $\mathscr{O}$ should include all observables in practice. They are assumed to satisfy the following three general postulates which are physically motivated.

(1) *Isotony*: If $\mathscr{O}_1$ contains $\mathscr{O}_2$, then $\mathfrak{A}(\mathscr{O}_1)$ contains $\mathfrak{A}(\mathscr{O}_2)$.

(2) *Poincaré covariance* (special relativity): Let $L = (\Lambda, a)$ denote the inhomogeneous Lorentz transformation $x \in M \to \Lambda x + a \in M$, where $a \in M$ is the amount of space-time translation and Λ is a homogeneous Lorentz transformation. We denote the transformed region by

$$L\mathscr{O} = \Lambda\mathscr{O} + a = \{\Lambda x + a; x \in \mathscr{O}\}. \tag{5.1}$$

Then the existence of an automorphism of α_L of $\mathfrak{A}$ satisfying

$$\alpha_L \mathfrak{A}(\mathscr{O}) = \mathfrak{A}(L\mathscr{O}) \tag{5.2}$$

is assumed.

(3) *Local commutativity* (causality): If two regions $\mathscr{O}_1$ and $\mathscr{O}_2$ are space-like separated in the sense that $x_1 - x_2$ is a space-like vector in the Minkowski metric for every $x_1 \in \mathscr{O}_1$ and $x_2 \in \mathscr{O}_2$, then $\mathfrak{A}(\mathscr{O}_1)$ and $\mathfrak{A}(\mathscr{O}_2)$ are assumed to commute, i.e.,

$$A_1 A_2 = A_2 A_1 \tag{5.3}$$

for any $A_1 \in \mathfrak{A}(\mathscr{O}_1)$ and $A_2 \in \mathfrak{A}(\mathscr{O}_2)$. Under the stated assumption on $\mathscr{O}_1$ and $\mathscr{O}_2$, no signals from one of them reaches to the other and a measurement in $\mathscr{O}_1$ and a measurement in $\mathscr{O}_2$ can be performed simultaneously without one influencing the other, according to the principle of relativity and causality. If two observables are simultaneously measurable, then they are supposed to commute with each other according to quantum mechanics.

If we go over to the algebra generated by each $\mathfrak{A}(\mathscr{O})$ or further to its closure, the properties assumed in (1), (2), and (3) persist. Hence we may assume

$\mathfrak{A}(\mathcal{O})$ to be a C^*-algebra, or more specifically a von Neumann algebra. $\mathfrak{A}$ will be taken to be the norm closure of the union of $\mathfrak{A}(\mathcal{O})$, i.e., $\mathfrak{A}$ will be the C^*-algebra generated by all $\mathfrak{A}(\mathcal{O})$. To avoid possible complication, one may restrict $\mathcal{O}$ to have a special form. Usually it is restricted to double cones:

$$DC(x,y) \equiv \{z \in M; x-z, z-y \text{ future time-like }\} \tag{5.4}$$

where $x, y \in M$ and $x - y$ is future time-like.

In quantum field theory, one is interested in a special class of states, i.e., either states with a finite number of particles, or states with finite energy. In particular, a vacuum state defined as a state with a minimal energy is of central interest.

If the Fourier transform of the translated observables $\alpha_{(1,a)}A$ as a function of $a \in M$ has a support in a subset S_A of M, then the change of the energy-momentum of a Hilbert space vector by A is limited to vectors in S_A. If a state vector Ω is annihilated by A whenever S_A is disjoint from a set S_Ω, then the energy momentum of the state obtained from Ω by a quasilocal perturbation (i.e., application of $A \in \mathfrak{A}$) is restricted to S_Ω. A state vector which has a minimal energy in any coordinate system of special relativity (i.e., a ground state in relativistic sense) is characterized by S_Ω which is contained in the closed future cone (the convex hull of the future light cone). This defines a vacuum vector Ω or a vacuum state

$$\omega_0(A) = (\pi_0(A)\Omega, \Omega), \tag{5.5}$$

where π_0 is the cyclic (GNS-)representation of $\mathfrak{A}$ associated with ω_0. It is invariant under translation $\alpha_{(1,a)}$, $a \in M$, and there exists a continuous unitary representation $U_0(a)$ $(a \in M)$ of $M = \mathbf{R}^4$ such that

$$\begin{aligned} &U_0(a)\Omega = \Omega, \\ &U_0(a)\pi(A)U_0(a)^* = \pi(\alpha_{(1,a)}A) \qquad (A \in \mathfrak{A}). \end{aligned} \tag{5.6}$$

The generators $P = (P^0, P^1, P^2, P^3)$ of

$$U(a) = \exp i(P, a) \tag{5.7}$$

are the energy-momentum operators in the representation space and the defining property of the vacuum state stated above amounts to the *positivity of energy*:

$$\mathrm{Spec}(P) \subset \text{ the closed future cone }. \tag{5.8}$$

This together with (1)–(3) are standard general assumptions of the theory of local observables. We add two more technical assumptions which are crucial in subsequent sections. As quoted earlier, the extremal vacuum state is pure, which we will assume below, i.e., π_0 is assumed to be irreducible.

For a region $\mathcal{O}$, $\mathcal{O}'$ denotes the causal complement of $\mathcal{O}$:

$$\mathcal{O}' = \{x \in M; x - y \text{ is space-like for any } y \in \mathcal{O}\}. \tag{5.9}$$

For unbounded regions such as $\mathcal{O}'$, we define $\mathfrak{A}(\mathcal{O}')$ to be the C^*-algebra generated by the union of $\mathfrak{A}(\mathcal{O}_1)$ over all $\mathcal{O}_1 \subset \mathcal{O}'$, i.e., all $\mathcal{O}_1$ space-like to $\mathcal{O}$. The following relation is known as *Duality* in theory of local observables:

$$\pi_0(\mathfrak{A}(\mathcal{O}'))' = \pi_0(\mathfrak{A}(\mathcal{O})). \tag{5.10}$$

Here R' denotes the commutant of the algebra R of operators. By the local commutativity, the left-hand side includes the right-hand side. This assumption strengthens it to equality and reflects (if $\mathcal{O}$ is a double cone) the physical idea that (expectation values of) observables at a fixed time (in any relativistic coordinate, i.e., on a space-like hyperplane) should determine the state completely (as an initial data) and hence should generate the whole algebra of operators. We note that this is a representation dependent property.

The quantum field is linear in test functions and hence has an additivity relative to a union of regions. The *Additivity Assumption* is its local algebra version:

$$\pi_0\left(\mathfrak{A}\left(\bigcup_{i=1}^{n}\mathcal{O}_i\right)\right) = \pi_0\left(\bigcup_{i=1}^{n}\mathfrak{A}(\mathcal{O}_i)\right)''. \tag{5.11}$$

We will use a weak version of this called *Weak Additivity*:

$$\pi_0(\mathfrak{A})'' = \left(\bigcup_{a\in M}\pi_0(\mathfrak{A}(\mathcal{O}+a))\right)''. \tag{5.12}$$

6. Theory of sectors. For a given vacuum state ω_0, we are interested (in quantum field theory) in those states ω corresponding to some "local" excitation of ω_0 in the sense that ω is close to ω_0 in spatial infinity, i.e.,

$$\text{(i)} \quad \|(\omega-\omega_0)|\mathfrak{A}(\mathcal{O}')\| \text{ is small for large } \mathcal{O} \tag{6.1}$$

(the difference $\omega - \omega_0$ restricted to the causally complementary region $\mathcal{O}'$ of $\mathcal{O}$ being measured by its norm) with an additional condition that the local excitation is transportable (by any inhomogeneous Lorentz transformation $L = (\Lambda, a)$), i.e.,

$$\text{(ii)} \quad \alpha_L\omega(A) \equiv \omega(\alpha_L A) \text{ is norm continuous in } L. \tag{6.2}$$

If a state ω satisfies (i) and (ii), then any normal states of the cyclic representation associated with ω also satisfy (i) and (ii). They are called a sector. So we are interested in sectors satisfying (i) and (ii). On the basis of the duality assumption for the vacuum sector, Doplicher, Haag, and Roberts developed the theory of sectors beginning in 1969 **[52–55]**.

From experience in quantum field theory, one knows that such a structure can be obtained from the field algebra $\mathcal{F}$ and the (compact) gauge group G as follows. To each $\mathcal{O}$, there is a von Neumann algebra $\mathcal{F}(\mathcal{O})$ and the field algebra $\mathcal{F}$ is the C^*-algebra generated by all $\mathcal{F}(\mathcal{O})$. There is a compact group G consisting of automorphisms which map each $\mathcal{F}(\mathcal{O})$ onto itself. The algebra $\mathfrak{A}(\mathscr{O})$ of local observables is the G-fixed point algebra of $\mathscr{F}(\mathscr{O})$ (the algebra of gauge invariant elements in $\mathscr{F}(\mathscr{O})$):

$$\mathfrak{A}(\mathscr{O}) = \mathscr{F}(\mathscr{O})^G. \tag{6.3}$$

The vacuum representation space of $\mathscr{F}$ (taken to be irreducible) as well as the algebra $\mathscr{F}$ itself splits into a sum of irreducible representations of G, labeled by the unitary equivalence class ξ of the representation. All sectors of $\mathfrak{A}$ appear as irreducible representations, labeled by ξ.

The local net $\mathscr{F}(\mathscr{O})$ of field algebras satisfies (i), (ii), and the normal commutation relations

$$F_1 F_2 = \pm F_2 F_1 \tag{6.4}$$

if $F_1 \in \mathscr{F}(\mathscr{O}_1)$ and $F_2 \in \mathscr{F}(\mathscr{O}_2)$, where $\mathscr{O}_1$ is space-like to $\mathscr{O}_2$. There is a homomorphism from $\{\xi\}$ to $Z_2 = \{1, -1\}$ denoted by sign ξ (the product of ξ is by the tensor product and then by decomposition to irreducibles). The sector ξ is called the Bose or Fermi sector according to sign $\xi = +$ or $-$. Suppose F_1 and F_2 belong to representations ξ_1 and ξ_2. If both are Fermi, then the $-$ sign is taken in (6.4). Otherwise the $+$ sign is taken.

It is a remarkable achievement of the theory of sectors that this is the most general situation. All sectors satisfying the selection criterion (i) and (ii) under the basic assumptions about the vacuum sector can be described by one field algebra $\mathscr{F}$ and a compact gauge group G as above **[56]**.

Note that only commutativity (of observables) at space-like separation is assumed at the start. This implies an alternative of commutation or anti-commutation for the field. In quantum field theory, there is a choice of $\pm$ in (6.4) which is different from the normal commutation relations but it is known there that the normal commutation relations can be recovered by the so-called Klein transformation **[57]**. This is included in this general result.

The so-called parastatistics corresponds to the case of a noncommutative gauge group G. Each sector corresponds to an irreducible representation of G and the dimension of this representation of G is the order of the parastatistics. The example of parafields obtained as the sum of identical fields exactly fits this description. The theory of sectors more or less says that there are no other possibilities for statistics of the (asymptotic in or out) particles.

Actually, a possibility of an "infinite" statistic appears in the Doplicher-Haag-Roberts analysis of sectors but it can be excluded if one assumes that the conjugate sector also satisfies the selection criterion (i) and (ii), or sectors are generated (as a semigroup) from those representations in which massive particles isolated from the continuum appear **[53, 58, 59]**.

The last assumption, without the selection criterion (i), selects those sectors which have a perturbation from the vacuum localized in a string-like region (i.e., the difference from vacuum is localized in a space-like cone). The theory of such sectors has been developed in **[58, 59]**.

7. Standard split inclusion and local current algebras. Buchholz [**60**] pointed out the existence of an intermediate type I factor N between local algebras $\mathfrak{A}(\mathcal{O}_1)$ and $\mathfrak{A}(\mathcal{O}_2)$ of free scalar fields when $\overline{\mathcal{O}}_1 \subset \mathcal{O}_2$:

$$\mathfrak{A}(\mathcal{O}_1) \subset N \subset \mathfrak{A}(\mathcal{O}_2). \tag{7.1}$$

This notion of an intermediate type I factor became a very powerful tool by the work of Doplicher and Longo [**61**] and found applications to mathematics in connection with the Stone-Weierstrass theorem [**62**] and subfactors [**63**] as well as to quantum field theory as described below.

For von Neumann algebras R_1 and R_2, the inclusion relation $R_1 \subset R_2$ is called standard split inclusion if

$$\text{(i)}\quad \overline{R_1\Omega} = \overline{R_2\Omega} = \overline{(R_2 \wedge R_1')\Omega} = H \quad \text{(total space)} \tag{7.2}$$

for some vector Ω, and

$$\text{(ii)}\quad \text{There exists a type I factor } N \text{ satisfying } R_1 \subset N \subset R_2. \tag{7.3}$$

The first property holds for an irreducible vacuum vector by the Reeh Schlieder theorem (the weak additivity is needed here) if $R_1 = \mathfrak{A}(\mathcal{O}_1), R_2 = \mathfrak{A}(\mathcal{O}_2)$, and $\mathcal{O}_1' \wedge \mathcal{O}_2$ has a nonempty interior. The second property is equivalent to the possibility of local preparation of states [**64, 65**] in the sense that for any normal state ω of $\mathfrak{A}(\mathcal{O}_1)$, there exists a projection operator E_ω in $\mathfrak{A}(\mathcal{O}_2)$ which prepares the given state ω from any normal state ψ of $\mathfrak{A}(\mathcal{O}_2)$, i.e., $\psi_E = \omega$ for $E = E_\omega$ and

$$\psi_E(A) = \psi(EAE)/\psi(E). \tag{7.4}$$

For other physical conditions leading to (ii), see [**66**].

Under the assumption of the standard split inclusion $\mathcal{F}(\mathcal{O}_2) \supset \mathcal{F}(\mathcal{O}_1)$ for the field algebra, there exists a unitary representation of the gauge group G by unitaries in $\mathcal{F}(\mathcal{O}_2)$ implementing the action α_G of G on $\mathcal{F}(\mathcal{O}_1)$. Going over to generators, one obtains a current algebra in $\mathcal{F}(\mathcal{O}_2)$, namely a map J from the Lie algebra of G to selfadjoint operators affiliated with $\mathcal{F}(\mathcal{O}_2)$ satisfying

$$\begin{aligned} &[J_u, J_v] = iJ_{[u,v]}, \\ &\alpha_g(J_u) = J_{g(u)}, \\ &[J_u, F] = \delta_u(F) \quad \text{for } F \in \mathcal{F}(\mathcal{O}_1), \end{aligned} \tag{7.5}$$

where α_g acts on J_u via its spectral projections which are in $\mathcal{F}(\mathcal{O}_2)$ and δ_u is the generator of α_g (acting on $\mathcal{F}(\mathcal{O})$) [**64, 67**].

The construction of the local current algebra extends to generators of the Poincaré group. In particular, the positive energy spectrum can be retained for local generators [**64**].

The convergence of local generator to global generator is also known [**68**].

8. Quantum statistical mechanics. A C^*-algebra $\mathfrak{A}$ together with a continuous one-parameter group α_t of automorphisms of $\mathfrak{A}$ ($t \in \mathbf{R} \to \alpha_t(A) \in \mathfrak{A}$ is continuous for each A) is called a C^*-dynamical system and is a convenient abstract setting appropriate for quantum statistical mechanics of spin lattice systems. (For example, see **[69]**.)

To give some idea about it, we describe a simple model of a spin lattice system called the one-dimensional XY-model. The sites of a one-dimensional lattice may be labeled by an integer $n = 0, \pm 1, \pm 2, \ldots$. For each lattice site n, the algebra of spin observables at the site n is isomorphic to 2×2 matrices and is linearly spanned by Pauli matrices $\sigma_x^n, \sigma_y^n, \sigma_z^n$, and the identity. For each subset I of $\mathbf{Z}$, $\mathfrak{A}(I)$ denotes the C^*-tensor product of $\mathfrak{A}_n$, $n \in I$. In particular, $\mathfrak{A} = \mathfrak{A}(\mathbf{Z})$ is the C^*-algebra of observables.

A model of the quantum system on a 1-dimensional spin lattice can be specified by writing the Hamiltonian $H(N)$ on a finite part $[-N, N]$ of the lattice. For the XY-model, we may take

$$(8.1) \qquad H(N) = -J \sum_{|n| \leq N} \{(1+\gamma)\sigma_x^n \sigma_x^{n+1} + (1-\gamma)\sigma_y^n \sigma_y^{n+1} + \lambda \sigma_z^n\},$$

where J, γ, and λ are real constants. The C^*-dynamical system is defined by

$$(8.2) \qquad \alpha_t(A) = \lim_{N \to \infty} e^{itH(N)} A e^{-itH(N)},$$

where the limit is shown to exist for each $A \in \mathfrak{A}$ and defines a continuous one-parameter group of automorphisms of $\mathfrak{A}$.

One of the aims of statistical mechanics is to find the thermal equilibrium state of the system when the dynamics is given. For a finite system ($N < \infty$), the dynamics is given by an inner automorphisms:

$$(8.3) \qquad \alpha_t^{(N)}(A) = e^{itH(N)} A e^{-itH(N)}, \qquad H(N) \in \mathfrak{A}.$$

For such a system, Gibbs obtained the following expression for equilibrium states at an absolute temperature T.

$$(8.4) \qquad \varphi_\beta(A) = Z^{-1} \operatorname{Tr}(e^{-\beta H} A), \qquad Z = \operatorname{Tr} e^{-\beta H},$$

where $\beta = 1/(kT)$, k is a (Boltzmann) constant, and Z is the normalization coefficient called the partition function.

Kubo **[70]** and Martin and Schwinger **[71]** used some property of the analytic continuation of (8.4), which was later formulated in the C^*-algebra approach by Haag, Hugenholtz, and Winnink **[72]** as the following KMS-condition:

As is easily proved, there is a dense set of A in $\mathfrak{A}$ such that $\alpha_t(A)$ has an analytic continuation from the function of a real variable t to an entire function $\alpha_z(A)$ of a complex variable z. The KMS-condition for a state φ of $\mathfrak{A}$ is the relation

$$(8.5) \qquad \varphi(A\alpha_{i\beta}(B)) = \varphi(BA)$$

which is required to hold for all α-entire B described above and for all $A \in \mathfrak{A}$. (For an alternative equivalent formulation, see, for example **[69]**.) This is now adopted as the definition of equilibrium states in the C^*-algebra approach. The Gibbs state (8.4) is the unique KMS state for the finite volume dynamics (8.3).

About the same time as [72] was written, Tomita developed a theory of generalized Hilbert algebras in order to deal with type III cases [73]. While expounding Tomita's theory, Takesaki [74] discovered that the modular automorphisms of a given state φ, which is the central object of the Tomita-Takesaki Theory, fulfills the KMS condition (with $\beta = -1$). This discovery made the powerful Tomita-Takesaki theory available to the equilibrium quantum statistical mechanics and also physical pictures and notions of the latter available to the theory of von Neumann algebras.

Later, in studying quantum field theory, Bisognano and Wichmann [75] found that the modular automorphisms of the vacuum state for the algebra $\mathfrak{A}(W)$ localized in the wedge region

$$W = \{x; |x^0| < x^1\} \tag{8.6}$$

coincide with the Lorentz boost transformation. Sewell [76] used this in his understanding of the Hawking temperature for the black hole radiation.

The C^*-algebra approach to quantum statistical mechanics of the spin lattice system has been fruitful, providing general theorems, such as the uniqueness of equilibrium states for one-dimensional systems (except for the case of extremely long-range interactions) [77, 78], and methods of computations such as in the case of the phase diagram for the ground states of the one-dimensional XY-model [79]: unique ground states for $|\lambda| \geq 1$ as well as for $\gamma = 0$; two pure ground states for $|\lambda| < 1, \gamma \neq 0$, $(\lambda, \gamma) \neq (0, \pm 1)$; and an infinite number of pure ground states (or four mutually disjoint ground state representations, two of which being soliton sectors) for $(\lambda, \gamma) = (0, \pm 1)$, with singularities of correlation functions at $|\lambda| = 1$ (strong external magnetic field) as well as at $\gamma = 0$, $|\lambda| < 1$ (XY symmetry). These concrete results are obtained with the help of abstract arguments about equivalence and disjointness of representations of the canonical commutation relations.

9. Solid state physics. We now briefly describe the use of the noncommutative Chern character (see the article of A. Connes in the present Proceedings) for an explanation of the quantum Hall effect in solid state physics [80].

Suppose a magnetic field B and an electric field E, which are mutually orthogonal, are applied to a thin strip of metal orthogonal to the magnetic field. Then the current j flows perpendicular to E and B. The relation $j = \sigma E$ defines the Hall conductivity σ. This Hall effect was discovered and discussed by E. H. Hall in 1879. One century later in 1980, K. von Klitzing, G. Dorda, and M. Pepper discovered that the Hall conductivity is quantized, i.e., $\sigma_{\text{Hall}} = me^2/h$ with an integer n, within an accuracy of 10^{-7}. After theoretical work, one can obtain σ_H via the Kubo formula as the noncommutative Chern character.

REFERENCES

1. J. von Neumann, *Mathematische Grundlagen der Quantenmechanik*, Springer, 1932; English transl., *Mathematical foundation of quantum mechanics*, Princeton Univ. Press, Princeton, N. J., 1955.
2. ——, *Continuous geometry*, Princeton Univ. Press, Princeton, N. J., 1960.
3. ——, *Die Eindeutigkeit der Schrödingerschen Operatoren*, Math. Ann. **104** (1931), 570–578.
4. F. J. Murray and J. von Neumann, *On rings of operators*, Ann. of Math. (2) **37** (1936), 116–229.
5. ——, *On rings of operators.* II, Trans. Amer. Math. Soc. **41** (1937), 208–248.
6. J. von Neumann, *On rings of operators.* III, Ann. of Math (2) **41** (1940), 94–161.
7. F. J. Murray and J. von Neumann, *On rings of operators.* IV, Ann. of Math. (2) **44** (1943), 716–808.
8. E. Schrödinger, *Die gegenwartige Situation in der Quantenmechanik*, Naturwissenschaften **23** (1935), 807–812, 823–828, 844–849.
9. E. P. Wigner, *Die Messung Quantenmechanischer Operatoren*, Z. Phys. **133** (1952), 101–108.
10. H. Araki and M. M. Yanase, *Measurement of quantum mechanical operators*, Phys. Rev. **120** (1960), 622–626.
11. H. Stein and A. Shimony, "Limitations on measurement," in *Foundation of Quantum Mechanics* (B. d'Espagnat, ed.), Academic Press, New York, 1971, pp. 56–76.
12. E. P. Wigner, *The problem of measurement*, Amer. J. Phys. **31** (1963), 6–15.
13. J. Earman and A. Shimony, *A note on measurement*, Nuovo Cimento B **54** (1968), 332–334.
14. A. Fine, *Insolubility of the quantum measurement problem*, Phys. Rev. D **2** (1970), 2783–2787.
15. S. Kamefuchi et al. (eds), *Foundations of quantum mechanics*, Proc. Internat. Sympos. Phys. Soc. Japan, Tokyo, 1984.
16. L. M. Ruth and A. Inomata (eds), *Fundamental questions in quantum mechanics*, Gordon and Breach, New York, 1986.
17. Daniel M. Greenberger (ed.), *New technique and ideas in quantum measurement theory* (Proc. Internat. Conf.), Ann. New York Acad. Sci. **480** (1980).
18. M. Namiki et al. (eds.), *Foundations of quantum mechanics*, Proc. Second Internat. Sympos., Phys. Soc. Japan, Tokyo, 1987.
19. G. Badurek, H. Ranch, and A. Zeilinger (eds.), *Matter wave interference*, (Proc. Internat. Workshop), Phys. B **151** (1988).
20. R. Penrose and C. J. Isham (eds.), *Quantum concepts in space and time*, Clarendon Press, Oxford, 1986.
21. Proc. Internat. Conf. on Quantum Statistics and Foundational Problems of Quantum Mechanics, Hadronic J. Suppl. **1** (1985), 193–693.
22. P. Lahti and P. Mittelstaedt (eds.), *Foundations of modern physics*, Proc. Internat. Sympos., World Scientific Press, Singapore, 1985.
23. V. Gorini and A. Frigerio (eds.), *Fundamental aspects of quantum theory*, Plenum Press, New York, 1986.
24. A. van der Merwe et al. (eds.), *Microphysical reality and quantum formalism*, Proc. Internat. Conf., Reidel, Dordrecht, 1988.
25. P. Lahti and P. Mittelstaedt (eds.), *Foundations of modern physics*, Proc. Internat. Sympos., World Scientific Press, Singapore, 1987.
26. L. Kostro, A. Posiewnick, J. Pykacz, and M. Żukowski (eds.), *Problems in quantum physics*, Proc. Internat. Sympos., World Scientific Press, Singapore, 1988.
27. G. Birkhoff and J. von Neumann, *The logic of quantum mechanics*, Ann. of Math. (2) **37** (1931), 823–843.
28. C. Piron, *Axiomatique quantique*, Helv. Phys. Acta **37** (1964), 439–468.
29. G. W. Mackey, *The mathematical foundations of quantum mechanics*, Benjamin, New York, 1963.
30. A. M. Gleason, *Measures on the closed subspaces of a Hilbert space*, J. Math. Mech. **6** (1957), 885–893.

31. E. Christensen, *Measures on projections and physical states*, Comm. Math. Phys. **80** (1982), 529–538.
32. F. J. Yeadon, *Measures on projections in W^*-algebras of type* II_1, Bull. London Math. Soc. **15** (1983), 139–145.
33. ——, *Finitely additive measures on projections in finite W^*-algebras*, Bull. London Math. Soc. **16** (1984), 601–625.
34. R. V. Kadison, *Isometries of operator algebras*, Ann. of Math. (2) **54** (1951), 325–338.
35. J. Feldman, Dissertation, Univ. of Chicago, 1954. (Also, *Isomorphisms of finite type* II *rings of operators*, Ann. of math. (2) **63** (1956), 565–571.)
36. H. A. Dye, *On the geometry of projections in certain operator algebras*, Ann. of Math. (2) **61** (1955), 73–89.
37. E. D. Wigner, *Group theory and its applications to the quantum theory of atomic spectra*, Academic Press, New York, 1959. (German original 1931.)
38. U. Uhlhorn, *Representation of symmetry transformations in quantum mechanics*, Ark. Fys. **23** (1963), 307–340.
39. R. V. Kadison, *Transformations of states in operator theory and dynamics*, Topology **3** (1965), 177–198.
40. H. Araki, *An application of Dye's theorem on projection lattices to orthogonally decomposable isomorphisms*, Pacific J. Math. **137** (1989), 1–13.
41. V. Bargmann, *Note on Wigner's theorem on symmetry operations*, J. Math. Phys. **5** (1964), 862–868.
42. H. Araki, *On the algebra of all local observables*, Prog. Theoret. Phys. **32** (1964), 844–854.
43. M. Takesaki, *Disjointness of the KMS states of different temperatures*, Comm. Math. Phys. **17** (1970), 33–41.
44. H. Araki, *Type of von Neumann algebra associated with free field*, Prog. Theoret. Phys. **32** (1964), 956–965.
45. W. Driessler, *Comments on lightlike translations and applications in relativistic quantum field theory*, Comm. Math. Phys. **44** (1975), 133–141.
46. ——, *On the type of local algebras in quantum field theory*, Comm. Math. Phys. **53** (1977), 295–297.
47. R. Longo, *Notes on algebraic invariants for noncommutative dynamical systems*, Comm. Math. Phys. **69** (1979), 195–207.
48. P. D. Hislop and R. Longo, *Modular structure of the local algebras associated with the free massless scalar field theory*, Comm. Math. Phys. **84** (1982), 71–85.
49. R. Longo, *Algebraic and modular structure of von Neumann algebras of physics*, Operator Algebras and Applications (R. V. Kadison, ed.), Proc. Sympos. Pure Math., vol. 38, part 2, Amer. Math. Soc., Providence, R.I., 1982, pp. 551–566.
50. H. Araki, *Remarks on spectra of modular operators of von Neumann algebras*, Comm. Math. Phys. **28** (1972), 267–277.
51. E. Størmer, *Spectra of states, and asymptotically abelian C^*-algebras*, Comm. Math. Phys. **28** (1972), 279–294.
52. S. Doplicher, R. Haag, and J. E. Roberts, *Field, observables and gauge transformations.* I; II, Comm. Math. Phys. **13** (1969), 1–23; **15** (1969), 173–200.
53. ——, *Local observables and particle statistics.* I; II, Comm. Math. Phys. **23** (1971), 199–230; **35** (1974), 49–85.
54. S. Doplicher and J. E. Roberts, *Field, statistics and non abelian gauge groups*, Comm. Math. Phys. **28** (1972), 331–348.
55. S. Doplicher, *Problems in quantum field theory and in operator algebras*, Lecture Notes in Math.. vol. 1132, Springer-Verlag, Berlin and New York, 1985, pp. 117–128.
56. S. Doplicher and J. E. Roberts, *C^*-algebras and duality for compact groups: why there is a compact group of internal symmetries*, Proc. Seventh Internat. Congr. Math. Phys. (M. Mebkhout and R. Seneor, eds.), World Scientific Press, Singapore, 1987, pp. 489–498.
57. H. Araki, *On the connection of spin and commutation relations between different fields*, J. Math. Phys. **2** (1961), 267–270.
58. D. Buchholz and K. Fredenhagen, *Locality and the structure of particle states*, Comm. Math. Phys. **84** (1982), 1–54.
59. K. Fredenhagen, *On the existence of antiparticles*, Comm. Math. Phys. **79** (1981), 141–151.

60. D. Buchholz, *Product states for local algebras*, Comm. Math. Phys. **36** (1974), 287–304.
61. S. Doplicher and R. Longo, *Standard and split inclusions of von Neumann algebras*, Invent. Math. **73** (1984), 493–536.
62. R. Longo, *Solution of the factorial Stone-Weierstrass conjecture. An application of the theory of standard split W^*-inclusions*, Invent. Math. **76** (1984), 145–155.
63. ——, *Simple injective subfactors*, Adv. in Math. **63** (1987), 152–171.
64. D. Buchholz, S. Doplicher, and R. Longo, *On Noether's theorem in quantum field theory*, Ann. Physics **170** (1986), 1–17.
65. S. Werner, *Local preparability of states and the split property in quantum field theory*, Lett. Math. Phys. **13** (1987), 325–329.
66. D. Buchholz and E. Wichmann, *Causal independence and energy-level density of localized states in quantum field theory*, Comm. Math. Phys. **106** (1986), 321–344.
67. S. Doplicher, "Current algebra and the local nature of symmetries in local quantum theory," in *Trends and Developments in the Eighties* (S. Albeverio and Ph. Blanchard eds.), World Scientific Press, Singapore 1985, pp. 139–155.
68. C. D'Antoni, S. Doplicher, K. Fredenhagen, and R. Longo, *Convergence of local charges and continuity properties of W^*-inclusions*, Comm. Math. Phys. **110** (1987), 325–348.
69. D. Bratteli and D. W. Robinson, *Operator algebras and quantum statistical mechanics*. II, Springer-Verlag, Berlin and New York, 1981.
70. R. Kubo, *Statistical-mechanical theory of irreversible processes*. I. *General theory and simple applications to magnetic and conduction problems*, J. Phys. Soc. Japan **12** (1957), 570–586.
71. P. C. Martin and J. Schwinger, *Theory of many-particle systems*. I, Phys. Rev. **115** (1959), 1342–1373.
72. R. Haag, M. N. Hugenholtz, and M. Winnink, *One the equilibrium states in quantum statistical mechanics*, Comm. Math. Phys. **16** (1967), 81–104.
73. M. Tomita, *Standard forms of von Neumann algebras*, Fifth Functional Analysis Symposium of the Math. Soc. of Japan, Sendai, 1967.
74. M. Takesaki, *Tomita's theory of modular Hilbert algebras and its applications*, Lecture Notes in Math., vol. 128, Springer-Verlag, Berlin and New York, 1970.
75. J. J. Bisognano and E. H. Wichmann, *On the duality condition for a Hermitian scalar field*, J. Math. Phys. **16** (1975), 985–1007; **17** (1976), 303–321.
76. G. Sewell, *Relativity of temperature and the Hawking effect*, Phys. Lett. A **79** (1980), 23–24.
77. H. Araki, *On uniqueness of KMS-states of one-dimensional quantum lattice*, Comm. Math. Phys. **44** (1975), 1–7.
78. A. Kishimoto, *On uniqueness of KMS states of one-dimensional quantum lattice systems*, Comm. Math. Phys. **47** (1976), 167–170.
79. H. Araki and T. Matsui, *Ground states of the XY model*, Comm. Math. Phys. **101** (1985), 213–245.
80. J. Bellissard, *K-theory of C^*-algebras in solid state physics*, Lecture Notes in Phys., vol. 257, Springer-Verlag, Berlin and New York, 1986, pp. 99–156.

KYOTO UNIVERSITY, JAPAN

Proceedings of Symposia in Pure Mathematics
Volume **50** (1990)

Mathematics Motivated by Physics

ARTHUR JAFFE

Introduction. John von Neumann had a deep interest in the natural world, and a major stimulus to his mathematical inspiration came from questions of quantum theory. Many of his papers directly concern problems of physics—such as his study of the representations of the canonical commutation relations, or his work on the theory of measurement. But in addition, much of von Neumann's mathematical work has evolved over the years into central tools for the study of physics. One example comes from von Neumann's work on operator algebras, which is now used in various modern formulations of the axioms for the theory of quantum fields. Another is the index for type II factors, which is related by the work of Jones to conformal field theory. In this spirit, I will describe a set of mathematical ideas growing out of work in physics, but which were very much inspired by the mathematics von Neumann developed. The focus of this paper is two topics: the analysis and the geometry of function spaces. Posed differently, this is the theory of analysis and geometry of infinite dimensional manifolds. Today, as in the past, physics remains an inspiration for analysis and geometry; one of the most active areas of current mathematical research centers on ideas evolving from the theory of quantum fields.

Von Neumann's theorem on the uniqueness of representations of the Weyl relations concerns quantum mechanics. More precisely, it is limited to systems with a finite number of degrees of freedom. In contrast, local quantum field theory involves systems with an infinite number of degrees of freedom and yields examples of unitarily inequivalent representations of the Weyl relations. Other sorts of infinite dimensional representations arise in the study of conformal field theory, especially representations of infinite dimensional algebras (Virasoro, Kac-Moody, etc.) which one finds in solutions to quantum

1980 *Mathematics Subject Classification* (1985 *Revision*). Primary 81-02, 81E08, 82A15, 46L60; Secondary 81C35.
This paper is in final form and will not be submitted for publication elsewhere.
Supported in part by the National Science Foundation.

field equations with special symmetries. One way to attempt to parameterize these examples — a way suggested by physics—is to investigate solutions of the nonlinear equations of motion.

I will describe such an investigation here. The work is mathematics, although the formulations arise from physics. Analysis on function spaces forms the basis for the existence theorems for nonlinear quantum fields and for the nonlinear equations which they satisfy. The geometry associated with these equations is closely tied to the theory of noncommutative differential geometry. In particular, the geometry of quantum field theory gives rise to noncommutative geometry in an infinite dimensional setting, and provides examples of the theory of entire cyclic cohomology.

1. **Nonlinear equations of quantum field theory.** Operator-valued solutions φ to nonlinear, hyperbolic partial differential equations are quantized fields. The simplest such case is a nonlinear, two-dimensional wave equation on a cylinder; then $\varphi = \varphi(x,t)$ is a function of two variables $(x,t) \in S^1 \times \mathbf{R}$. A typical such equation is

$$\Box\,\varphi + \varphi + \varphi^3 = 0\,, \tag{1}$$

where $\Box\,\varphi = \varphi_{tt} - \varphi_{xx}$ denotes the action of the wave operator. This is called the "φ^4 equation," as it arises from an energy functional which is quartic when expressed in terms of φ. What is special to quantum field theory is that the initial data satisfy constraints called the "canonical commutation relations." This is just an infinite dimensional Heisenberg algebra, which is defined as follows: let $\varphi(f,t) = \int \varphi(x,t)f(x)dx$, where f is a real, C^∞ function on the circle. The Heisenberg algebra is generated by these $\varphi(f,t)$'s, which are required to satisfy

$$[\varphi(f,0), \varphi_t(g,0)] = i\langle f, g\rangle_{\mathscr{L}^2(S^1)}\,. \tag{2}$$

Here $\varphi(f,t)$ and $\varphi_t(f,0)$ are linear, selfadjoint operators on a Hilbert space $\mathscr{H}$. Equations such as (1–2) must be interpreted as equations for an operator-valued distribution φ. Thus the nonlinearity in φ must be defined, independently of the equation itself.

Similar examples, the $\mathscr{P}(\varphi)$-equations, arise from replacing $\varphi + \varphi^3$ by $\mathscr{P}'(\varphi)$, where $\mathscr{P}$ is a polynomial which is bounded from below. This class of $\mathscr{P}(\varphi)$ equations on a cylinder arise from energy densities $\frac{1}{2}\,(\varphi_t^2 + \varphi_x^2) + \mathscr{P}(\varphi)$. The equations are

$$\Box\,\varphi + \mathscr{P}'(\varphi) = 0\,. \tag{3}$$

The first problem in understanding these equations is to define the nonlinearities and to show the existence of solutions. The positive solution to this question is a classic, fundamental result of constructive quantum field theory. General references can be found in **[GJ4]** for the following two theorems:

THEOREM 1. *A precise definition of polynomials in φ can be given. With this definition, and with $\mathscr{P}$ as above, solutions exist to the $\mathscr{P}(\varphi)$ equations* (3) *with initial data satisfying* (2).

The definition of $\mathscr{P}(\varphi)$, consistent with the nonlinear equation, involves a solution to the physics problem of renormalization. The procedure includes three steps: Step 1. First truncate the Fourier series of φ, $\varphi = \sum_{j=-N}^{N} \varphi_j e^{ijx}$. Step 2. Define the modified monomial φ^k by replacing φ^k with a polynomial of the form $\varphi^k + \mathscr{R}_k$, where $\mathscr{R}_k$ is a $(k-2)$-degree polynomial of the form $\sum_{j=1}^{[\frac{k}{2}]} c^j \varphi^{k-2j}$. Here $c = c(N)$ diverges for N tending to infinity as $O(\log N)$. The addition of $\mathscr{R}_k$ to φ^k is called the additive renormalization of φ^k. Step 3. Finally, show that a suitable limit $N \to \infty$ exists and defines a solution to the equations (2-3), as well as to the standard axioms for quantum fields. In the language of physics, this means that we have given a nonperturbative proof of renormalization for these equations, as well as providing a proof of existence. The choice of $c(N)$ is determined by perturbation theory, see **[GJ1-4]**. This classic route is somewhat *ad hoc*, because it is necessary to introduce $c(N)$.

The second problem is to replace the cylinder by Minkowski space:

THEOREM 2. *The Minkowski space limits of these models, $S^1 \times \mathbf{R} \to \mathbf{R} \times \mathbf{R}$, exist and satisfy the standard axioms of quantum field theory* (*Wightman axioms, Haag-Kastler axioms*).

I do not wish to elaborate the definition of $\mathscr{P}(\varphi)$ or its derivative, or on the existence proof for these equations. Rather, I will describe here some recent results of a surprising nature concerning certain systems of nonlinear quantum field equations which depend on both scalar and spinor fields. These equations are chosen in a special way so that they possess a symmetry, called supersymmetry, which relates the scalar fields to the spinor fields. In the case of these systems, it is actually possible to solve the quantum field equations without any renormalization! In other words, we can omit Step 2 above; it is unnecessary to introduce artificially the $O(\log N)^j$ coefficients c^j into the equations.

For this reason I call such quantum field equations *a priori* quantum field equations; these *a priori* equations do not require ad hoc modification in the definition of the nonlinear terms. They are close in spirit to ordinary nonlinear equations. From the aesthetic point of view, this certainly is preferable to infinite renormalization, but it remains to be seen whether such principles (such as supersymmetry) are actually relevant for equations which describe nature.

Previous arguments that renormalization is always necessary were based on the assumption of scalar equations—not on systems of scalars and spinors. The fact that the equations have solutions without renormalization is related to a cancellation of the putative divergences for scalars with those for spinors.

Not every supersymmetric equation has this property; but on a cylinder equations with "$N = 2$ type" of supersymmetry do behave nicely. The resulting equations are well-behaved; one can understand from perturbation theory how different divergent terms arise, but cancel. Such behavior was suggested in perturbation theory in the physics literature in 1974, but only recently has it been understood as a mathematical theory **[JLWie1-2]**. This theory is intimately tied to the geometry of loop space.

Let me describe these examples in a bit more detail. Consider a complex scalar field $\varphi = \frac{1}{\sqrt{2}}(\varphi_1 + i\varphi_2)$, where φ_1 and φ_2 are selfadjoint; we couple this field to a 2-component Dirac field ψ. The interaction is specified by a polynomial of the form

$$V(\varphi) = \sum_{j=1}^{n} a_j \varphi^j , \qquad a_j \in \mathbf{C} .$$

At a classical level, the equations themselves are variational equations of a standard functional, which corresponds to the energy function,

$$H = \frac{1}{2} \int_{S^1} \left\{ |\varphi_t|^2 + |\varphi_x|^2 + |V'|^2 + \bar{\psi}(i\gamma^1 \partial)\psi + \bar{\psi} W \psi \right\} dx . \tag{4}$$

In this expression we use the Dirac matrices

$$\gamma_0 = \begin{pmatrix} 0 & 1 \\ 1 & 0 \end{pmatrix} , \qquad \gamma_1 = \begin{pmatrix} 0 & -1 \\ 1 & 0 \end{pmatrix} . \tag{5}$$

We define the adjoint $\bar{\psi} = \psi^{\mathrm{tr}} \gamma_0$, and the interaction term $\bar{\psi} W \psi$ is

$$W = V'' \begin{pmatrix} 1 & 0 \\ 0 & 0 \end{pmatrix} + (V'')^* \begin{pmatrix} 0 & 0 \\ 0 & 1 \end{pmatrix} .$$

The interactions arising from (4) could be called generalized $\mathscr{P}(\varphi)$-Yukawa models; they have polynomial interactions given by $\mathscr{P} = |V'|^2$, and boson-fermion couplings determined by W, which in the quadratic case would be the standard Yukawa coupling of physics.

THEOREM 3. *The equations arising from the interaction* (4) *on a cylinder have solutions without renormalization. The components φ of the initial data satisfy the canonical commutation relations* (2) *while the components of the Fermi field satisfy canonical anticommutation relations*

$$\{\psi_\mu(x), \psi_\nu(y)\} = \{\bar{\psi}_\mu(x), \psi_\mu(y)\} = 0,$$
$$\{\bar{\psi}_1(x), \psi_2(y)\} = \{\bar{\psi}_2(x), \psi_1(y)\} = \delta(x - y)I .$$

One technical tool in the analysis of these problems is the use of Pfaffians on Hilbert space. The theory of Fredholm determinants is well developed as the perturbations of the identity, $\mathrm{Det}\,(I - T) = 1 + f(T)$, where $f(T)$ is given by a Fredholm series for $T \in \mathscr{L}_1$, the Banach space of trace class operators. On the other hand, the corresponding theory of Pfaffians is relatively new. Pfaffians of skew-symmetric, finite dimensional matrices are a standard invariant. The Pfaffian of an infinite-dimensional, skew matrix, however, does not exist. But as in the case of the determinant, the relevant object is not the Pfaffian of one matrix but the ratio of two Pfaffians. In the finite dimensional case, with A invertible, define the *relative Pfaffian* of A and B by $\mathrm{Pf}(A,B) = \frac{\mathrm{Pf}(A^{-1}-B)}{\mathrm{Pf}(A^{-1})}$. It has the property

$$\mathrm{Pf}(A,B)^2 = \mathrm{Det}(I - AB)\,. \tag{6}$$

Actually the relative Pfaffian can be defined both when A is not invertible and in the infinite dimensional setting. One only requires that A and B are Hilbert-Schmidt. This Pfaffian is the limit of finite rank approximations and so it has the property (6). An analytic theory of such Pfaffians on Hilbert space has been developed in **[JLWe3]**.

The second ingredient in the proof is the use of functional integral representations of the heat kernels e^{-tH}, namely Feynman-Kac type representations of the form

$$\langle \chi, e^{-tH}\psi\rangle = \int \bar{\chi}\psi\, d\mu_H\,.$$

Here $d\mu_H$ is a generalization of the measures on the space of distributions $\mathscr{D}'$ (measures for the $\mathscr{P}$-theories) to the setting of functionals on a Grassmann algebra (needed for the inclusion of spinors). The required technical framework takes a lengthy development; in the case of the work cited here, the delicate *a priori* cancellations are established in **[JL1, JLWe2, JLWiel-2]**, and are based on a refinement of techniques in **[GJ4, SS]**.

The basis for the analysis of functional integrals (for problems with an infinite number of degrees of freedom) is the phase-cell localization method and the cluster expansion method developed in **[GJ1-4]**. These expansions are very similar in spirit to renormalization group methods, but they are inductively defined expansions rather than iteratively defined ones. Furthermore, these methods lead to theorems in the mathematical sense. Related ideas occur both in such mathematical work as Carleson's theorem on the convergence of Fourier series **[Ca]** or Fefferman's analysis of the spectrum of the Laplace operator **[F]**.

Work on the above supersymmetric examples gives a spectral theory for certain Dirac operators Q and Laplace operators H on function spaces. The Laplace operator, for example, has the general form

$$\Delta + \mathscr{P},$$

where $\mathscr{P}$ is a potential (possibly involving a coupling to fermions). The detailed analysis of the spectrum of such operators is within the scope of the methods of constructive quantum field theory.

2. **Supersymmetry and geometry.** The variational equations corresponding to this energy function are supersymmetric. The generator of the supersymmetry is a selfadjoint operator Q which has three basic properties:

(i) It is selfadjoint and local,

$$Q = Q^* = \frac{1}{\sqrt{2}} \int_{S^1} Q(x)dx\,.$$

The last equality holds as a form equality on a suitable domain, and the operator Q is the closure of this form.

(ii) It is a square root of the Hamiltonian,

$$H = Q^2\,.$$

(iii) There is a $\mathbf{Z}_2$-grading γ, which anticommutes with Q.

Thus γ is a selfadjoint, unitary operator, $\gamma = \gamma^* = \gamma^2$, and

$$Q\gamma = -\gamma Q\,.$$

More generally, let $A \to A^\gamma = \gamma A \gamma$ denote the action of the grading. Every operator A can be decomposed into even and odd parts under the grading, $A = A_+ + A_-$, where $A^\gamma_\pm = \pm A_\pm$. Then $Q^\gamma = -Q$. In the representation for which γ is diagonal, the operators Q and H have the form

$$Q = \begin{pmatrix} 0 & Q_- \\ Q_+ & 0 \end{pmatrix}\,, \qquad H = \begin{pmatrix} Q_-Q_+ & 0 \\ 0 & Q_+Q_- \end{pmatrix}\,. \tag{7}$$

THEOREM 4 **[JL1-2, JLWe1-2]**. *The operator Q has an explicit representation in terms of fields. Namely*

$$Q(x) = \psi_1(\varphi_t - \varphi_x^* - iV'(\varphi)) + \psi_2(\varphi_t^* - \varphi_x - iV'(\varphi)^*) + \textit{hermitian conjugate}\,.$$

With this definition, the operator Q is selfadjoint and Fredholm. It satisfies (i-iii) *above, and furthermore the heat kernel e^{-tH} is trace class for all $t > 0$.*

With Q in hand, we can discuss the infinitesimal supersymmetry transformation which it generates. An infinitesimal Lie symmetry is a derivation; such a symmetry is generated by a commutator. An infinitesimal supersymmetry is a superderivation; it is generated by a supercommutator. This infinitesimal dA is defined by

$$dA = QA - A^\gamma Q\,.$$

In the language of physics, the even elements under the grading are called bosonic, while the odd elements are fermionic. If A is bosonic, then $dA = QA - AQ$, while if A is fermionic, then $dA = QA + AQ$. Since Q anticommutes with γ, a bosonic operator is transformed into a fermionic one, and a fermionic operator is transformed into a bosonic one. Explicitly, the field operators φ and the components ψ_j of ψ have the infinitesimal transformation properties

$$d\varphi = \frac{-i}{\sqrt{2}}(\psi_1+\psi_2) \quad d\psi_1 = \frac{1}{\sqrt{2}}(\varphi_t^*-\varphi_x+iV'(\varphi)^*) \quad d\psi_2 = \frac{1}{\sqrt{2}}(\varphi_t-\varphi_x^*+iV'(\varphi))\,.$$

The superderivation property is

$$d(AB) = (dA)B + A^\gamma(dB)\,.$$

The square of any superderivation is a derivation,

$$d^2(AB) = (d^2A)B + A(d^2B) .$$

In fact d^2 is the derivation given by commutation with H, namely $d^2A = [H, A]$; it is the infinitesimal generator of the time translation group.

3. **Index theory and noncommutative geometry.** The form of Q given in Theorem 4 is relevant to a standard index problem for a Dirac-type operator (see [**W**, **JL2**]). The index of Q_+ is given by the graded trace of the heat kernel, and we have

$$\text{Index}(Q_+) = \text{Tr}\,(\gamma e^{-H}) .$$

The index of a Dirac operator is a fundamental geometric invariant. The index is stable under continuous deformations of Q, and in many cases it has been computed for quantum field theory by a homotopy argument. Of course, one relies on the existence theory of Theorem 4 to justify the statement that the index is a mathematically well-defined object.

The analysis of Dirac operators on compact manifolds is one of the now standard points of departure for index theory. The fact that a Dirac operator has a square equal to a Laplace operator is fundamental. This phenomenon occurs in the infinite dimensional case, for example in the study of a Dirac-type operator Q on loop space. In the simplest case of that example, one is led to Dirac operators such as the ones of Theorem 4. One finds [**JLWe1**, **JL2**] in these examples, the following result.

THEOREM 5. $\text{Index}(Q_+) = \text{Degree}(V')$.

The number of ground states of H is greater or equal to magnitude of this index, $|\,\text{Index}(Q_+)|$. In the quantum mechanical case, it is known that the two numbers are equal [**JLL**]. We conjecture that this vanishing theorem also holds in the general case described by Theorem 5, but a better understanding of the phenomenon is necessary [**JL2**]. Atiyah has suggested that this phenomonon may be interpreted in terms of Floer cohomology [**A**].

The fact that quantum field theory naturally fits into index theory yields one framework for the geometry of infinite dimensions. Supersymmetric quantum field theory also provides examples of the noncommutative geometry of Alain Connes. In fact, Connes' geometry is a natural setting for index theory. One clear fact is that the cohomological nature of field theory is beginning to emerge, and it is being studied by this and other points of view. Infinite dimensional spaces have entered geometry in several ways, but the generalization inspired by quantum field theory is very natural. A fundamental starting point for the investigation of the geometry of infinite dimensions is to study noncommutative geometry of infinite dimensions.

The basic reference for noncommutative geometry is [**C1**], with natural generalizations to the infinite dimensional setup [**C2**, **JLO1**]. Cyclic homology and cohomology have been studied by other workers, especially Feigen,

Tsygan, and Quillen, but the framework of Connes naturally lends itself to a description of physics and to a connection with quantum field theory. The reason is that the basic objects studied by Connes are the cochains $f_n(a_0, \cdots, a_n)$, which are $(n+1)$-linear functional on an algebra $\mathfrak{A}$. These functionals play the same role in noncommutative geometry as the Wightman functions (or the Euclidean Green's functions) do in quantum field theory. In the field theory case, the operators a_j are field operators or products of field operators, and $\mathfrak{A}$ is an "algebra of field operators."

There is a complex on the space $\mathscr{C}(\mathfrak{A})$ of cochains, given by the coboundary operator ∂ of entire cyclic cohomology,

$$\cdots \longrightarrow \mathscr{C}(\mathfrak{A}) \xrightarrow{\partial} \mathscr{C}(\mathfrak{A}) \xrightarrow{\partial} \mathscr{C}(\mathfrak{A}) \longrightarrow \cdots,$$

where $\partial^2 = 0$. The pertinent problem in noncommutative geometry is to find a cocycle $f_n = \tau_n$ for the coboundary operator; the cocycle condition is $\partial\tau = 0$. The cocycle encodes the cohomological information of the Wightman functions. We have given a simple formula for τ **[JLO1]**: Let $a(t) = e^{-tH} a e^{tH}$. Then

THEOREM 6. *The functionals*

$$\tau_n(a_0, \dots, a_n) = \int_{0 \le t_1 \le \cdots \le t_n \le 1} \operatorname{Tr}(\gamma a_0 da_1(t_1) da_2(t_2) \cdots da_n(t_n) e^{-H})\, dt_1 \cdots dt_n \tag{8}$$

are a cocycle for the coboundary operator ∂ of entire cyclic cohomology.

This cocycle has an intimate connection with index theory. In the case that $n = 0$, the cocycle $\tau_0(I)$ is just the index of Q_+. More generally, if e is an idempotent in the algebra generated by field operators, then the vector bundle determined by e has an index problem given by τ

$$\operatorname{Index}((eQe)_+) = \sum_{n=0}^{\infty} (-1)^n \frac{(2n)!}{n!} \hat{\tau}_n(e, e, \cdots, e), \tag{9}$$

where $\hat{\tau}$ denotes a normalized form of τ (see **[C1, C2, JL3]**). This can be extended to the case of the superderivation d itself. In the examples above, the function (9) is stable under variation of the lower order coefficients in the potential V, and captures the cohomology. The fact that τ has such a simple expression in terms of the elements of the algebra $\mathfrak{A}$ yields the possibility to use the formula (8) for computations. This has been done for the index, but more general computations are clearly needed.

4. **Operator algebras and statistical mechanics.** The connection between geometric invariants of noncommutative geometry and statistical mechanics may seem far-fetched. But the modular operator in operator algebras arises from the theory of Gibbs states and the "KMS condition." This condition is a central feature of the abstract treatment of statistical mechanics on the one hand, and of Tomita's theory of modular automorphisms on the other,

especially as exploited by Connes. Here we study a superversion of this condition.

Let $\mathfrak{A}_\alpha$ denote the subspace of $\mathfrak{A}$ such that the automorphism $\alpha_t : a \to e^{itH}ae^{-itH}$ extends to an entire function of t. If $\mathfrak{A}$ is a C^*-algebra, then $\mathfrak{A}_\alpha$ is always dense in $\mathfrak{A}$. We define a super-KMS functional as a continuous linear functional ω on $\mathfrak{A}$, such that for all elements of $\mathfrak{A}_\alpha$,

(10) $\qquad \omega(da) = 0, \qquad$ super-translation invariance

and

(11) $\qquad \omega(ab) = \omega(b^\gamma \alpha_i(a)), \qquad$ super-KMS property.

The super-KMS property is reflected in the action of the grading, and for this reason ω is not in general positive. A standard example of a super-KMS functional is given by a supertrace ω, namely

(12) $$\omega(a) = \mathrm{Str}(ae^{-H}) = \mathrm{Tr}(\gamma ae^{-H}).$$

We can also study cases where (12) is not valid, such as when H may have continuous spectrum, or when H may not even exist. We can construct such functionals as limits of functionals of the form (12). A standard example is the infinite volume limit of quantum field theory. This has been studied in a simple case [We], and is now under investigation in a realm of more general cases.

PROPOSITION 7. *If ω is a super-KMS functional on $\mathfrak{A}$, then the function*

(13) $$f(t_1, \dots, t_n) = \omega(a_0, \alpha_{t_1}(a_1), \dots, \alpha_{t_n}(a_n))$$

extends to an analytic function in the strip $0 \le \mathrm{Im}\, t_1 \le \cdots \le \mathrm{Im}\, t_n \le 1$. *Furthermore, for t in the strip,*

(14) $$|f| \le \mathrm{const.} \prod_{j=0}^{j=n} \|a_j\| \, .$$

PROOF. One starts on the subalgebra $\mathfrak{A}_\alpha$. Using identity (11) and a Phragmén-Lindelöf type of theorem, one can establish (14) inductively in n. The passage to analyticity on the full algebra $\mathfrak{A}$ follows by the Weierstrass approximation theorem.

THEOREM 8. *The functionals τ_n defined by*

(15) $$\tau_n(a_0, \dots, a_n) = \int_{0 \le t_1 \le \cdots \le t_n \le 1} \omega(a_0, \alpha_{it_1}(da_1), \dots, \alpha_{it_n}(da_n))dt$$

are a cocycle for the entire cyclic cohomology.

This is a natural extension of Theorem 6, and similar questions have been studied by Kastler [**K**]. An exponential form of studying the infinite dimensional (entire) cyclic cohomology has been proposed by Quillen [**Q**] and developed in [**JO**]. It gives an elegant way to verify the algebraic properties of τ by reducing the computations to those of generating functions (exponential series in the universal algebra generated by the fields).

5. **The bad influence of physics on mathematics.** This provocative title is meant to draw attention to some comments on the modern development of a relation close to von Neumann's heart. I believe that the interaction over the past ten years between mathematics and physics has sped the course of mathematical development. Once again physics has become a central partner to mathematical discovery. Striking examples can be found in gauge theory or in analysis in infinite dimensions. Physics has provided mathematical inspiration, pointing to new ideas and opening new avenues for mathematical research. Along the way, ideas from physics have led to interesting conjectures and to some new mathematical theorems. Recently there has been a flurry of activity oriented toward conjectures and coming from particle physicists. Physics has also become a very fashionable field for mathematicians to study, so it is natural that stimulating new points of view have developed.

Having written enthusiastic accounts of this phenomenon in the past [**J1–3**], I would now like to play the devil's advocate. As mathematicians and physicists come in closer contact, attitudes and ideas from one subject are communicated to workers in the other subject. What bothers me is a certain "sociological" gulf between physics and mathematics; this is partly a matter of language, but more a matter of attitude, working habits, and an evolved tradition. In particular, the center of this question revolves about how one regards mathematical understanding, including the status of a mathematical piece of work, and what weight or importance one attaches to work at various stages. I contend that a seed has been planted that could eventually change the character of mathematics as we know it today, with a change for the worse.

One aspect of this, but certainly not the only one, concerns the concept of a "mathematical proof." It is easiest to illustrate details of my argument in terms of a mathematical proof, but the reader should keep in mind the fact that this discussion is really symbolic for a set of more subjective criteria concerning insight and understanding in mathematics. So for the sake of being specific, let me consider proofs.

The notion of a mathematical proof certainly makes the subject of mathematics very special when compared with all other intellectual pursuits, such as the study of history, or of biology, or even of physics. These subjects appeal to the natural world for empirical evidence. Mathematics, on the other hand, does not have such a mirror; it must create a set of internal values and standards.

Most mathematical statements begin life as conjectures. A conjecture becomes a mathematical theorem when it is proved by commonly accepted logic. The standards of what constitutes a proof have evolved over hundreds of years; there is no doubt in the mind of traditional mathematicians what a proof means.

In comparison, since physicists can appeal to nature, they generally work in a very different style. For example, it is not always necessary to make a clear distinction between a conjecture and a theorem. This may sound radical, but the notion of a proof in much of the physics literature is what a mathematician would call a plausibility argument. This is evident when one hears a physicist refer to a "nonrigorous proof of a statement." In mathematics, there is no such thing as a nonrigorous proof: one can give a proof; or one can give an intuitive outline of a proof, or of a possible proof. One can also give a proof, modulo some stated but unproved assertions. In the last case, one reduces a proof to a well-posed conjecture. A "nonrigorous proof" is simply not a proof, but lies in some other category.

On the other hand, the typical worker in the physics community is not trained to use the logic which is commonly accepted by mathematicians; hence it is difficult for the physicist to pass from an illuminating outline to a proof. Such an outline may give great insight—possibly even into a well-known theorem—without necessarily providing a technical avenue to its proof. However, one must distinguish between insight into a known result (which has already been proved in the traditional sense) and a conjecture (which may be suggested by the outline but which may not survive after precise formulation). New points of view can be very interesting, but they become important when they lead to new results.

I like to use the term "mathematical physicist" to describe the small number of persons familiar with physics who actually prove mathematical theorems. This mathematical physicist is really a mathematician who gets inspiration and motivation from problems of physics. While this modern meaning is not universally accepted, clearly some distinction is necessary to separate various working styles.

Over the past ten years, making this distinction has become more and more difficult because of the appearance on the scene of a new generation of young theoretical physicists who have learned large amounts of modern mathematics, but who are not mathematical physicists in the sense above. This new generation does not use their mathematical knowledge to prove new mathematical theorems. Rather they give beautiful outlines of complicated arguments, arguments which often yield extremely interesting mathematical speculations—some of them far reaching— but nevertheless mathematical conjectures. Members of this new generation do not take the next step to try to refine the conjectures, or to elevate them to mathematical results.

Of course, some mathematical conjectures are of tremendous importance: they may provide new insights uniting diverse fields of mathematics. Or a speculation may distill years of thought into a precious nutshell. Still other pioneering work can lead to entirely new areas of mathematics, such as Riemann's contributions to geometry and analysis. Most speculations, however, are less glorious; and mathematics is full of them. It is certainly unreasonable for a substantial fraction of the mathematical community to work on conjectures without refining their understanding to shape precise statements, but a full plate of conjectures can be a pretty heady meal.

A problem arises if conjectures are represented as true understanding. Mathematicians know that technical snags can cloud the path to any result; famous conjectures were worked on over the course of many years—and sometimes even over centuries. In the case one of these famous problems is solved—i.e., a proof is given—major mathematical credit also goes to the person who gives the first proof, not just to the originator of the conjecture. Some well-known mathematicians have even devoted their entire careers to working toward the proof of one famous theorem. Even partial results, such as Falting's proof of Mordell's conjecture—a step toward finiteness of the number of solutions to Diophantine equations—are greeted with wide acclaim.

Speculation, conjecture, or incomplete understanding should not be presented as a finished product; rather it is a tentative and unstable state in need of thorough and disciplined support. This is not a philosophical conclusion; other approaches have been tried in the past, with poor consequences. A striking example of these effects can be seen in the collapse of the Italian school of algebraic geometry in the 1930s, after a generation of brilliant speculation.

There is a sharp contrast between these values and those of physics. In physics, the various stages of an argument are given very different weights. Richard Feynman has boasted about his acumen at reasoning with mathematicians, because he understands mathematical statements in terms of testing some special cases. Physicists are often satisfied to test a number of cases and conclude that the hypothesis is true. The four color problem, for example, could be regarded solved by map makers. This is consistent with the philosophy that physics is an inductive science. Furthermore, in physics a much higher value is placed on an everyday —as opposed to revolutionary— conjecture of a result, than on the verification of whether it is actually true or false.

I am sometimes asked how to describe the difference between the way mathematicians and physicists think. My answer is that in the initial stages of discovery I believe that they think in exactly the same way. Both mathematicians and physicists appeal to intuition, try out examples, and propose outlines of arguments. The difference really comes when mathematicians carry this procedure on for many more iterations than physicists do. In the end, mathematicians formulate hypotheses precisely and turn outlines into proofs. They eventually eliminate the false starts, and they often do not publish their results until their conclusions are fairly complete. Physicists are more prone to rush into print and to be satisfied with partial arguments or tentative conclusions. The difference between the two cultures is very much a matter of degree.

One subject that has been degraded by its use in the formal physics literature is the area of function space integrals, or path integrals. While there is a well-developed mathematical literature on this subject, many first rate mathematicians have come to believe that any argument based on functional integrals is mathematically suspect. That is not the case. A great amount of interesting mathematics is based on *mathematically well-defined* functional integrals. Perhaps this qualifier should be used in the future to identify such in the mathematical literature, but I am sorry to say that it appears necessary!

Of course, what distinguishes physics most from mathematics is the fact that physics is an explanation of nature. The connection between nature and the equations of physics may indeed involve some assumptions which are difficult to understand. For this reason physics is not necessarily bound to mathematical logic, but this can be justified because physics is partially an empirical science.

Another characteristic of physics is its relatively short attention span for a given set of problems, compared with the long time-scale of mathematics. At the present time, there is a different trend in theoretical physics, which I alluded to above. Large numbers of people study string theory or other mathematically oriented areas. They provide a rich structure, full of ideas and potential connections between different branches of physics (such as gravitation and strong interactions) and mathematics. But the recent developments of these workers do not necessarily fit the picture of an empirical science: while there is little evidence at this point that string theory is a correct description of nature, many people don't seem to care.

On the other hand, physicists have helped provide brilliant new insights into some branches of mathematics. Physics has again become one of the most fascinating sources of mathematical speculations. However, awaiting the proper formulations of these ideas for the description of nature, one could ask whether string theorists, and other related workers, should be regarded as physicists or as mathematicians.

This is at the heart of the problem. Many physicists regard this type of work as mathematics, because it has little contact with experience. It is not physics in the sense of the word as commonly accepted twenty years ago. On the other hand, most mathematicians currently say that what most of these scientists do is physics; what this statement really means is that it is something *other* than mathematics. In fact, the work may be *neither* physics *nor* mathematics as we know them today.

REFERENCES

[A] M. F. Atiyah, *New invariants for 3- and 4-dimensional manifolds, The mathematical heritage of Hermann Weyl*, R. O. Wells, Jr., Editor, American Mathematical Society, Providence, R.I., 1988, pp. 285–289.

[Ca] Lennart Carleson, *On convergence and growth of partial sums of Fourier series*, Acta Math. **116** (1966), 135-157.

[C1] A. Connes, *Noncommutative differential geometry*, Publ. Math. IHES **62** (1985), 257–360.

[C2] ——, *Entire cyclic cohomology of Banach algebras and characters of θ-summable Fredholm modules*, K-Theory **1** (1988), 519–548.

[CM] Alain Connes and Henri Moscovici, *Cyclic cohomology and the Novikov conjecture*, preprint.

[EFJL] K. Ernst, P. Feng, A. Jaffe, and A. Lesniewski, *Quantum K-Theory II. Homotopy invariance of the Chern character*, J. Funct. Anal., to appear.

[F] C. Fefferman, *The Heisenberg uncertainly principle*, Bull. Amer. Math. Soc., **9** (1983), 125–322.

[GJ1] J. Glimm and A. Jaffe, *A $\lambda\phi^4$ quantum field theory without cut-offs.* I, Phys. Rev. **176** (1968), 1945–1961.

[GJ2] ——, *The $\lambda(\phi^4)_2$ quantum field theory without cut-offs.* IV. *Perturbations of the Hamiltonian*, J. Math. Phys. **13** (1972), 1568–1584.

[GJ3] ——, *Positivity of the ϕ^4_3 Hamiltonian*, Fortschritte der Physik **21** (1973), 327–376.

[GJ4] ——, *Quantum physics*, Second Edition, Springer-Verlag, New York, 1987.

[GJS] J. Glimm, A. Jaffe, and T. Spencer, *A convergent expansion about mean field theory, Part* I. *The expansion*, Ann. Phys. **101** (1976), 610–630; *Part* II. *Convergence of the expansion*, Ann. Phys. **101** (1976), 631–669.

[J1] A. Jaffe, "Ordering the universe: the role of mathematics," in *Renewing U.S. mathematics*, D. H. Van Osdol, Editor, National Academy Press (1984) pp. 1–19.

[J2] ——, "Mathematics at the heart of science," *Balomenos lectures*, University of New Hampshire Press, Durham, N. H., 1988, pp. 117–162.

[J3] ——, *The importance of mathematics*, SUNY Research **9** (1989), 4–7.

[JL1] A. Jaffe and Lesniewski, *A priori estimates for $N = 2$ Wess-Zumino models on a cylinder*, Commun. Math. Phys. **114** (1988), 553–575.

[JL2] ——, *Supersymmetric field theory and infinite dimensional analysis*, in "Nonperturbative Quantum Field Theory," Proceedings of the 1987 Cargèse Summer School, G. 't Hooft *et. al.*, Editors, Plenum Press, New York, 1988.

[JL3] ——, *An index theorem for super-derivations*, Commun. Math. Phys. **125** (1989), 147–152.

[JL4] ——, *Geometry of Supersymmetry*, in "1988 Erice Lectures," A. Wightman and G. Velo, Editors, Plenum Press, New York, to appear.

[JLL] A. Jaffe, A. Lesniewski, and M. Lewenstein, *Ground state structure in supersymmetric quantum mechanics*, Ann. Phys. **178** (1987), 313–329.

[JLO1] A. Jaffe, A. Lesniewski, and K. Osterwalder, *Quantum K-Theory I: the Chern character*, Commun. Math. Phys. **118** (1988), 1–14.

[JLO2] ——, *On Super-KMS Functionals and Entire Cyclic Cohomology*, K-Theory **2** (1989), 675–682.

[JLWe1] A. Jaffe, A. Lesniewski, and J. Weitsman, *Index of a family of Dirac operators on loop space*, Comm. Math. Phys. **112** (1987), 75–88.

[JLWe2] ——, *The two-dimensional, $N = 2$ Wess-Zumino model on a cylinder*, Commun. Math. Phys. **114** (1988), 147–165.

[JLWe3] ——, *Pfaffians on Hilbert space*, J. Funct. Anal., **83** (1989), 348–363.

[JLWe4] ——, *The loop space $S^1 \to \mathbf{R}$ and supersymmetric quantum fields*, Annals of Physics **183** (1988), 337–351.

[JLWie1] A. Jaffe, A. Lesniewski, and C. Wieczerkowski, *Heat kernel regularization of quantum fields*, Commun. Math. Phys. **121** (1989), 337–344.

[JLWie2] ——, *A Priori Quantum Field Equations*, Annals of Physics **192** (1989), 2–20.

[JLWis] A. Jaffe, A. Lesniewski, and M. Wisniowski *Deformation of super-KMS functionals*, Commun. Math. Phys. **121** (1989), 527–540.

[JO] A. Jaffe and K. Osterwalder, *Ward identities for noncommutative geometry*, preprint.

[K] D. Kastler, *Cyclic cocycles from graded KMS functionals*, Commun. Math. Phys. **121** (1989), 345–350.

[Q] D. Quillen, *Algebra cochains and cyclic cohomology*, Publ. Math. I.H.E.S. **68** (1989), 139–174.

[SS] E. Seiler and B. Simon, *Bounds in the Yukawa$_2$ quantum field theory: upper bound on the pressure, Hamiltonian bound and linear lower bound*, Commun. Math. Phys. **45** (1975), 99–114.

[We] J. Weitsman, *A supersymmetric field theory in infinite volume*, Ph.D. Thesis, Harvard University, 1988.

[W] E. Witten, *Supersymmetry and Morse Theory*, J. Diff. Geom. **17** (1982), 661–692.

HARVARD UNIVERSITY

Proceedings of Symposia in Pure Mathematics
Volume **50** (1990)

The Mathematical Implications of Fundamental Physical Principles

I. E. SEGAL

Introduction. Nothing was more important in von Neumann's scientific career than the interplay between mathematics and external scientific issues. And probably no one in the quantum mechanical era has contributed as much to this interplay, and especially as it involved fundamental physics. He was above all the father of rigorous quantum phenomenology, but he also made major contributions to the mathematical elucidation of the classic general principles of physics: causality, symmetry, and stability. This work reacted back on theoretical physics, and this in turn on mathematical developments, in a complex interactive pattern.

The mathematical innovations of von Neumann and of like-minded others of his generation, together with experimental progress in physics since von Neumann left the scene and the maturity of the work that he and his mathematical colleagues bequeathed, have drastically altered the state of fundamental theory, both mathematical and physical. Indeed, it looks as if mathematical physics may be on the verge of a kind of settlement of basic issues. Such a possible settlement would not mean that progress on the foundations would stop or diminish, any more than the work of Euclid and later on non-Euclidean geometry had such an effect, or that of Kolmogoroff and others in mathematical probability has this effect in that area. It would mean only that a mathematically secure and physically lucid basis for further research would be established that incorporates the central principles and formalism of fundamental theoretical physics.

Although von Neumann was outstanding in the quantum era, crucial contributions were made by other mathematicians—Leray, Stone, Veblen, Weyl,

---

1980 *Mathematics Subject Classification* (1985 *Revision*). Primary 81E08, 81E20, 83F05, 85A40; Secondary 35B40, 43A85, 47C15, 81G20.

Research supported in part by the National Science Foundation under grants DMS 8706094 and DMS 8706119.

This paper is in final form and no version of it will be submitted for publication elsewhere.

0082-0717/90 $1.00 + $.25 per page

and Wiener, among others. What these men had in common was a profound inspiration from fundamental physics, captured in essence in a rigorous and beautiful mathematical form. At the time they worked there could be no assurance of a 'practical' impact on the physical issues that contributed to their motivation. Their work was presented primarily as pure mathematics, and justified by its naturalness, intrinsic interest, and technical challenge, and—but not at all exclusively—its potential physical applicability. Each of the men cited is the father of a distinctive and crucial line of research for mathematical physics, which connected closely with von Neumann's ideational thrust. Thus, Leray founded the modern theory of nonlinear partial differential equations; Stone developed algebraic representation theory and correlated it with spectral theory in Hilbert space; Veblen's work on generalized space-times and invariant wave equations, in part with von Neumann, brought modern geometry to bear on theoretical physics in a lucid and fruitful way; Weyl founded modern Lie group theory, applied it to quantum theory, and expounded enormously stimulating ideas about the nature of space-time; and Wiener invented functional integration theory.

Many others made significant contributions, but this is not the place to go into the long and complex history of the interaction between mathematics and fundamental physical theory. It will suffice here to describe briefly the four general principles that had emerged by World War II, and that had rather direct physical interpretations.

Causality has several forms: nonrelativistic (Newtonian), involving the specification of the state of a system at a given time, together with dynamics determining the state at later times; and relativistic (Einsteinian), in which the finite velocity of propagation of physical effects is the basic constraint in particular. Relativistic causality is directly experienced, both empirically and intuitively, and is a very strong theoretical constraint.

Symmetry is the modern form of conservation laws, and in part represents relativity, or Einstein's law for the addition of large velocities. The basic space-time symmetries in particle and quantum field theories preserve causality. These symmetries form a group of transformations of space-time that maps the future of any point into the future of the transformed points; the notion of 'future' being ordained by the causal structure of space-time.

Stability is obviously essential for observational physics, at least relative to the time required for measurement. Theoretically, it is connected with the general principle that physical systems tend to evolve toward states of lower energy; this implies that only systems whose energy is bounded below can be stable. The energy is the infinitesimal generator of temporal evolution, which is the most important subgroup of the symmetry group, connecting symmetry with stability, and ultimately with causality.

Phenomenology refers to the description of states, observables, the probability distributions of observables in given states, and the like. Classically, states are represented by points of phase space; observables are functions on phase space; and temporal evolution is represented by the action of a one-parameter group of transformations on phase space that are 'symplectic', or preserve a canonical differential two-form given in phase space. The classical 'Hamiltonian' is the function on phase space, unique within an additive constant, that generates the temporal evolution group in accordance with symplectic analysis, as the modern form of Lie's theory of contact transformations is known. With the advent of quantum mechanics, which was made necessary by the production of particles and the apparently inherently random character, in part, of this production, states were represented by vectors in Hilbert space (with some semantic confusion, avoided in a later definition, cf. below). Observables were represented by selfadjoint operators in this Hilbert space, and temporal evolution by a one-parameter unitary group, whose selfadjoint generator, given by Stone's theorem, was the Hamiltonian. This somewhat overstates the degree of mathematical specification in the naive form of quantum phenomenology proposed by Heisenberg, Dirac, and others in the 1920s. The 'correspondence principle' states that the quantum Hamiltonian is obtainable from the classical Hamiltonian by judicious substitution of quantum variables (or 'canonical observables') for the classical ones. The most crucial contributions of von Neumann to the mathematical analysis of fundamental physics were in the area of quantum phenomenology, and are closely connected with his work in operator theory and operator algebra, which will be discussed below.

Besides the mathematical treatment of these four fundamental principles, the correlation between higher mathematics and empirical physics that has emerged in recent years depends also on some crucial postwar experiments and observations, at the extremes of distance. In astronomy, the 1960s saw the discovery of the cosmic microwave background radiation by Penzias and Wilson, and of quasars, primarily by Schmidt, providing important new constraints for fundamental physical theory. At the other distance extreme, particle experiments revealed a vast proliferation of quasistable particles, whose lifetimes are large compared to an expected basic scale of the order of 10^{-23} sec., but short compared to that of the neutron, the only classical particle that is unstable (via radioactivity). Theory was shaken also by the observation of violations of discrete symmetries—parity P, charge-conjugation C, and even in a strangely unique instance, of CP. There were two distinct kinds of neutrino, when one seemed plenty, and mysterious pseudo-electrons, which were identical to ordinary electrons except that they were hundreds of times heavier; and so on.

On the table, if not actively pursued in recent times, were the theoretical ideas represented by Mach's principle, Minkowski's group theoretic analysis of special relativity, and Einstein's Equivalence Principle. Together with the developments indicated above, they have led to striking progress at the interface between pure mathematics and fundamental physics. From this has emerged a tentative synthesis of mathematics of the von Neumann-Wiener variety—emphasizing operator theory, harmonic analysis, and functional integration—with the physics of extreme distances. One of the important byproducts of high-speed computing, in which von Neumann led the way, was to computer-intensive statistics, which has made it possible to dispense with the older, less objective and certain parametric statistics; this has greatly facilitated the empirical validation of the astronomical part of this synthesis.

This synthesis may be called 'chronometric' theory, inasmuch as it originates in a study of the nature of time and duration. The central idea of the theory applies to the general class of simple Lie groups whose Lie algebras contain a nontrivial invariant convex cone (physically interpretable as infinitesimal displacement into the future), or equivalently, that admit a positive-energy unitary representation (as required for a stable particle-theoretic interpretation). As yet, however, the mathematical development has been largely confined to the case of the group SU(2, 2), which also figures prominently in the work of Veblen and von Neumann, and is intrepretable physically as the group of all local one-to-one transformations that preserve causality in a neighborhood of an arbitrary point in Minkowski space. The universal cover of this group acts as the group of global causality-preserving transformations on the Einstein Universe, which will be denoted **M**, and which will play a fundamental role as a space-time model in the following. The mathematical work summarized below includes contributions from many collaborators and former students, and crucially S. M. Paneitz and J. F. Nicoll, but also essentially J. C. Baez, R. W. Goodman, L. Gross, H. P. Jakobsen, A. Klein, M. Kon, B. Kostant, E. Nelson, N. S. Poulsen, D. A. Vogan, Jr., M. Weinless, and Z.-f. Zhou.

Aspects of von Neumann's scientific initiatives. I attended some of von Neumann's lectures as an undergraduate more than fifty years ago, and saw him in a variety of capacities thereafter. He seemed a passionate lecturer, urbanely rational as a colleague, personally considerate, and almost inhumanly powerful technically. More relevantly, he was a broad integrated personality whose work cannot be neatly categorized or linearized without misrepresenting its overall character. Much as he enjoyed and rose quickly to technical challenges, and resolved much more than his share of them, he seemed primarily concerned with the clear and viable formulation and development of fundamental scientific questions and issues, and only rather incidentally concerned with the solution of mathematical puzzles. The mathematical novelty implicit in quantum mechanics, and the divergence of its most fundamental form, i.e., quantum field theory, seemed outside the scope of traditional mathematics, spurring him toward fundamental mathematical innovation. At the same time the Wedderburn structure theory had a revival in various forms, and its extension to infinite-dimensional algebras ultimately provided an issue along purely mathematical lines that connected with quantum phenomenology. A decade later, the war focused his interests on nonlinear partial differential equations, whose intractability motivated his program in high-speed computing. He worked in a total scientific way, in which the questions, which he formulated himself, are no less important than the answers. But at the same time he did provide a great many answers and created powerful mathematical instruments not less vital for the development of quantum physics than the new accelerators.

I will begin with quantum phenomenology, and describe how von Neumann's initiative here has developed into constructive quantum field theory—the solution of nonlinear partial differential equations exemplifying the physical constraints indicated earlier. This initiative correlates closely with his work in rings of operators, which was discussed at length by other contributors to this Symposium, particularly as regards 'local' theory (factors), and I will only deal briefly with the global theory and its relation to the local theory. In a curious way, quantum field theory has led into a less well-known phase of von Neumann's work, that on the "Geometry of Complex Domains", with Veblen. This work together with the three general constraints from classical fundamental physics suggested a very natural theory of the nature of the cosmological redshift, which computer-intensive statistical studies show to be in remarkably close agreement with systematic astronomical observation, in great contrast to the Doppler ('expanding universe') theory. This in turn has led to a kind of cosmological mathematics, in which the same idea is applied to constructive quantum field theory in four dimensions, to the temporal asymptotics of the Yang-Mills equations, and to a useful reinterpretation of scattering theory. This work has finally reacted back on fundamental microphysics, suggesting a unified theory of the fundamental forces. This in turn creates new scientific issues in the areas that von Neumann did much to initiate.

Quantum phenomenology. A simple but important early work of von Neumann was the axiomatization of the concept of 'expectation value in a state'. This was a linear functional E on an algebra of observables—to begin with, he dealt with the Hermitian operators in a finite-dimensional space as a model—that was 'positive', i.e., nonnegative on nonnegative observables, and 'normalized', i.e., one on the unit observable. I thought it was in keeping with von Neumann's apparent philosophy of operationalism (the variant of positivism espoused by Bridgman) to call E simply a state. Physically important concepts such as pure state, density matrix, probability distribution of an observable in a given state, etc., were then definable in a natural and effective way. Later von Neumann searched for the right algebra of observables. A long paper in 1936 established an interesting direction, but was inconclusive.

In the meantime the important issue of the relation between the Heisenberg and Schrödinger forms of quantum mechanics was settled by the Stone-von

Neumann theorem, which showed their essential equivalence, by virtue of the unicity of the irreducible representation of the Heisenberg relations, with natural regularity conditions. Independently of its application to quantum mechanics, this theorem can be regarded as the central one for L_2-harmonic analysis, and can be extended to arbitrary locally compact abelian groups, and beyond, as shown by Mackey. In the case of quantum fields, involving an infinite number of canonical p's and q's, the unicity fails, but at the time the concept of quantum field was too loosely formulated for this to be a demonstrable theorem, and physical practice assumed essential unicity. Thus the journalistically excellent but mathematically heuristic treatment of light theory in von Neumann's 1932 book found it convenient to make a 'symplectic' transformation of the p's and q's, which is not implementable by a unitary transformation—ending up nevertheless with a divergent Hamiltonian operator. This divergence was present in the original formulation by Dirac, and remains to this day, forming the central foundational problem of quantum field theory.

'Rings', as von Neumann called them, or weakly closed selfadjoint operator algebras on Hilbert space (abbreviated to W^*-algebras, and also known, more lengthily, as von Neumann algebras), were the focus of von Neumann's research (in part with F. J. Murray) for almost a decade. His interest in them came from pure algebra, with the Wedderburn theory as prototype, from his intensive concern with quantum phenomenology, and from the needs of infinite-dimensional representation theory for groups, and for groups of transformations, which can be regarded as including ergodic theory. In particular, quantum field theory was an important motivation, and using his quite original theory of infinite tensor products of Hilbert spaces, he constructed some interesting examples, which however lacked nontrivial symmetry features. His construction of the approximately finite II_1 factor—a central simple W^*-algebra distinct from that of all bounded operators on Hilbert space—was an exciting algebraic triumph, and at the same time this factor turned out to arise naturally in connection with the free fermion quantum field.

But strikingly original and beautiful as the theory of rings was, its relevance to the more physical side of quantum phenomenology is in a way limited, as exemplified in the complications of the 1936 paper, also known as the 'onion' paper because of the difficult penetrability of its succession of brackets (I am informed by S. Kakutani that the paper acquired this name in an early Tokyo seminar on it). In retrospect it is clear that the problem with von Neumann's approach was that it is *spatial* rather than *algebraic*, i.e., it depends on the way in which the observables are represented as acting on the vectors in a linear space in which they are represented as operators, which may be unphysical. A decade after the 1936 paper I modified the von Neumann axioms, retaining a Jordan algebra of observables, but eliminating unphysical topological aspects, using instead simple axioms for the supremum of a (bounded) observable. The result was a purely algebraic treatment of states, probability, simultaneous observability, etc.—the basic paraphernalia of quantum phenomenology. This applies in particular to the case of a C^*-algebra, which is the same as a W^*-algebra, except that its topology was that of uniform convergence, which does not require any specific representation by operators on a Hilbert space, and is definable purely algebraically, on the basis of ideas going back to Stone, Gelfand, and Neumark.

The abstract characterization of C^*-algebras published during the war by Gelfand and Neumark was unexpected by von Neumann, and was quite influential in spurring the algebraic approach. The approach to quantum phenomenology proposed soon after the war, in which observables are selfadjoint elements of an abstract C^*-algebra, states are normalized positive linear functionals on the algebra, and motions are represented by automorphisms of the algebra rather than unitary transformations on an underlying Hilbert space, seems now to have become generally adopted in rigorous theoretical work.

There were two principal 'practical' payoffs for the C^*-algebraic approach. First, it made possible a rigorous treatment of the states corresponding to points in the continuous spectrum of a basic operator, such as the energy, which in Dirac's treatment had purely symbolic representation as vectors of infinite norm. Second, it permitted the extension of the Stone-von Neumann theorem to the infinite-dimensional case, with the modification that unitary equivalence is replaced by an algebraic *-isomorphism, which in the finite-dimensional case implies unitary equivalence. This in turn made it possible to represent symplectic actions (such as that used by von Neumann) in a simple and rigorous way, as automorphisms of the 'Weyl' C^*-algebra generated by the bounded functions of arbitrary finite sets of the infinite set of p's and q's. An important if immediate byproduct of the investigation was the projective unitary representation of the symplectic group that implemented this action in the finite-dimensional case. This remarkable 'harmonic' representation is the natural counterpart for the symplectic group to the spin representation of the orthogonal group, but had remained undiscovered because of its infinite-dimensionality; it has since been instrumental in pure mathematical investigations from group and number theory to asymptotic analysis. In the same way, the spin representation of the orthogonal group was extended to the infinite-dimensional case, as an automorphic representation of the 'Clifford' C^*-algebra, which is correspondingly generated by functions of arbitrary finite sets of an infinite set of p's and q's satisfying anticommutation rather than Heisenberg commutation relations, or Clifford relations.

But the algebraic formalism presented a problem in the definition of the vacuum or ground-state. In elementary quantum mechanics this is definable as the lowest proper vector of the Hamiltonian, but this does not apply algebraically, there being no given vector space, or for that matter Hamiltonian, except as a (in general, unbounded) derivation of the algebra of observables. Instead it seemed reasonable to define a vacuum for a one-parameter group of automorphisms of a C^*-algebra $\mathbf{A}$ as a state E such that $E(A^tB)$ is a positive-frequency function of t for arbitrary A and B in $\mathbf{A}$, where A^t denotes the image of A under the automorphism of parameter t; i.e., the Fourier transform of $E(A^tB)$ is supported by the nonnegative reals. This was validated by application to the Weyl algebra, where it showed the unicity of the usual mode of quantization purely on the basis of symmetry and stability. This could also be interpreted as a uniqueness theorem for the solution of *quantized* linear equations, such as $\Box\phi + V\phi = 0$, where V is a given nonnegative bounded measurable function on space.

Alternatively, the free boson and fermion fields are uniquely determined by temporal invariance together with stability; the rather ad hoc prescriptions used in the physical literature are indeed the only possibilities, within unitary equivalence. Each type of field has three quite different but unitarily equivalent representations, depending on which operators are diagonalized or pseudodiagonalized. The original Fock-Cook representations (heuristically treated by Fock in a Euclidean space setting, and rigorously developed by Cook in a fully covariant formalism that has been universally adopted) diagonalize the particle numbers. The wave properties of the field are expressed by the diagonalization of the fields themselves; or in the case of fermion fields, pseudodiagonalizations in which the p's are left multiplications on a Clifford algebra and the q's are right multiplications twisted by a basic automorphism of the Clifford algebra. In either case the field Hilbert space $\mathbf{K}$ is represented as a space of square-integrable operators over a *real part* $\mathbf{H}'$ of the underlying single-particle complex Hilbert space $\mathbf{H}$; in the boson case, the measure is the isotropic centered normal distribution over $\mathbf{H}'$, while in the fermion case a 'noncommutative' measure is involved. More specifically, $\mathbf{K}$ may then be represented as the L_2-closure of the Clifford algebra over $\mathbf{H}'$, relative to the unique normalized trace, determined as the usual matrix trace normalized to unity on 1, on restriction to the Clifford algebra over any finite even-dimensional subspace, which is isomorphic to a complete matrix algebra. This is perhaps the simplest construction for the approximately finite II_1, to which the ring generated by the p's (or that generated by the q's) is isomorphic, as well as the most essential occurrence of a non-type I ring outside the subject of operator algebra per se. In a third representation, $\mathbf{K}$ is represented as the space of antiholomorphic square-integrable functions or spinors over $\mathbf{H}$, in the respective boson and fermion cases, and the creation operators are pseudodiagonalized, by their representation as multiplication operators; being nonnormal, they cannot be strictly diagonalized.

These rigorous and invariant formulations of free fields made possible clearcut and simple answers to a variety of natural questions. The infinite-dimensional analogue of the question treated by the Stone-von Neumann theorem in the finite-dimensional case could be studied, and it was found that even apparently very closely related Weyl systems were not unitarily equivalent. For example, if all the p's are multiplied by two and all the

q's are divided by two, a new Weyl system was obtained that was not unitarily equivalent to the original one. This in turn was found to be related to the phenomenon discovered by Cameron and Martin, of the nonabsolute continuity of simple transformations on Wiener space, for example simple multiplication by two in $C_0[0, 1]$. Such results could be considered special cases of a sharp formulation in Hilbert space, to the effect that the linear transformation T on the real Hilbert space $\mathbf{H}'$ is absolutely continuous with respect to the isotropic normal distribution in $\mathbf{H}'$ if and only if T^*T differs from I by a Hilbert-Schmidt operator. This implies that if T is any symplectic transformation on $\mathbf{H}$, i.e., continuous real-linear invertible transformation leaving invariant the imaginary part of the inner product, and if $(\mathbf{K}, W)$ denotes the Weyl system of the free boson field over $\mathbf{H}$, then the Weyl system $(\mathbf{K}, W_T)$, where $W_T(z) = W(Tz)$, is unitarily equivalent to $(\mathbf{K}, W)$ if and only if $[T, i]$ is Hilbert-Schmidt on $\mathbf{H}$ as a real space, by virtue of a factorization for symplectics established by Shale. A similar development applies to fermion fields; in both cases, von Neumann's infinite tensor products of Hilbert spaces were an essential ingredient in the proof.

The Hilbert space approach is both more invariant and more powerful than the Wiener space approach to functional integration, but virtually all of the transformations T used in practice (e.g., that used in von Neumann's treatment of light theory) are excluded by the absolute continuity limitation. They are readily incorporated however as automorphisms of the C^*-Weyl algebra; there is a unique automorphism that carries $W(z)$ into $W(Tz)$, for all z in $\mathbf{H}$, in the context just indicated, and indeed this is valid for arbitrary Weyl systems over $\mathbf{H}$, not merely that for the free field.

The next stage in development was the treatment of nonlinear quantized field equations, but before going into that it has to be noted that with the determination of a vacuum for an automorphism group of C^*-algebra we are back in the W^*-algebraic context, in part. For the vacuum state determines canonically a concrete Hilbert space on which the C^*-algebra is represented, and in which the automorphism group is unitarily implementable. The problem is that this ultimate Hilbert space cannot be given a priori, but must be solved for, somewhat like the boundary in a free-boundary nonlinear p.d.e. problem. In addition, a so-called 'broken symmetry' arises from the circumstance that the vacuum is in general not invariant under the full symmetry group acting as automorphisms of the C^*-algebra, but only under a subgroup. Thus the analysis cannot be conducted wholly in either a C^*- or W^*-algebraic framework, and in some later developments, even the * is not given a priori and must be solved for.

Global vs. local theory of rings. Previous speakers at this Symposium have concentrated effectively on the local theory of rings, but it is the global theory that implies most directly to external issues, and provides a natural conjunction of operator algebra and measure theory, which were two of von

Neumann's favorite subjects. The *global theory* treats rings directly, without recourse to a presentation as a direct integral of factors. The *local theory* attempts to reduce questions to two parts: (a) the corresponding issue for factors; and (b) conventional Lebesgue integration theory, over the space of factors.

There is a completely explicit structure theory, in spatial terms, for rings that are abelian, or commutors of such, or more generally of type I. The unitary invariants of such rings are simply describable in terms of spectra, cardinal number multiplicities, and measure space invariants deriving from Maharam's classification. Thus in this case there is a complete counterpart to the Wedderburn theory, but the hope for such a theory in the case of rings of other types was dashed with the discovery of an uncountable proliferation of factors of these types.

In practice, 'exotic' factors rarely occur, except in situations lacking in smoothness (infinite discrete groups) or local issues in free quantum field theory, where they disappear when a more physical formulation of localization (so-called 'Newton-Wigner' or 'quantum' localization) is used.

Overall, the study of exotic factors is fascinating per se but has at the same time branched off from the most basic issues that apparently motivated von Neumann—quantum field theory, the Wedderburn theory, representation theory for continuous groups, and groups of transformations. The global theory avoids measurability problems that are ideationally irrelevant and technically messy, and seems to suffice for 'practical' needs, at the same time that it unifies abstract integration theory with the operator ring context. The global multiplicity theory for rings that derives from noncommutative integration theory and work of Griffin and others provides for type II and III rings an analog to the multiplicity theory for type I rings (based on the equivalence between maximal abelian *-algebras of operators on Hilbert space and the L_∞-multiplication algebras of measure space) establishing the basic case of unit multiplicity. The noncommutative integration theory also serves to establish the generalization of the Plancherel theorem to arbitrary locally compact unimodular groups, and to show for example that the L_2 convolution algebra of an open simple Lie group is *-algebraically and measure-theoretically equivalent to the algebra of functions from the group dual as a measure space to the Hilbert-Schmidt operators; but here a combination of global and local theories is involved.

The relation between the theories may be illustrated by consideration of von Neumann's theorem that the 'ring' generated by a selfadjoint operator on a separable Hilbert space consists only of its bounded Baire functions. The local approach—pure direct integral theory—gives no insight here; the ring is a direct integral of type I_1 rings, whose structure could not be more trivial, but this does not help. Rather the structure theorem for abelian rings shows that such a ring is equivalent algebraically and as regards the operational calculus to L_∞ on a measure space, which by separability may be taken as

[0, 1] with an appropriate Borel measure, on which every measurable function is Baire. But I will say more later about the reduction and noncommutative integration theory, after discussing theoretical physical applications.

Constructive quantum field theory. The classical (numerically-valued) and quantized solutions of the wave or similar relativistic equations are similar in many respects, but there is a fundamental difference in that the initial (or Cauchy) data for the quantized field are not at all arbitrary, but are fixed (within an inessential unitary equivalence) by the stability constraint together with the commutation relations (the form of the latter is in turn uniquely determined by relativistic invariance). The free field at a fixed time turns out not to be an actual function on space, but only a distribution; for example, for the wave equation itself, it takes the form $(-\Delta)^{-1/4}w(x)$, where $w(x)$ denotes the white noise in space. This means that powers of the field are purely symbolic, a priori. This in turn means that the Heisenberg ansatz for derivation of the quantum Hamiltonian, involving substitution of the canonical quantum operators for the classical ones, is only symbolic. Thus for a classical nonlinear wave equation such as $\Box\phi + g^2\phi^3 = 0$—which has perfectly well-defined global solutions throughout Minkowski space-time—there is a priori no meaning to the nonlinear term after substitution of a putative operator-valued solution satisfying the canonical commutation relations.

There is no question that a simple limiting approach yields a meaningless result, identically $+\infty$ in the case of the symbolic Hamiltonian for the equation given. A variety of 'renormalization' methods for dropping infinities were in use when in 1950 Wick proposed a partial standardization, applicable to the convolution of the field with a smooth function, rather than to the field itself. This appeared to be only slightly less opportunistic than earlier theoretical physical practice, but Nelson had the idea to investigate the result of using this definition, applying the Heisenberg ansatz not to the interacting field, which was unknown but to the free field, in one dimension, with periodic boundary conditions. It developed that the formal expression provided by the Heisenberg ansatz could indeed be interpreted as a semibounded self-adjoint operator in the free Hilbert space. To be more specific, consider the prototypical wave equation $\Box\phi + p'(\phi)$ in $R^1 \times S^1$, where $\Box$ denotes the wave operator $\partial_t^2 - \partial_x^2$ ($t \in R^1$, x in the reals mod 2π), and p is a given real polynomial that is bounded below (to model the constraint of physical stability). As a classical equation this has the energy, or Hamiltonian function

$$\mathbf{H}(\phi) = \int \left[\tfrac{1}{2}(\partial_t\phi)^2 + \tfrac{1}{2}(\partial_x\phi)^2 + p(\phi)\right] dx$$

which is 'conserved', i.e., independent of t. According to conventional doctrine, quantization is effected by replacing the 'classical', real-valued, field ϕ by a 'quantized', or operator-valued field ϕ, satisfying the canonical commutation relations, which for fixed t take the form (setting $\dot\phi(t,x) = \partial_t\phi(t,x)$)

$$[\phi(x), \phi(y)] = 0 = [\dot\phi(x), \dot\phi(y)]; \qquad [\phi(x), \dot\phi(y)] = i\delta(x-y).$$

The operator-valued field $\phi(x)$ that is inserted in order to obtain the conserved quantum field Hamiltonian $\mathbf{H}(\phi)$ is of course the interacting field, which however is unknown a priori, being in fact essentially the objective of the theoretical analysis. Nelson's ansatz substitutes in the place of this unknown field ϕ_{int} the known free field ϕ_{free}, at the fixed time t_0 in question; both ϕ_{int} and ϕ_{free} satisfy the canonical commutation relations. This in effect converts the a priori nonlinear equation $\Box\phi + p'(\phi) = 0$ into a linear one that is mathematically much more tractable; but there is no reason for the solutions of these two equations to be identical, or even unitarily equivalent. The solution of the linear equation is however an interesting and useful object of study, and in fact a surprising thing develops; the solution of the linear equation which has been made mathematically unexceptionable *satisfies a local nonlinear wave equation.* But the good fortune is not at all unmitigated; the equation satisfied by the field ϕ is not the original one, but rather of the form $\Box\phi + :: q'(\phi) :: = 0$, where q is a polynomial that agrees with p only in its term of highest degree, in general; and here the generalized powers have vanishing expectation value with respect to the *physical* vacuum, whose state vector is the lowest eigenvector of $\mathbf{H}(\phi_{\text{free}})$, which distinction from earlier is indicated by the $:: \cdots ::$ in place of the $: \cdots :$. Thus the Nelson ansatz produced a strong salient into the solution of quantized nonlinear wave equations, but it by no means went all the way—still less could it be applied directly in the four-dimensional case.

I felt at this point that the deeper development of constructive quantum field theory depended on a better understanding of space-time and its symmetries, in continuation of lines growing out of the work of Veblen and von Neumann, rather than in the technical refinement of the functional analytic lines that had broken into the two-dimensional case. Cosmology has provided a proving ground for this geometrical line, and we seem now to be back at the point where functional analytic initiatives need to be resumed, and can be more incisive. Let me digress, therefore, to describe how mathematical cosmology evolved into a kind of cosmological mathematics that provides a powerful approach to the theory of nonlinear wave equations in four dimensions, both classical and quantized.

The geometry of complex domains and space-time in the large. The seminar on the Geometry of Complex Domains that von Neumann conducted in the 1930s jointly with Oswald Veblen had both a mathematical and a physical origin. Veblen proceeded from finite-dimensional, and even finite geometries, while von Neumann took off from the infinite-dimensional, in the form of his continuous geometries, but they were both concerned with the mathematical explication of the new theoretical physics. In their day the empirical base for application of their ideas was too rudimentary to confirm their insights, and for the most part they were too modest to set into print more than the unexceptionable mathematics. But I recall very well von Neumann's casual but pointed remarks on the schizoid character of time in quantum mechanics, as both a coordinate and the parameter in the evolutionary group, and Veblen's strong feeling that his 'conformal space-time' could be a truer model for space-time than the conventional Minkowski space to which it was locally essentially identical. In conjunction with the notes of their seminar, this raised the question of how one could, or whether one could, distinguish by physical observation between the conformal space-time model (more precisely, its universal cover, required to enforce global causality, as shown by H. Rudberg), and Minkowski space.

At first glance the question of establishing such a distinction appears almost philosophical and perhaps devoid of empirical significance, but von Neumann's comment about the theoretical modelling of physical time provided a clue. The natural times in Minkowski space $\mathbf{M}_0$ and in the universal cover of $\mathbf{M}$ of conformal space (represented most simply as a real projective quadric of hyperbolic signature) were locally almost indistinguishable, but fundamentally different in the large—notwithstanding the local identity of $\mathbf{M}_0$ and $\mathbf{M}$ as acted on by the local group of all causality-preserving transformations; in particular, their effective identity as regards Maxwell's equations, which govern light, which is basically all that is observable in large scale astronomy. Since energy is represented as the infinitesimal generator of temporal evolution, this implied a difference between the natural energies in $\mathbf{M}_0$ and $\mathbf{M}$, and indeed if one energy was conserved, the other could not be, as a consequence of the noncommutativity of the two temporal evolution groups. The same two energies also figured in an exemplification of Minkowski's idea, abstracting the replacement of the Galilean by the Lorentz group, that an apparent physical symmetry group may be a limiting case of a mathematically more attractive and empirically more relevant group.

It turned out that the energy difference $E - E_0$ between the energies in $\mathbf{M}$ and $\mathbf{M}_0$ is always nonnegative, and increases from 0 at the time of emission for a localized source, monotonely. This implies that if the conventional $\mathbf{M}_0$-energy is what is actually measured observationally—and only the $\mathbf{M}_0$-energy is determined by the local form of the wave function, the $\mathbf{M}$-energy including in addition a component that depends on its global form—then a free particle, in particular a photon, will experience a redshift when propagated over a very long period of time. The effect here is similar to the difference between distances as measured on a sphere S^2 and as measured in a plane on which S^2 has been stereographically projected; the conformality of this map is replaced by the preservation of causality in the relations between $\mathbf{M}$ and $\mathbf{M}_0$. It appears as a more natural origin for the cosmological redshift than the 'Expanding Universe', and is free of the adjustable parameters that intervene in that theory and of the need for ad hoc scenarios required in it (such as 'Evolution' for quasars and radio sources, and 'Inflation' at the hypothetical beginning of the universe), but there is one apparent problem with it. The relation between $\mathbf{M}$ and $\mathbf{M}_0$ is locally precisely one of stereographic projection, and as in the case of distances on S^2, the effect locally varies with the square of the distance, rather than linearly as stated by the Hubble Law.

The distances, however, are not observed, but estimated in a complex model-dependent manner; what are observed rather, apart from the redshifts, are the apparent brightness and diameter. Careful investigation shows that the reported confirmation of Hubble's Law has been based on large components of wishful thinking and self-fulfilling prophecy. Objective and efficient statistical methods, applied to galaxy samples much larger and more precisely specified than the samples used by Hubble and his successors, show that the observable consequences of a linear redshift-distance law are more deviant from actual measurements than even a cubic law, and that a quadratic law provides the optimal, and indeed an excellent fit. It should be remembered that Hubble himself was one of the most explicit in expressing skepticism regarding the Doppler interpretation of the observed redshifts, but that there was then no apparent alternative explanation. Moreover the directly observed quantities reported by Hubble in his original small sample more closely follow a quadratic or higher power law than a linear one. The basic samples shown in textbooks, to follow the linear law are those of 'cluster' galaxies, whose selection criteria has been persistently obscure, apart from the major cluster galaxy catalog of Abell, whose definition is based explicitly on the assumption of the Hubble Law. Cluster observations are moreover subject to many model-dependent corrections, and the definition of clusters suffers from the lack of a direct objective means to distinguish foreground and background galaxies from 'true' cluster members.

In any event, the bulk of galaxy redshift measurements are at redshifts of the order of 0.05 or less, whereas today there are observations on quasars of redshifts almost 100 times greater. From early on, quasars have appeared highly deviant from the Hubble Law or its large-redshift extension in the form of Friedman cosmology; but systematic observations on quasars are in excellent agreement with the hypothesis that the redshift is represented by $E - E_0$. In addition to such statistical features of large samples of extragalactic objects, there are many problems that require elaborate ad hoc scenarios in the Friedman cosmology, that disappear in the framework of the other cosmology, which has been called 'chronometric' because it originated in the question raised by von Neumann of the proper theoretical description of time in quantum mechanics. For example, apparent velocities of jets in radio sources well in excess of that of light are a striking phenomenon that is difficult to deal with in the frame of Friedman cosmology, but is straightforwardly intelligible in the chronometric theory; and in fact provides a direct estimate of the extragalactic distance scale, which some consider the most important physical parameter in astronomy, and has been actively disputed for decades. In chronometric cosmology this parameter R may be described as the radius of space, which is represented as a sphere S^3, in laboratory units; together with h and c, R provides the three fundamental units required for physics, in a physically natural way that is invariant under the full 15-parameter conformal group of all causality-preserving transformations in **M**.

Natural as the chronometric cosmology appears from a mathematical standpoint, the question of its empirical validity is basically statistical, and would hardly be tractable without the high-speed computers whose development von Neumann sponsored. Older statistical methods were dependent on parametrizations that developed from plausibility and convenience more than direct observation. Rapid computer simulations now make possible efficient nonparametric methods that provide objective probabilities of deviations of specified observations from theoretical prediction, in circumstances far more complex and realistic than would be possible by theoretical estimation of probabilities. These methods must be used to the hilt to make efficient use of astronomical data in cosmological testing because of the inherent truncation involved (only the objects apparently brighter than some limit are observable, leading to distance-dependent sampling) and also because of the absence of objective a priori theoretical parametrizations of distributions such as that of intrinsic luminosities of galaxies. They unequivocally reject the Hubble Law, but accept the Lundmark (quadratic) Law, which moreover fits so accurately that it predicts precisely the statistical characteristics of the deviations of the Hubble Law from observation, serving to indicate that the attribution of these deviations to hypothetical perturbations incapable of direct observation (in the nature of epicycles, but of lesser predictive power) is unscientific. Figure 1 shows an example of the predictions of the Hubble and Lundmark Laws concerning the results to be expected for analyses predicated on the other law; the Hubble Law predicts that the Lundmark Law will fit badly, with a large dispersion and a large correlation of intrinsic magnitude with redshift, and thus is quite unable to explain its actual excellent fit, while the Lundmark Law predicts that the Hubble Law analyses will show large deviations of precisely the statistical character that is observed. Another example, dealing with directly observed data, treats the apparent magnitude-redshift relation in the four largest published complete galaxy samples, as observed, and as predicted by redshift-distance power laws, using the observations in question to estimate in a statistically optimal way the distribution of intrinsic magnitudes of the galaxies, permitting the elimination of the a priori bias deriving from the fixed apparent magnitude limit of the sample, by subtraction of the conditional expectation of the intrinsic magnitude for given redshift, in a sample having the given limiting magnitude (cf. Figure 2). Figure 3 shows that in the case of the largest available complete quasar sample, the Friedman cosmology is unable to predict its own distribution of intrinsic luminosities of quasars consistently; but that the deviations, postulated by defenders of this cosmology to arise from effects incapable of direct observational substantiation, known as 'evolution', are identical to those predicted by the chronometric cosmology, for the results of analyses predicated on the

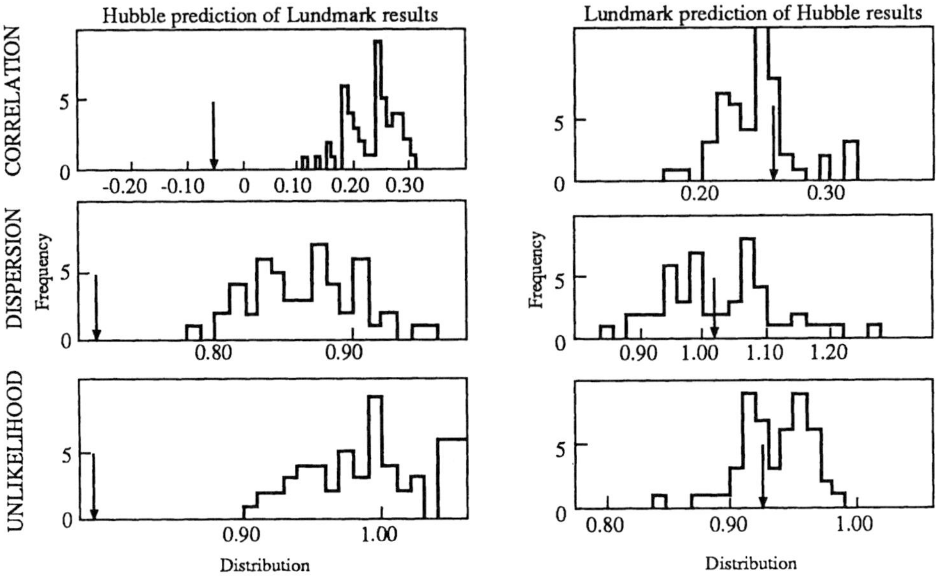

FIGURE 1. Mutual predictions of two redshift-distance laws for the results of analyses predicated on the other law. The arrows show the actual results; the histograms show the predicted distributions of results, obtained in 100 computer simulations. The data are the measurements reported by N. Visvanathan for the sample of all galaxies of a specified type in the whole sky brighter than a specified limit (290 galaxies).

Friedman cosmology; and that the distribution of intrinsic magnitudes derived from the chronometric cosmology is, in contrast to that derived from the Friedman cosmology, entirely self-consistent.

These results have not been scientifically rebutted in the astronomical literature. The rigorous and optimal statistical techniques for estimation of the distributions of intrinsic luminosities are ignored and appeals are made to results for 'cluster' galaxies notwithstanding the explicit assumption of the Hubble Law in the definition used in the chief catalog of clusters (that of Abell); subjective selection procedures, from this catalog and of the cluster members (including on occasion the deletion of objects from a putative sample on the basis of observations of their magnitudes and redshifts, in part, with the stated objective of eliminating so called '*cD*' galaxies); and many model-dependent and observationally uncertain but substantial corrections, some not fully reported in the literature (for aperture, spectral shape, background cluster luminosity, cluster statistics, etc.).

On this empirical basis it is appropriate to conclude—necessarily tentatively, as always in physical science—that large-scale space-time is actually better represented by the universal cover **M** of the conformal space emphasized by Veblen and von Neumann. This conclusion is strengthened on the

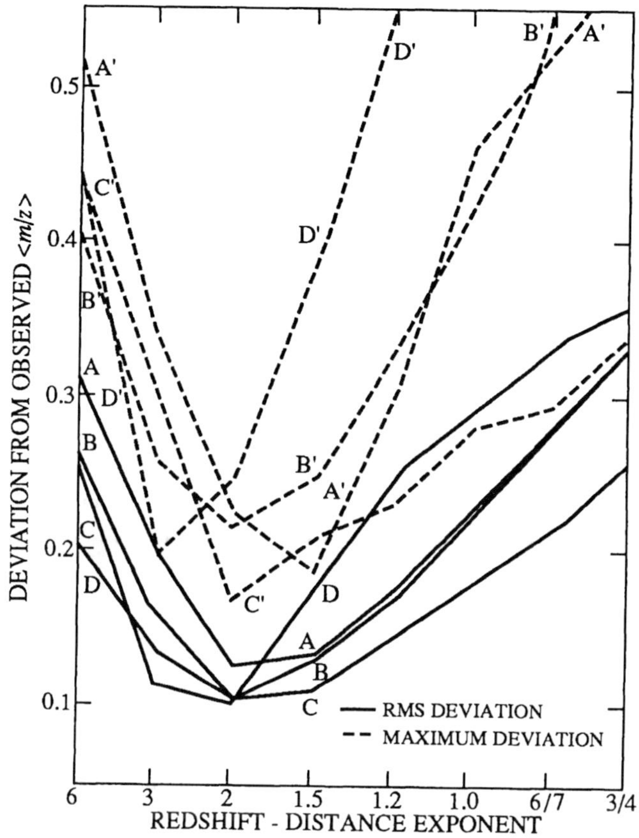

FIGURE 2. The deviations between theoretical predictions based on redshift-distance power laws and direct observations, in four large complete galaxy samples. The rms and maximum deviations are shown as functions of the redshift-distance exponent p. The truncation effect resulting from the use of a fixed completeness limit has been removed by a statistically optimal technique.

theoretical side by the fact that **M** is in a sense the universal space-time, satisfying the extremely natural conditions of causal structure and spatial and temporal isotropy; all other that are four-dimensional ($\mathbf{M}_0$ and the universal covers of the de Sitter spaces) are open submanifolds that are orbits under subgroups of the causal group **G** of **M**. A schematic drawing of the invariant imbedding (under the Poincaré group, extended by scaling transformations) of $\mathbf{M}_0$ into **M** is shown in Figure 4. Moreover, **M** is apart from $\mathbf{M}_0$ the only such space-time that can be factored as time $\times$ space. In a formal mathematical way it might appear as a deficiency in **M** that it admits no invariant metric under **G**, but only an eight-parameter family of metrics invariant under the seven-dimensional maximal essentially compact subgroup **K** of **G**; but this is actually physically highly explanatory. The decomposition of **M** as $R^1 \times S^3$

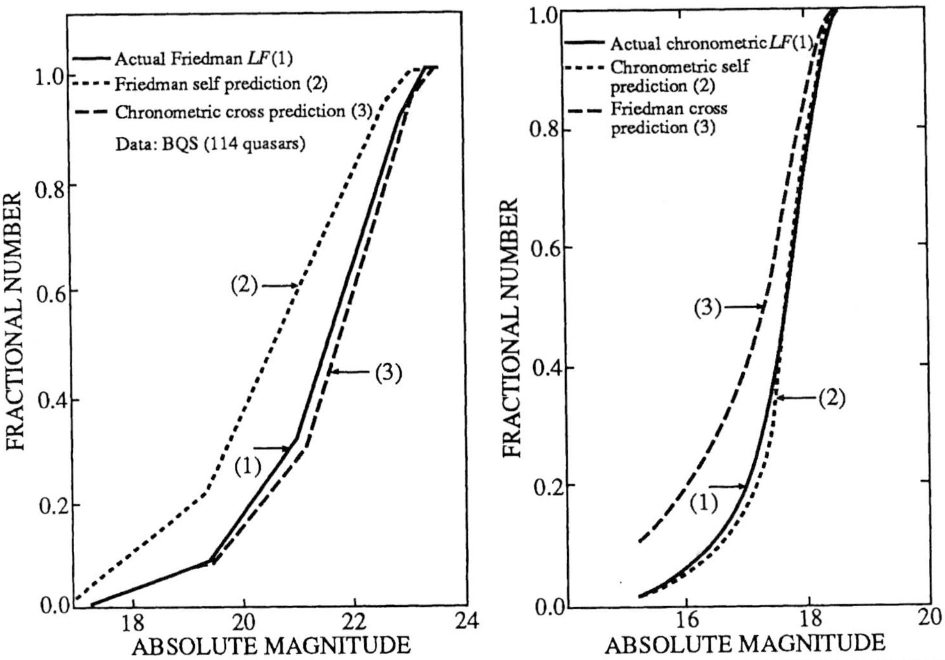

ACTUAL VS. PREDICTED ROBUST *LFs* FOR THE FRIEDMAN COSMOLOGY ($q_0 = 0$)

Cumulative normalized Friedman (nonevolutionary) luminosity functions for the BQS

ACTUAL VS. PREDICTED ROBUST *LFs* FOR THE CHRONOMETRIC COSMOLOGY

Cumulative normalized chronometric luminosity functions for the BQS

FIGURE 3. Estimates and predictions of the chronometric and Friedman cosmologies for the quasar LF(='luminosity function', or distribution of intrinsic luminosities). The Friedman cosmology is unable to consistently predict its own LF (cf. curves (1) and (2) on the left), without introducing ad hoc hypotheses incapable of direct observational substantiation. However, the chronometric cosmology predicts precisely the Friedman LF (cf. curves (1) and (3) on the left) as well as its own. The data are the measurements reported by M. Schmidt and R. F. Green for a complete sample of quasars in a specified region (approximately one quarter of the sky; 114 quasars).

represents space-time as the Einstein Universe, which has long appeared as an excellent model for the large-scale gravitational structure of the universe; and **K** is then represented as the isometry group of the Einstein Universe. The geometry of **M** in conjunction with Mach's principle determines an inertial frame that fixes the specific decomposition of **M** that designates cosmic space and cosmic time; in other terms, suitably interpreted, the frame in which the

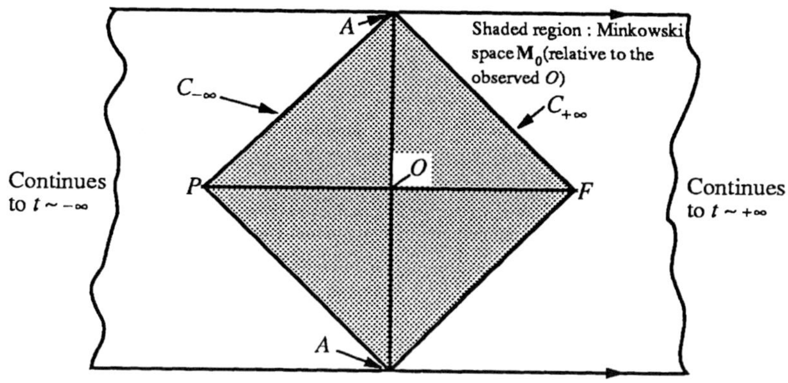

$O =$ Observer position $t = 0, \rho = 0$
$A =$ Spatial antipode of observed position $(t = 0, \rho = \pi)$
F (resp. P) $=$ infinite future (resp. past) of any point in
$M_0[F't = +\pi, \rho = O$(resp. $P't = -\pi, \rho = 0]$
$C_{\pm\infty} =$ infinite future (resp. past) of the space-like surfaces $x_0 = s$ in M_0 as s tends to $\pm\infty(C_{\pm\infty}, \rho \pm t = \pi)$

The embedding of Minkowski into chronometric space-time. Spacetime translations, Lorentz transformations, and scaling in M_0 extend uniquely and continuously to causality-preserving transformations defined in all of M. Under this (Poincaré) group, M splits up into enumerably many subregions similar to M_0, obtainable from it by successive powers of the transformation ζ, together with the boundaries of these regions. The time axis in M is to be identified with the same axis below (space is closed).

FIGURE 4. The Poincaré-invariant imbedding of Minkowski space into the universal cover **M** of conformal space-time. $C_{-\infty}$ and $C_{+\infty}$ are distinct in this imbedding, although they are identified in the conformal compactification.

energy of the energetic contents of the universe is minimal. It is an observational fact, as interpreted chronometrically, that there exists a frame in which these contents—luminous matter (galaxies, quasars, etc.) and background radiation—appear close to being stationary, with typical velocities less than $10^{-4}c$.

It must be admitted that nothing in the way of deep mathematics is required to derive observable predications from the geometry of the Einstein Universe, in conjunction with Maxwell's equations and quantum mechanics. A nontrivial, but essentially straightforward computation, shows that the redshift z (defined so that $1 + z$ is the factor by which wave length λ is increased) varies with distance r (measured in natural units, i.e., radians on

S^3) and takes the form $z = \tan^2(r/2) + O(\lambda^{-1/2})$ in which the $O(\lambda^{-1/2})$ term dependent on the Maxwell wave function in question is negligible in the observable frequency range. Since apparent luminosity varies as $(\sin r)^{-2}$, which is diminished by the factor $(1+z)^{-1}$ by redshifting, the resulting prediction is that L_{obs} varies as $(1+z)z^{-1}$. In a similar way, the volume of space out to the redshift z is expressible in closed form, for comparison with the numbers of objects of a designated type observed in appropriate samples, under a theoretical assumption of a large scale homogeneous spatial distribution. Geometrically, the apparent diameter varies as $(\sin r)^{-1}$, or as $(1+z)z^{-1/2}$; but the quantum-mechanical complementarity between time and position coordinates on the one hand, and energy-momentum on the other, requires a factor of $(1+z)^{-1}$ to compensate for the factor of $1+z$ applicable to the wave length (and thereby preserve the commutation relations between space-time coordinates and energy-momentum).

But the application of these cosmic ideas to mathematical issues arising in part in microphysics does lead to nontrivial results that do not seem obtainable in any other essentially different way. Physically, there is no reason why the same time, energy, and symmetry group that is applicable cosmically should not also apply microscopically; and to all types of fields and particles, not merely photons. I first discuss some mathematical aspects.

Cosmological mathematics. I will give three examples of how the cosmological idea of going to $\mathbf{M}$ can resolve or clarify questions that are a priori in $\mathbf{M}_0$—one classical, one quantum-mechanical, and one about scattering in either context.

One of the most interesting nonlinear wave equations is that of Yang-Mills; it is invariant not only under the conformal group but under the infinite-dimensional group of gauge-transformations. But it is difficult; like the Einstein equation, it involves nonlinear constraints on the Cauchy data at an initial fixed time, for example, and in addition its nonlinearity is relatively singular. These difficulties have prompted its study in a Euclidean context, in which the physical Lorentzian metric is replaced by a positive definite one, and in which considerable progress has been made. One of the invariants associated with the equation, and one of central importance in connection with its quantization along functional integration lines, is the integrated Lagrangian, or action. When finite in the Euclidean regime, an important class of solutions admit extension to the conformal compactification S^4 of Euclidean four-space, making possible their exhaustive analysis.

In the Lorentzian case, the powerful theory of elliptic equations does not apply and the difficulties would appear to be much greater. At first glance it appeared unlikely that the equations would admit global solutions in $\mathbf{M}_0$, at least generically, and it required estimates not only in different function spaces, as usual in the treatment of nonlinear equations but also in two different carefully chosen gauges, to establish this.

However, the action integral over $\mathbf{M}_0$ involves an infinite time integration, which has no reason to exist, from a conventional relativistic position. But taking the viewpoint that $\mathbf{M}$ is the proper global space-time arena, and noting that $\mathbf{M}_0$ has compact closure in $\mathbf{M}$, and that the equations are conformally invariant, as is the Lagrangian density, the action integral appears formally as one over a compact region in $\mathbf{M}$, and thus plausibly finite. In fact this idea works, in conjunction with a carefully detailed and nontrivial study of the relations between solutions of the equations in $\mathbf{M}_0$, and in $\mathbf{M}$, requiring three different gauges. The action integral in $\mathbf{M}_0$ is absolutely convergent provided a Cauchy datum is in a simple Sobolev space, and the solution of the equations in $\mathbf{M}_0$ extends uniquely to a global solution in $\mathbf{M}$, as the universal cover of the conformal compactification of $\mathbf{M}_0$.

In quantum field theory a similar relation between conformally invariant fields in $\mathbf{M}_0$ and $\mathbf{M}$ exists. To take the simplest example, the quantized field ϕ_0 for the wave equation $\mathbf{M}_0$ is related by unitary equivalence to the restriction to $\mathbf{M}_0$ of the quantization of the wave equation in $\mathbf{M}$. This makes possible an extension to four dimensions of a modified form of the Nelson ansatz, giving rise to an expression for the total Hamiltonian that becomes a well-defined selfadjoint operator when the putative interacting quantum field is replaced by the known free field, for the conformally invariant nonlinear equation $\Box\phi + g\phi^3 = 0$, which is prototypical. Purely in terms of $\mathbf{M}_0$, it follows in particular that the integral over $\mathbf{M}_0$ of the 'Wick-ordered' power, $:\ \phi(x)^4\ :$, has a natural and unique interpretation as a selfadjoint operator—a result that is far outside the range of conventional techniques. Corresponding results are applicable to arbitrary conformally invariant nonlinear equations modelling interactions between particles represented as the quanta of positive-energy conformally-invariant free fields. This includes, for example, the case of quantum electrodynamics, with a vanishing mass for the electron. The general idea is that temporal evolution of the quantized field may as well be expressed in terms of the Einstein time t in $\mathbf{M}$ as in terms of the Minkowski time x_0 in $\mathbf{M}_0$, and that by virtue of conservation of energy, there is a formal equivalence between $\mathbf{H}(\phi_{\text{free}})$ and its integral with respect to t, from $-\pi$ to $+\pi$, multiplied by $(2\pi)^{-1}$. The latter integral is then manifestly invariant under Einstein temporal evolution in addition to being a densely defined sesquilinear form on the C^∞ vectors for the Einstein Hamiltonian, and hence by a theorem of N. S. Poulsen, an essentially selfadjoint operator.

In both classical and quantum theory, the scattering transformation for a nonlinear wave equation plays a basic role. This is interpreted as carrying the putative incoming free field at times approaching $-\infty$ into the outgoing free field at times approaching $+\infty$, and hence involves infinite limits. Such limits do not at all exist generically for perturbations of one-parameter evolutionary groups, and it has seemed remarkable, as well as technically quite difficult to establish, that they do in fact exist for the perturbations modelling local interactions that occur in physical theory. For conformally invariant equations whose solutions are extendable from $\mathbf{M}_0$ to $\mathbf{M}$, the infinite limit in question is essentially equivalent to the entirely finite propagation from the trace of the solution on the part C_- of the boundary of $\mathbf{M}_0$ in $\mathbf{M}$ that is in the past, to the corresponding part in the future, C_+. This mapping is implemented by the generator of the infinite cyclic central discrete subgroup of $\mathbf{G}$, which maps t into $t+\pi$, and each point of space into its antipode. It applies in particular to the Yang-Mills equation, and defines thereby an effective scattering transformation, even though there is no corresponding free field in such cases. $C_\pm$ are lightcones, and what is involved from a partial differential equations standpoint is a Goursat problem, now solved in conjunction with J. Baez and Z. Zhou, for equations such as the conformally invariant scalar equation earlier indicated. This explains simply why the scattering transformation exists, and gives a sharp and satisfactory extension to the earlier work of Strauss, who initiated the study of this prototypical equation; the scattering transformation applies to any solution having a Cauchy datum of finite Einstein energy, and acts continuously in the corresponding norm.

In summary, ideas going back to Veblen and von Neumann provide a natural and mathematically productive model for space-time, and indicate a non-Doppler nature for the cosmological redshift, providing predictions virtually without adjustable parameters that are in excellent agreement with objectively delineated observations. But what about particle phenomena, gravity, and the large-small connection? Mathematically these involve at the foundational level quantum fields and thereby operator algebra, for the treatment of quantum phenomenology, and harmonic analysis, to deal with symmetries and quantum kinematics.

Rings and harmonic analysis. The logical starting point for ring theory was the reduction, or direct integral theory, but for essentially accidental reasons this was not published until after von Neumann's work with Murray on the local theory (factors) was published. By this time (1947), C^*-algebras and their states had been studied, and perhaps the simplest way to approach reduction theory is as the decomposition theory for a given state E of a ring $\mathbf{A}$ relative to a given abelian C^*-algebra $\mathbf{C}$ of the commuting ring $\mathbf{A}'$ of $\mathbf{A}$. E may be expressed then as an integral $\int_S E_x \, dm(x)$ of states E_x, where x ranges over the spectrum S of $\mathbf{C}$. The canonical association of representations with states then associates with each point x the components $\mathbf{A}(x)$ and $\mathbf{H}(x)$. The $\mathbf{H}(x)$ are 'infinitesimal' rather than finite subspaces of $\mathbf{H}$, but the decomposition can be effected uniquely and smoothly if, as is no essential loss of generality, S is represented as a 'Stone' space.

There are two main cases as regards the choice of $\mathbf{C}$. Assuming for simplicity that $\mathbf{H}$ is separable, if $\mathbf{C}$ is the center of $\mathbf{A}$, then the components $\mathbf{A}(x)$ are almost everywhere factors, and the decomposition of $\mathbf{A}$ as a direct integral of factors is unique. If however $\mathbf{C}$ is chosen as a maximal abelian subalgebra of $\mathbf{A}'$, then the components are almost everywhere irreducible, but unicity is in general lacking. In the case of a group of measure-preserving transformations, ergodicity is equivalent to irreducibility for the naturally associated C^*-algebra, and the representation of the measure as a direct integral of ergodic ones is obtained.

In case E is central, or a trace, $E(AB) = E(BA)$, a more incisive theory applies, and the condition that E be normalized can be relaxed so that E need be defined only on an appropriately dense subspace of $\mathbf{A}$. Indeed, such a linear functional E is substantially the linear extension of what von Neumann called a 'weight': a countably additive, nonnegative, unitarily invariant function on the projections in the ring $\mathbf{A}$. Noncommutative integration theory provides for such a structure $(\mathbf{A}, E)$ a generalization of Lebesgue integration theory, which via spectral theory can be regarded as the theory for the special case in which $\mathbf{A}$ is abelian. In particular, there is a unitarily- and $*$-invariant notion of 'measurable operator', generalizing that of measurable function, closed under sequential convergence, etc. There is a well-defined space $L_p(\mathbf{A}, E)$ of measurable operators extending the usual notion of a Lebesgue space, and extensions of the Fubini, Radon-Nikodym, Riesz-Fischer, and Riesz representation theorems can be established. The action by left multiplication of $\mathbf{A}$ on $L_2(\mathbf{A}, E)$ defines a standard form for $\mathbf{A}$, describable as of 'unit multiplicity', in terms of which the multiplicity theory for type I and II rings attains a simple and natural form.

The noncommutative integration theory was instrumental in the development of the generalization of the Plancherel theorem for R^n, and the Peter-Weyl theorem for compact groups, to general unimodular locally compact groups. The corresponding 'Plancherel' measure on the spectrum of the center of the group ring is unique, and its precise specification has served in the work of Harish-Chandra as a central goal and guidepost in the now most imposing edifice of the infinite-dimensional representation theory for semisimple Lie groups.

In practice, problems of explicit quantum physics seem to have a fairly adequate operator-algebraic foundation, on the basis of the developments indicated above. In particular, for example, the analogs of the Schrödinger representation in the infinite-dimensional cases, which exists for fermions as well as bosons, are simply expressed in terms of spaces of the form $L_2(\mathbf{A}, E)$, where $\mathbf{A}$ is generated by certain of the quantized field operators, and E is the vacuum state. However, some conspicuous key problems remain. One is the quantization of nonunitary representations of transformation groups; since only three of the many observed quasi-elementary particles are stable, the practice in much of the physical literature of modelling all of them by unitary representations may be an oversimplification. Another is the construction of appropriately invariant measures in function space; apart from Gaussian and related measures, the measures on which heuristic quantum field theory is currently quite dependent are presently almost wholly nebulous.

Issues in infinite-dimensional cohomology and the explication of quantum-field-theoretic anomalies are less simply formulated but quite topical and are in part dependent on operator-algebraic development.

Fundamental particles as representations of a transformation group. It is natural to ask, in relation to chronometric cosmology, what happens to the redshifted energy; and what is its analog for massive particles? The theory uses the representation of the Einstein energy H as the sum H_0+H_1 of the conventional relativistic, or Minkowski energy H_0, and the super-relativistic energy H_1. The decomposition is uniquely determined by transformation properties under scaling, under which the Minkowski coordinates x_j $(j = 0, 1, 2, 3)$ are transformed into λx_j, for some constant λ. More specifically, H_0 and H_1 are respectively scale-covariant and scale-contravariant, i.e., transform like r and $1/r$, respectively. Because scaling commutes with the Lorentz group, the decomposition $H = H_0 + H_1$ is Lorentz-covariant, relative to any given point of observation. Locally, in Minkowski space, H, H_0, and H_1 are represented by vector fields that generate local one-parameter groups of causality-preserving transformations. Setting $\partial_0 = \partial/\partial x_0$, H_0 is represented by ∂_0 and H_1 by $-\frac{1}{4}x^2\partial_0 + \frac{1}{2}x_0(x_0\partial_0 + x_1\partial_1 + x_2\partial_2 + x_3\partial_3)$, which is the transform of ∂_0 under (conformal) inversion. The vector field for H is most simply expressed as ∂_t, where t is the Einstein time coordinate. For a localized state, H_1 substantially vanishes, and is otherwise positive, indicating that it represents an attractive force. By virtue of their scaling characteristics and invariance under rotations, H_0 and H_1 correspond to effective potentials of the form λr and $-G/r$, where r is the Euclidean distance. λr has been established phenomenologically as the primary subnuclear potential, by experiments in the past two decades—an analytic form that has appeared puzzling within the frame of conventional relativistic theory.

These and other considerations are quite suggestive of the proposal that H_1 represents gravity, in the case of a massive particle, and a generalized form of gravity, in the case of photons, so that the chronometric redshift can be regarded as gravitational in an extended sense. As for the physical role of the redshift energy, it may represent the origin of the energy dissipated by the large mass loss from galaxies, which has otherwise been persistently difficult to explain.

The classic principle of Mach may be formulated more specifically as the statement that stable mass represents the interaction energy of the object in question with the cosmic background. The Einstein Equivalence Principle asserts a kind of local conformal invariance, inasmuch as transformation to an accelerated frame is infinitesimally precisely a conformal transformation. It is natural to integrate this to a global principle of conformal invariance of the fundamental interactions. But isn't this inconsistent with the non-scale-invariance of mass? Not at all, in terms of the foregoing specification of

Mach's principle. According to this, the gravitational mass is only as invariant as the cosmic background from which it derives, by the attainment of an equilibrium over a very long period of time. Taking the Einstein Universe as an approximation to the large-scale gravitational structure of the universe, this background is invariant only under the seven-parameter Einstein isometry group, and hence gravitational mass would be expected to have similarly restricted invariance.

But is there an intrinsic component to the inertial mass, above and beyond the gravitational mass derived from the interaction energy of the object with the background? If so, wouldn't that violate conformal invariance? Not at all, in the chronometric framework; even photons have nonvanishing 'Einstein' mass, even though their Minkowski mass vanishes; here mass is defined as the infimum of the spectrum of the corresponding energy operators. In fact, the Einstein mass of the photon is precisely two, in chronometric units, in which the mass of the proton is of the order of 10^{40}, so that this intrinsic mass is unobservably small. It is nevertheless important theoretically, for one reason because it eliminates the so-called infrared divergences that exist in the Minkowski framework, and for another because it is indicative of a natural physical interpretation that reconciles conformal invariance with physical mass for massive particles, such as the electron. These have intrinsic Einstein mass $\frac{5}{2}$, if modelled as spinor fields of weight $\frac{3}{2}$ as usual; but their physical mass is much greater, of the order of 10^{37}.

Thus the lack of scaling invariance of mass is an important clue to the origin of the physical (observed, inertial) mass, as composed primarily of a conformally-invariant intrinsic mass, which however is unobservably small, and a Machian or gravitational mass that is invariant only under the subgroup leaving invariant the large-scale gravitational structure of the universe—the Einstein isometry group in chronometric theory. There is in addition a quantum-field-theoretic selfmass, in the case of the electron, e.g., deriving from its interaction with the field it creates, but while this may be large compared to the 'bare' mass of $\frac{5}{2}$, it is plausibly small compared to the electron's interaction energy with the entire cosmic background.

To be sure, there is a clear mathematical difference between the conventional massless particles such as the photon and the neutrino on the one hand and the massive ones such as the electron, nucleon, and mesons. The former are represented by 'small' conformally invariant subspaces or factors of the generic class of fields in which they are included; thus their wave functions are determined by their restrictions to a fixed time, and as abstract irreducible representations of **G** are of Gelfand-Kirillov dimension three. In the massive case, the specification of the wave function requires in addition the gravitational mass, besides the restriction to a fixed time, and the Gelfand-Kirillov dimension of the corresponding representation is four. But this is not to say that m_p and m_e may not yet be estimated from dynamical equilibrium considerations and cosmic parameters of a statistical character, even though the chronometric model indicates that it is unlikely that they can be determined a priori in a group-theoretic way. It seems significant that all nonlinear wave equations found renormalizable in the physical literature are conformally invariant, apart from a linear term that can be interpreted in line with the foregoing as describing the gravitational masses of the particles involved.

For a mathematical description of the elementary particles it seems necessary as usual in modern physical theory to distinguish between the 'bare' particles, which have conceptually not yet begun to interact, and the observable 'physical' particles, which have been 'clothed' by the interaction. The foregoing considerations suggest that the fundamental bare elementary particles should be represented by the stable **G**-invariant factors of representations of the *transformation group* $(\mathbf{G}, \mathbf{M})$; here **G** is the universal, and in fact two-fold, cover of the connected group of all causal transformations on **M**. Such representations are, in physical terms, simply general conformally invariant fields, unrestricted in mass. From a geometrical point of view this says that the basic fields inhabit the section spaces of vector bundles induced from representations of the isotropy subgroup of **G** in its action on **M**—which subgroup turns out to be the scaling-extended Poincaré group (more precisely, its two-fold universal cover). But which representations? Traditionally, fermions have been assumed to be spinor particles, and the fundamental bosons to be vector, scalar, etc. particles, all of which transform trivially internally under the space-time translation subgroup T of the Poincaré group. If **M** is the proper space-time arena rather than $\mathbf{M}_0$, there is no special reason for the action of T on physical fields to be trivial, and the spinor and form bundles seem correspondingly dubiously basic.

There are more natural bundles, whose factors coincide with those of the spinor and form bundles, group-theoretically, but which transform nontrivially under T. It will suffice here to discuss primarily the fermion case. The *spannor* representation of the universal cover of the scaling-extended Poincaré group $\mathbf{P}^e$ differs from the usual spin representation only in an off-diagonal term that deforms into 0 as the cosmic distance scale R becomes infinite (in a sense similar to the deformation of the Poincaré into the Galilean groups as c becomes infinite). The **G**-invariant factors of the spannor bundle are identical to those of the usual spinor bundle, but they are assembled in a more subtle way, evidenced, e.g., by the scalar character of the casimir of **G** in its action on the spinor fields, and by its action as a Dirac (first-order differential) operator on the spannor fields. The *plyor* bundle is similarly related to the form bundle, and the Casimir again acts nontrivially, this time giving rise to the Maxwell rather than Dirac equation. Thus there is no need to postulate the Dirac and Maxwell equations; they arise purely group-theoretically, and there is a natural local interaction between spannors and plyors that is patterned after and includes the conventional quantum electrodynamic interaction between electrons and photons. These fields and their interaction are moreover $U(1)$ gauge-invariant, in a manner interpretable physically as lepton and baryon number conservation, as well as conservation of charge. There is a plausible effective division between leptons and baryons, as of respective approximate conformal weights $\frac{3}{2}$ and $\frac{5}{2}$, and their trilinear interaction with the bosons as represented by plyors unifies all known types of microscopic interaction and as represented by the H_1 component of the total energy, includes also the gravitational interaction. The inducing representations for the spannors and plyors are real, and in the corresponding real bundles there is a natural invariant complex structure that differs from the conventional one in a way that effectively replaces P by CP, when interpreted in terms of the invariant factors, which can explain why parity as conventionally defined is not conserved, but only CP (approximately).

In any event, the hypothesis that the fundamental physical forces are conformally invariant—although the distribution of energy in the universe is only invariant (in principle, only approximately) under the maximal subgroup **K** of **G**, producing the 'broken symmetry' represented by the conformal non-invariance of mass—is an attractive one theoretically that is moreover consistent with general physical principles, and indicated by systematic redshift and related observations. At the same time it is a prodigious challenge to mathematical development, and in the specific form of chronometric theory, a synthesis of theoretical physical development made possible only by the increasing sophistication and abstraction of mathematics along with its concrete and computer-intensive specification and elaboration. Von Neumann's work has been ideationally as well as technically crucial in establishing this direction and making it concretely accessible.

None of the section spaces of conformally invariant bundles that have any likelihood of relevance is fully decomposable under **G**, or even under the Poincaré subgroup, implying limitations on the interpretation of observed laboratory states as assemblies of free particles in the relativistic classification due to Wigner. There are inherent off-diagonal terms that are not directly observable, and may in part be responsible for 'missing-energy' experiments presently interpreted in terms of unobserved neutrinos. The quantization of these indecomposable section spaces, or of nonunitary group representations in general, is an important mathematical problem that exemplifies the unending interplay between experiment and theory. More specifically, the C^*-algebraic quantum phenomenology cannot be applied directly because the * is not given a priori, in the case of fermions especially; rather it must be determined from stability and symmetry considerations. Some recent developments in operator algebra relating to the introduction of an appropriate * may be useful here, but the detailed quantization of induced vector bundles along empirically relevant lines may stretch computers to the utmost, into symbolic group theory and explicit harmonic analysis. The spirit of the total von Neumann will be needed and live on in enterprises of this type.

References

Y. Choquet-Bruhat, S. M. Paneitz, and I. E. Segal, *The Yang-Mills equations on the universal cosmos*, J. Funct. Anal. **53** (1983), 112–150.

I. M. Gelfand and M. Neumark, *On the imbedding of normed rings into the ring of operators in Hilbert space*, Rec. Math. (N.S.) **12** (1943), 197–213.

S. M. Paneitz, I. E. Segal, and D. A. Vogan, Jr., *Analysis in space-time bundles.* I–IV, J. Funct. Anal. **47** (1982), 78–142; **49** (1982), 335–414; **54** (1983), 18–112; **75** (1987), 1–57.

I. E. Segal, *Postulates for general quantum mechanics*, Ann. of Math. (2) **4** (1947), 930–948.

_____, *Mathematical cosmology and extragalactic astronomy*, Academic Press, New York, 1976.

O. Veblen and J. von Neumann, *Geometry of complex domains*, (mimeographed notes), Seminar, Institute for Advanced Study, 1935–1936.

J. von Neumann, *Uber Funktionen von Funktionaloperatoren*, Ann. of Math. **32** (1931), 191–222.

_____, *Mathematische Grundlagen der Quantenmechanik*, Springer, Berlin, 1932.

_____, *On an algebraic generalization of the quantum mechanical formalism.* I, Rec. Math. (N.S). **1** (1936), 415–484.

_____, *On infinite direct products*, Compositio Math **6** (1938), 1–77.

Massachusetts Institute of Technology

Proceedings of Symposia in Pure Mathematics
Volume **50** (1990)

On the Principles of Large Scale Computing Machines

HERMAN H. GOLDSTINE AND JOHN VON NEUMANN

1. Introduction. During the recent war years a very considerable impetus has been given to applied mathematics in general, and in particular to mathematical physics, particularly in certain important fields which have not been, in the past, in the focus of most theoreticians' interest. Typical of these fields are various forms of continuum dynamics, classical electrodynamics through hydrodynamics to the theories of elasticity and plasticity. One might also mention various involved problems of statistics and of the significance of statistics, but the examples could be multiplied in many other directions. Again, partly under the influence of wartime necessities, but partly also as a natural outgrowth of normal industrial development, which is turning increasingly towards automatic scanning and control procedures, the methods of automatic perception, association, organization, and direction have been greatly advanced. These methods were in most cases of the high speed electromechanical, or of the extremely high speed electronic type. Modern radar, fire control, and television techniques are good examples of this.

These two streams of evolution have produced both an increased need for large scale, high speed, automatic computing, and the means, or the potential means, to develop the devices to satisfy this need. Accordingly a very extensive large scale renascence of interest in automatic computing machines has come about.

In this article we attempt to discuss such machines from the viewpoint not only of the mathematician but also of the engineer and the logician, i.e., of the more or less (we hope "less") hypothetical person or group of persons

---

1980 *Mathematics Subject Classification* (1985 *Revision*). Primary 68Q99.

This paper was never published. It contains material given by von Neumann in a number of lectures, in particular one at a meeting on May 15, 1946, of the Mathematical Computing Advisory Panel, Office of Research and Inventions, Navy Department, Washington, D.C. The manuscript from which this paper was taken also contained material (not published here) which was published in the Report, "Planning and Coding of Problems for an Electronic Computing Instrument".

really fitted to plan scientific tools. We shall, in other words, inquire into what phases of pure and applied mathematics can be furthered by the use of large scale, automatic computing instruments and into what the characteristics of a computing device must be in order that it can be useful in the pertinent phases of mathematics.

Since our aim is not the exceedingly difficult, unsafe, and partly contentious one of describing and comparing computing instruments, we do not attempt to give anything like a complete account of the remarkable developments that are at present taking place in numerous organizations. It is unavoidable that our account will be considerably biased by our own actual efforts in this field, and we cannot pretend to give a truly balanced picture of the "state of the art". We will, nevertheless, aim to deviate from this balance no more than is subjectively unavoidable. At any rate we shall from time to time make mention of such attributes of existing or proposed machines as are relevant to our subject.

As pointed out above, our discussion must center around two viewpoints: Where lie the main mathematical needs for high speed, automatic computing, and what characteristics of a computing device are effective in the various pertinent phase of mathematics? In attempting such a doubly oriented discussion, we find it impossible to give separate and consecutive accounts for each one of its two underlying viewpoints. It would, in fact, be desirable to give but one discussion based upon the broader problem: To what extent can human reasoning in the sciences be more efficiently replaced by mechanisms? A discussion of this question would, however, carry us too far afield. Instead we proceed in the following oscillating fashion. We do not fix the order of the discussion but move from one to the other viewpoint as frequently as seems desirable until a sufficient sense of the interconnections between the two problems has been established to conclude the paper with a joint discussion of both problems.

2. Importance to mathematics. Our present analytical methods seem unsuitable for the solution of the important problems arising in connection with nonlinear partial differential equations and, in fact, with virtually all types of nonlinear problems in pure mathematics. The truth of this statement is particularly striking in the field. Only the most elementary problems have been solved analytically in this field. Furthermore, it seems that in almost all cases where limited successes were obtained with analytical methods, these were purely fortuitous, and not due to any intrinsic suitability of the method to the milieu. This accidental character of such successes becomes particularly plausible if one realizes that changes in the physical definition of the problem, which are physically quite irrelevant and minor, usually suffice to make the previously successful analytical approach quite inapplicable. A typical example for this phenomenon: The introduction of a small nonconstancy of entropy or of a curvature (spherical or cylindrical symmetry) in the one-dimensional transient ("Riemann") or two-dimensional stationary ("Hodograph") situations in compressible, nonviscous, nonconductive fluid dynamics. Compare this "rigidity" of the nonlinear problem with the ease and elegance with which "perturbations" are handled in the linear calculus of quantum mechanics.

To continue this line of thought: A brief survey of almost any of the really elegant or widely applicable work, and indeed of most of the successful work in both pure and applied mathematics suffices to show that it deals in the main with linear problems. In pure mathematics we need only look at the theories of partial differential and integral equations, while in applied mathematics we may refer to acoustics, electrodynamics, and quantum mechanics. The advance of analysis is, at this moment, stagnant along the entire front of nonlinear problems. That this phenomenon is not of a transient nature but that we are up against an important conceptual difficulty is clear from the fact that, although the main mathematical difficulties in fluid dynamics have been known since the time of Riemann and of Reynolds, and although as brilliant a mathematical physicist as Rayleigh has spent the major part of his life's effort in combating them, yet no decisive progress has been made against them—indeed hardly any progress which could be rated as important by the criteria that are applied in other, more successful (linear!) parts of mathematical physics.

It is, nevertheless, equally clear that the difficulties of these subjects tend to obscure the great physical and mathematical regularities that do exist. To name one example: The emergence of shocks in compressible, nonviscous, nonconductive fluids shows that nonlinear partial differential equations tend to produce discontinuities, that their theory does not form a harmonic whole without these discontinuities, that the nature of the "characteristic curves" is probably seriously affected by them. It seems, furthermore, that the "proper" way to introduce these discontinuities necessarily violates Haekel's otherwise well established principle of the "permanency of formal laws", since shocks cause entropy changes in the nominally still, nonviscous, nonconductive flow. It also, and in connection with this, somehow impairs the "reversibility" of the flow. Yet, our present information about these phenomena, or rather about their deeper mathematical meaning, as well as about the details of the formation, interaction, and dissolution of these discontinuities, is worse than sketchy. Another example: The emergence of turbulence in incompressible, viscous hydrodynamics indicates that for nonlinear, partial differential equations of that mixed (parabolic-elliptic) type it is not always the knowledge of the simplest, most symmetric (laminar) individual solutions which matters, but rather information of certain large, connected families of solutions—where each of these "turbulent" solutions is hard to characterize individually but where the common statistical characteristics of the entire family contain the really important insights. Again our properly mathematical information on these turbulent solutions is practically nil, and even the united (semiphysical, semimathematical) information is most tenuous. Even the analysis of the situations in which they originate, the (linear!) perturbation-type stability discussion of the laminar flow, has been carried out only in rare cases, and with methods of great apparent difficulty.

It is important to avoid a misunderstanding at this point. One may be tempted to qualify these problems as problems in physics, rather than in applied mathematics, or even pure mathematics. We wish to emphasize that it is our conviction that such an interpretation is wholly erroneous. It is perfectly true that all these phenomena are important to the physicist and are usually mainly appreciated by him. Yet this should not detract from their importance to the mathematician. Indeed, we believe that one should ascribe to them the greatest significance from the purely mathematical point of view as well. They give us the first indication regarding the conditions that we must expect to find in the field of nonlinear partial differential equations, when a mathematical penetration into this area, that is so difficult to access, will at least succeed. Without understanding them and assimilating them to one's thinking even from the strictly mathematical point of view, it seems futile to attempt that penetration.

That the first, and occasionally the most important, heuristic pointers for new mathematical advances should originate in physics, is not a new or a surprising occurrence. The calculus itself originated in physics. The great advances in the theory of elliptic differential equations (potential theory, conformal mapping, minimal surfaces) originated in physical equivalent insights (Riemann, Plateau). This applies even to the heuristic approach to the correct formulations of their "uniqueness theorems" and of their "natural boundary conditions". Such advances as have been made in the theory of nonlinear partial differential equations are also covered by this principle, just in what seem to us to be the most decisive instances. Thus, although shock waves were discovered mathematically, their precise formulation and place in the theory and their true significance has been appreciated primarily by the modern fluid dynamicists. The phenomenon of turbulence was discovered physically and is still largely unexplored by mathematical techniques. At the same time, it is noteworthy that the physical experimentation which leads to these and similar discoveries is a quite peculiar form of experimentation; it is very different from what is characteristic in other parts of physics. Indeed, to a great extent, experimentation in fluid dynamics is carried out under conditions where the underlying physical principles are not in doubt and where the quantities to be observed are completely determined by known equations. The purpose of the experiment is not to verify a proposed theory but to replace a computation from an unquestioned theory by direct measurements. Thus wind tunnels are, for example, used at present, at least in large part, as computing devices of the so-called analogy type (or, to use a less widely used, but more suggestive, expression proposed by Wiener and Caldwell: of the measurement type) to integrate the nonlinear partial differential equations of fluid dynamics.

Thus it was to a considerable extent a somewhat recondite form of computation which provided, and is still providing, the decisive mathematical ideas in the field of fluid dynamics. It is an analogy (i.e., measurement) method, to be sure. It seems clear, however, that digital (in the Wiener-Caldwell terminology: counting) devices have more flexibility and more accuracy, and could be made much faster under present conditions. We believe, therefore, that it is now time to concentrate on effecting the transition to such devices, and that this will increase the power of the approach in question to an unprecedented extent.

We could, of course, continue to mention still other examples to justify our contention that many branches of both pure and applied mathematics are in great need of computing instruments to break the present stalemate created by the failure of the purely analytical approach to nonlinear problems. Instead we conclude by remarking that really efficient high-speed computing devices may, in the field of nonlinear partial differential equations as well as in many other fields which are now difficult to access or are entirely denied of access, provide us with those heuristic hints which are needed in all parts of mathematics for genuine progress. In the specific case of fluid dynamics these hints have not been forthcoming for the last two generations from the pure intuition of mathematicians, although a great deal of first-class mathematical effort has been expended in attempts to break the deadlock in that field. To the extent to which such hints arose at all (and that was much less than one might desire), they originated in a type of physical experimentation which is really computing. We can now make computing so much more efficient, fast, and flexible that it should be possible to use the new computers to supply the needed heuristic hints. This should ultimately lead to important analytical advances.

AMERICAN PHILOSOPHICAL SOCIETY

Proceedings of Symposia in Pure Mathematics
Volume **50** (1990)

Scientific Computing: Von Neumann's Vision, Today's Realities, and the Promise of the Future

JAMES GLIMM

1. Von Neumann's vision. Science was transformed by the invention of calculus. The impact of computers upon science will be at least as large.

Across a broad front, computers will allow the transition from qualitative to quantitative and from descriptive to predictive. John von Neumann foresaw that the equations describing scientific phenomena, once expressed in mathematical terms, could be solved numerically, without recourse to routine or repetitive experiment.

This vision is nothing less than the second half of the scientific revolution. Throughout four centuries we have expected that a successful scientific theory would have its major concepts expressed quantitatively as numbers and its major relationships expressed as mathematical equations; the truth of this theory was settled by experimental tests and hand calculations, often in idealized situations. The second half of the scientific revolution is no less sweeping in its goals. The solutions of the equations are also to be obtained on mathematical grounds, by numerical computation, without restriction to idealized cases.

Von Neumann is the founding father of modern scientific computing. His work in the major areas of this field—numerical analysis, numerical algorithms, computations, mathematical modeling, and asymptotic analysis—stands today as vital and seminal. In this lecture, we will trace several of

---

1980 *Mathematics Subject Classification* (1985 *Revision*). Primary 76N10, 65M05, 35L65.

Key words and phrases. conservation laws, artificial viscosity, shock waves, chaos.

Supported in part by the National Science Foundation, grant DMS-8619856; the Applied Mathematical Sciences subprogram of the Office of Energy Research, U. S. Department of Energy, under contract DE-FG02-88ER25053; the Army Research Office, grant DAAG29-85-0188; and the Air Force Office of Scientific Research AFOSR-88-0025.

This paper is in final form and no version of it will be submitted for publication elsewhere.

these topics, from von Neumann's work and times to the present, and then conclude with an assessment of von Neumann's vision for the future.

Our main conclusion is that von Neumann's vision remains attainable, though more difficult than was foreseen by him. The intrinsic difficulties of science have delayed the attainment of the goal. The coupling between very short and very long length scales within a single problem, and the proliferation of degrees of freedom are examples typical of these difficulties. Their importance was not fully appreciated by von Neumann. However, von Neumann did realize the potential for dramatic progress in computing technology, even where he did not foresee the specific routes this technology would follow. A combination of improvements in hardware, algorithms, and scientific theory and understanding still seems likely to fulfill the vision of von Neumann.

2. Numerical analysis and numerical algorithms. Von Neumann's work in numerical analysis included numerical stability, precision (i.e., estimation of round off error), inversion of matrices, and the computation of solutions containing discontinuities. We discuss the latter topic here. A system of equations of the form

$$U_t + \nabla \cdot F(U) = 0 \tag{2.1}$$

expresses conservation of the quantity U, with flux F, and for this reason such equations, called systems of conservation laws, arise very frequently in physics. Assuming that the Jacobian matrix

$$A(U) = \partial F(U)/\partial U \tag{2.2}$$

has real eigenvalues $\lambda_1, \ldots, \lambda_n$, (2.1) is said to be hyperbolic. These eigenvalues are the propagation speeds for small disturbances in (2.1), i.e., they govern the linear theory, defined by fixing some state U_0 and considering $A = A(U_0)$ as a constant matrix.

It is well known that such systems develop discontinuities (shock waves) spontaneously, even in the case of smooth initial data, in which case (2.1) must be interpreted in the weak sense as a distribution equation, with U belonging, e.g., to the space of functions of bounded total variation.

The problem was the following: What numerical algorithm would allow one to compute distributional derivatives effectively? In practical terms, the problem presents itself in the form of overshoots and unacceptable post shock oscillations. The solution found by von Neumann and Richtmyer **[37]** was to modify the equations slightly, so that distribution derivatives become ordinary derivatives, which can be approximated by finite differences. The parabolic regularization

$$U_t + \nabla \cdot F(U) = \varepsilon \Delta U \tag{2.3}$$

has the desired property. Its solutions are always smooth. This modification is too strong, but a nonlinear variant of it is what von Neumann and Richtmyer proposed.

For compressible gas dynamics, one of the equations is a balance of forces,

$$m_t + \nabla \cdot \left(\frac{m^2}{\rho} + p \right) = 0, \tag{2.4}$$

where m is the momentum, ρ the density, and p the pressure. The artifical viscosity of von Neumann and Richtmyer is obtained by the substitution

$$p \to p + q \tag{2.5}$$

in the momentum equation (2.4) and in the energy equation as well. Here q is a pseudopressure,

$$q = \begin{cases} \text{const}\, \rho (\partial U / \partial x)^2 & \text{in compression,} \\ 0 & \text{in expansion.} \end{cases}$$

One can check that q does add a diffusion term as in (2.3), but with a non-linear diffusion constant ε.

A very elegant method for achieving the same goal: second order accuracy and control of overshoots (the Gibbs phenomena of post shock oscillations) via artificial viscosity was found by Lax and Wendroff [**29**], in a formulation which was independent of the specific form of the equations (2.4) and (2.5), and was valid for general conservation laws. The notion of artificial viscosity dominated thinking in the computation of discontinuous solutions of conservation laws for two decades and many improvements were made to refine this idea.

In these methods, about five mesh blocks were used internal to a shock wave. As a result, considerable loss of detail and resolution prevented accurate computation of even moderately complicated shock wave interaction problems. This state of affairs was disturbed by three challenges to the finite difference community.

The first challenge came from very high resolution shock computations, in one space dimension, based on the random choice method (RCM) [**12**]. The method was introduced for theoretical reasons: it was used to prove the first existence theorem for hyperbolic conservation law systems valid for all time $t > 0$. The first computations with this method were given by Moler and Smoller [**35**] and it was later used by Chorin [**7**], who also advanced the idea of Lax [**30**] to use equidistributed rather than truly random numbers to reduce the stochastic variance in the method. On this basis RCM is essentially first order accurate.

The basic building block in RCM, as well as in earlier work of Godunov, was the solution of the Riemann problem, which is the Cauchy problem for (2.1) with jump discontinuous data:

$$U(x, t = 0) = \begin{cases} U_L, & x < 0, \\ U_R, & x > 0. \end{cases}$$

A second challenge came from the test problem of Sod, who compared all of the then available methods for the numerical solution of gas dynamics.

His test problem was the simple Riemann problem [**40**]. The comparison demonstrated that numerical resolution for both shock waves and especially for contact discontinuities was a genuine problem.

The third and definitive challenge came from Ben-Dor and I. I. Glass [**1**], who pointed out that for the simple Mach reflection problem, existing numerical methods failed to reproduce experimental density contours. The interest in this problem stems in part from the observation of von Neumann that there is a potential source of nonuniqueness in this problem, a matter which is not resolved to this day.

These challenges led to three responses. Chorin recognized that Riemann solutions were a promising avenue to improve multidimensional computations. His proposal to accomplish this goal [7] was to extend RCM to two dimensions by operator splitting (i.e., by the successive solution of one-dimensional problems in the x and y directions separately). Although this specific plan did not succeed, it did lead to two other approaches which were successful. About the same time, a variety of improved and related finite difference algorithms were proposed. We discuss one version of these ideas, explicitly dependent on the use of Riemann solutions. Van Leer [**31–33**] showed how to use the Riemann solvers in an improved (higher order) version of the Godunov method. Colella and Woodward developed and extended van Leer's ideas [**8**]. The main result of this line of work has been to reduce the computational shock width from five zones in the Lax-Wendroff scheme to about two to three mesh zones, or a factor of about two per linear dimension.

The basic idea of Godunov is to regard (2.1) as a flux balance equation and to use Riemann solutions to compute this flux balance. Godunov's version was only first order accurate, and consequently very diffusive (giving widely smeared out shocks). Van Leer computed the states going into the Riemann solution and the flux computation on the basis of higher order interpolation, with a limiter based on monotonicity constraints to avoid overshoots or undershoots near discontinuities. The Riemann solution itself was also higher order, using higher order data of the form linear + constant rather than constant left and right states.

The van Leer approach has two separate aspects: limiters to replace artificial viscosity and to prevent post shock oscillations and Riemann solvers to compute fluxes more accurately. Some version of van Leer's ideas use only the limiters in a finite difference context. Contributions by Boris, Book, Harten, Osher, and Roe should also be mentioned in connection with the improved difference schemes.

The third response to these challenges was a separate, and more fundamental, generalization of RCM to higher space dimensions. In two dimensions, the discontinuities are families of curves. The method of front tracking [**6, 13–15**] recognizes the most important of these discontinuities as independent computational degrees of freedom. In this method, the fronts are explicitly located and propagated dynamically. Riemann problems are used to find the (one-dimensional) wave speed in the direction normal to the front, while the smooth part of the solution is computed by any convenient finite difference method.

Front tracking was developed and a series of obstacles were overcome, in particular the bifurcation of front topology which results from colliding fronts was solved in a number of cases [**16**, **19**, **20**, **24**, **25**]. It has been tested on a considerable range of problems and appears to give increased resolution factors in the range three to five and occasionally 50 per linear dimension (27 to 125 and occasionally 10^5 per space time grid block) in comparison to the van Leer-Colella-Woodward methods discussed above [**17**]. On this basis front tracking has achieved one of its major scientific goals: a proof of scientific principle for this method.

Not all computational difficulties are associated with discontinuities, of course, and not all discontinuities should be tracked. For this reason a hybrid approach, combining the van Leer, front tracking, and local mesh refinement ideas with an asymptotic analysis of complex systems, has been proposed [**18**] as a promising approach to fluid computations. There are a number of computational ideas (spectral, multigrid, parallel algorithms, etc.) beyond those discussed here, and it is safe to say that the choice of the optimal method depends strongly on the problem to be solved.

3. Modeling of discontinuous waves. We pass from convergence rates to a more serious question: the fundamental accuracy and validity of the computation.

The pseudophysics of artificial viscosity was sufficient to allow the computation of shock waves. For these problems, the wave speed does not depend upon the value of physical (or artificial) viscosity. Since the wave speed is not altered in the artificial viscosity method, this method gives acceptable answers, at some expense in terms of convergence and resolution.

This plan fails for more complicated problems, for which the wave speeds or wave interactions depend in an essential fashion on the physical and chemical processes which occur within the idealized discontinuity. We consider two cases of increased difficulty. In the first case, the physics, and in particular the wave speeds, depend continuously on the viscosity and have unique limits as the viscosity tends to zero. This case requires the numerical resolution of internal layers in order that artificial viscosity does not give wrong viscosities and hence wrong wave speeds. For example the computations of detonations using van Leer methods [**9**] show that some twelve mesh blocks are needed inside the reaction zone to achieve correct wave speeds. A second, and more difficult, case can give rise to nonunique physics for the conservation law, as we now explain.

The conservation law (2.1), and in fact every law of physics, is an approximation to some more complicated theory. In simple cases, the limiting process which yields (2.1) from the more exact theories is continuous, and in these cases (2.1) defines a uniquely determined physics and has a mathematical formulation which gives (as far as we know) unique solutions. However, (2.1) has no length scales in it. Consequently, it is invariant under the change of scale transformations $x \to ax, t \to at, 0 < a$. If the more exact physics from which (2.1) is derived has two or more length scales, which have been set to zero to obtain (2.1), then the ratio of these length scales is a dimensionless number which must be given in order to supplement (2.1). In other words (2.1) can be essentially nonunique, or underspecified, both physically as well as mathematically.

Such is the case for combustion. Zeldovich, von Neumann, and Doring (ZND) [**10**, **36**, **42**] analyzed the wave structure associated with detonations and thereby established the scientific principles which determine the detonation wave speed. For most detonations, the wave speed is in the first case of depending on the viscosity, but having a unique zero viscosity limit. For most deflagrations (flames) the situation is the reverse; the zero length scale limit depends on a dimensionless number which is the ratio of two length scales.

A shock wave for the conservation law (2.1) is characterized by the relation

$$s[U] = [F], \tag{3.1}$$

where s is the wave speed, $[U] = U^{+} - U^{-}$ is the jump in U, and similarly $[F] = F(U^{+}) - F(U^{-})$. This relation is the statement that the jump discontinuity is a weak solution of (2.1).

Specializing to the equations of compressible gas dynamics, the relations (3.1) define curves in the p, $V = \frac{1}{\rho}$ plane. These curves are called Hugoniot curves. The points on the curve are the pressures p and specific volumes V of post-shock states which can be connected to a given (preshock) state. A line joining two states on the Hugoniot curve is called a Rayleigh line. The slope of the Rayleigh line joining two states on the Hugoniot curve is $-m^2$, where m is the mass flux across the shock. Thus in effect, the slope of the chord, or Rayleigh line, determines the wave speed.

For chemically reactive flow, the picture changes only slightly. The post combustion state lies on a distinct Hugoniot curve from the precombustion state. The ZND solution of the detonation wave structure problem was to pass through a nonreactive shock wave to a point on the nonreacting Hugoniot. From there, an extremum on the reacted Hugoniot defines the path to the post combustion state. Given the chemistry, there is a unique speed for this unsupported detonation wave, but in contrast to the previous case of inert gas shock waves, physical or artificial viscosity or wave front curvature will modify this wave speed.

Using equations derived by Jones [**2**, **27**, **28**], corrections to the detonation speed due to the wave front curvature for the ZND theory have been derived for reactions of arbitrary order, with Arrhenius kinetics, and a polytropic equation of state [**5**, **27**]. The same conclusions were also reached, also on the basis of Jones's equations, by Stewart and Bdzil [**41**] for the special case of a reaction order $\delta \leq 1$, and with a reaction rate which has no dependence on the state (i.e., on temperature or pressure). Furthermore, the same equations have been solved numerically [**3**, **4**] for a realistic equation of state (HOM, see [**34**]) and more general kinetics.

Conservation laws also rise in the description of fluid flow in petroleum reservoirs, and motivated by this application, Isaacson, Marchesin, and Plohr developed a numerical algorithm to solve Riemann problems for fairly general 2×2 systems of conservation laws [**26**]. This tool for computational experiments, combined with insights from geometry and bifurcation theory, has led to fundamental revisions in our understanding of conservation laws. Major contributors to this body of work include F. Furtado, E. Gomes, H. Holden, Eli Isaacson, D. Marchesin, P. Paes-Leme, F. Palmeira, B. Plohr, D. Schaeffer, M. Shearer, and B. Temple.

Specifically, shock waves and wave interactions of a qualitatively new character have been found and basic notions of stability and uniqueness have been shown to be inadequate. The detailed structure of wave interactions in a neighborhood of a point U_0 where eigenvalues coincide, $\lambda_i(U_0) = \lambda_{i+1}(U_0)$, has been obtained. Moreover, the shock theory has been related to a range of examples, including oil reservoir models, reactive fluid flow, elasticity, and phase transitions, with new conclusions being reached. The reader is referred to survey articles of the author [**21–23**] for details and further references. A major conclusion of these articles is that novel wave structures can be expected in typical one-dimensional systems of size 2×2 or higher. For conservation laws describing mechanical systems, with a phase space which is a direct sum of a configuration space and a momentum space, there is a pairing of eigenvalues, which makes the typical size for novel structures to be 4×4 or larger. The equations of elasticity, MFD, and tertiary oil recovery are all of this type. For wave interaction problems in two dimensions, similar and more serious nonuniqueness phenomena are known for smaller systems, including those of compressible gas dynamics.

Computations of discontinuous waves, in the nonunique, or underspecified case, pose very serious problems. Each numerical algorithm or choice of grid resolution is a model for physics and will select its own, generally incorrect, answer. The problem is almost as serious in the case (as occurred above) where the conservation law has a unique physical solution, but the wave speeds depend on the magnitude of the viscous terms in the equation, as in finite difference computations these numerically determined viscous terms are usually orders of magnitude larger than their physical values.

There are only two reliable strategies for computation in either of the two cases discussed in this section. A full numerical resolution of all essential length scales can be attempted, (this is normally prohibitively expensive and time consuming, even with van Leer type higher order Godunov algorithms), or frontal velocities must be an input to the computation. The second route is the one normally followed. If the frontal velocities are given explicitly, front tracking or closely related algorithms are obtained. Otherwise the finite difference algorithm must be tuned to give the correct wave speeds.

4. Fluid chaos and mixing. Still more difficult is the problem of chaos, turbulence, and fluid mixing. Here the difficulty is the large number of fluid degrees of freedom which play an essential role in the problem. Von Neumann considered the problem of chaotic fluid mixing. The problem was too large for computers at his disposal, for the algorithms he was using, and no progress was made. We show below a recent computation due to X.-L. Li (unpublished) for the two fluids compressible Rayleigh-Taylor fluid mixing problem. The computation is by the front tracking method and was performed on a Sun 4, which is only slightly more powerful than the supercomputers of von Neumann's day. We mention this comparison explicitly, not only to illustrate the progress made in the hardware side, but also to illustrate the progress which has been made with the algorithms; in particular, from a hardware point of view, the computations shown below could have been done in the years shortly after von Neumann's death.

The mixing or entrainment of one fluid into another is a problem in nonequilibrium thermodynamics, as it involves the determination of a diffusion constant. We are here interested in the situation when molecular diffusion is irrelevant, and the mixing occurs on a macroscopic level due to the chaotic flow associated with an unstable interface. The instability is caused by a light fluid being accelerated into a heavy one, and the result is that bubbles of light fluid and spikes and drops of heavy fluid are each entrained in the other. In Figure 1, a computation showing several successive time frames in the development of this instability is shown. Initially the heavy fluid is below and the light fluid is above, and the acceleration ("gravity") is directed upwards, so that the light fluid is pushed downwards into the heavy fluid. The initial interface has a small amplitude random disturbance, and at late time, a chaotic mixing zone develops. This computation is part of a systematic study of fluid mixing by the author and coworkers. The main purpose of this study is to develop an understanding of the chaotic mixing process. Among the results obtained in this line of work, we mention the computation of a single mode (bubble and spike) for late time, for a range of density ratios and compressibilities [**11**], the formulation of a simple model ordinary differential equation which duplicates the bubble and spike motion, again for a range of density ratios and compressibilities [**11**, **43**], the validation [**20**] of a statistical model [**38**, **39**] due to Sharp and Wheeler, for the chaotic interaction of many bubbles, and the observation that the single bubble terminal velocity does not coincide with bubble velocities in chaotic flow [**20**]. A distinct statistical model introduced by Zufiria [**44**] shares the property of giving reasonable agreement with gross experimental quantities, while the details of the bubble-bubble correlations are wrong in both models. The Zufiria model is restricted to a single point $A = 1, M^2 = 0$ in the Atwood number—compressibility plane. Improved statistical models which will better represent the bubble-bubble correlations are currently under development.

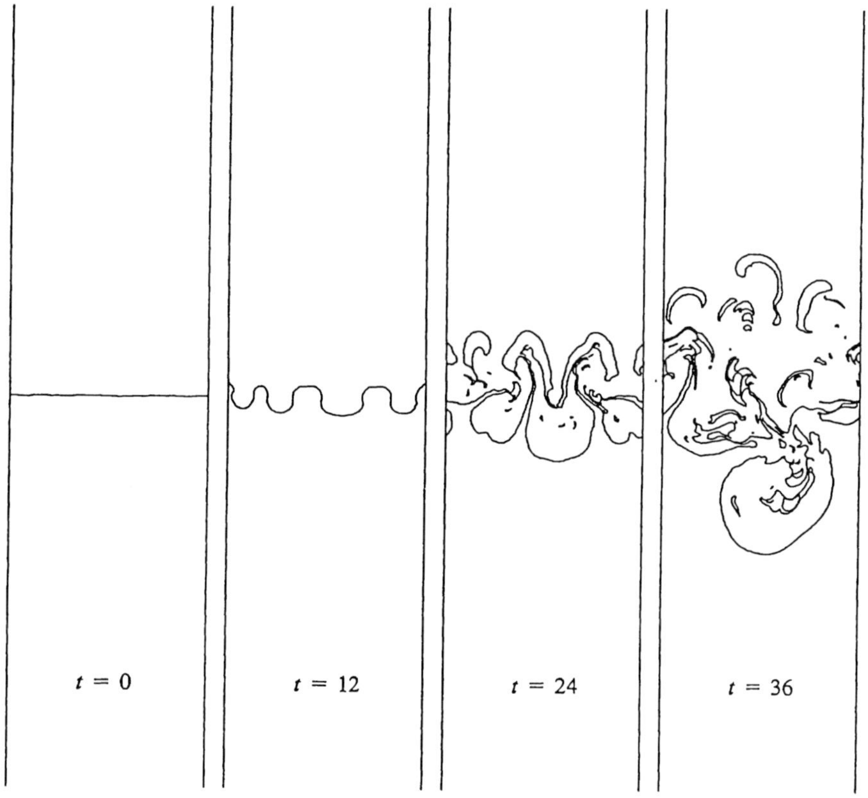

FIGURE 1. Plots of a material interface for successive time steps. The initial front has four bubbles with different radii. The compressibility M^2 for this case is about 0.5 and the density ratio is equal to 2. Picture courtesy of X.-L. Li.

5. The promise of the future. Scientific computing has become established as a third branch of science, along with theory and experiment. We can expect rapid progress on a broad front in science and technology. The progress will be driven by better hardware, better algorithms, and better theory. The time scale for the operation of these trends is not two or five years but is more likely to exceed fifty years. Bill Joy, of Sun Microsystems, speaks of "factors of two as far as the eye can see" in reference to prospects for improved computer hardware. This same trend has a strong historical basis, and has been documented in terms of the growth of computer power over several decades. It is unique across all of science and technology. There is no other area in which capabilities are increasing by factors of two, and which have put as many as four of these factors together in a short time. There are no other areas with a credible claim to do so in the future. The factors of two are nearly unique across all of history. It is a very unusual era in which any technology achieves sustained progress at this rate for an extended period. On this basis we project that computer technology will usher in a new historical era, just as the technologies for stone, copper, iron, steel, and industrial machinery each gave rise to their own historical eras.

Algorithmic developments in the utilization of computers have progressed at a comparable rate. The best efforts to quantify this trend have concluded that the improvements in overall computational effectiveness have derived about equally from hardware and from improved algorithms. We project comparable advances in scientific theory. Theory uses the computer in the same way it uses experiments: as a window of truth on the universe. It is easy to believe that a sharp increase in experimental knowledge will result in a flowering of theory. It will make little difference whether the experiments are laboratory measurements or numerical computations, once they are established to be scientifically correct. Laboratory science will retain its importance, however, both as a validation test for the computations and because on human time scales, the computer revolution will develop gradually.

We are still in the beginning phase of the computer revolution. This revolution can be regarded as the second half of the scientific-industrial revolution. Equations are not only to be posed mathematically; they will be solved on that basis also. Machinery will not only relieve the drudgery of our aching backs, but of our minds as well. The computer revolution could be dated as beginning in the 1950s, and so it is now nearly forty years old. It appears to be as full of life as ever, and it could easily be another sixty years before this technology matures.

Extremely difficult problems of science remain to be solved. The difficulties of large and small length scales can be illustrated within the problem of weather forecasting. The size of a rain drop is about 10^{-3} meters, while that of a large scale weather system is about 10^6 meters. These two interact. Specifically, the small length scales influence the large ones. The formation of rain drops affects the moisture content of the air, hence the cloud cover, reflectivity, temperature gradients, and finally the large scale dynamics of the atmosphere. The difficulties associated with a proliferation of degrees of freedom can be illustrated by the combustion of methane, certainly the simplest of all organic substances. Computational chemists have determined that a correct description of the combustion process over a range of temperatures and pressures requires a consideration of over ninety intermediate species (free radicals). Science is difficult and complex. It is clear that the broad progress projected above will not derive from hardware advances alone, but will require the expected progress in algorithms and theory as well.

There will be a trend toward the better integration of theory, algorithms, and hardware. This synergism results from the fact that combining the progress in two or in all three of these rapidly moving areas gives added power to the scientific enterprise. The intractable problems referred to above seem unlikely to yield to short term progress in only one of these areas.

Partial differential, integral, and stochastic equations have long been the universal language for much of science. As computers increasingly allow the routine solution of these equations, the power of this language increases. The reappearance of the universal scientist and the weakening of disciplinary boundaries appear to be coupled to the growth of scientific computing.

In summary, we believe that von Neumann's vision for scientific computing is undiminished, both in its present state and in its promise for the future.

REFERENCES

1. Ben-Dor and I. I. Glass, *Nonstationary oblique shock-wave reflections: Actual isopycnics and numerical experiments*, AIAA J. **16** (1978), 1146–1153.
2. B. Bukiet and J. Jones, *The competition between curvilinear geometry and chemistry in reactive fluid flow*, in Proc. VIII Internat. Congr. of Mathematical Physics (R. Seneor, ed.), World Scientific, Singapore, (1987), pp. 699–701.
3. B. Bukiet, *The effect of curvature on detonation speed*, SIAM J. Appl. Math. **49** (1989), 1433–1446.
4. ——, *Density profiles for diverging detonations*, Contemp. Math. **100** (1989), 63–76.
5. B. Bukiet and J. Jones, *The competition between curvature and chemistry in a spherically expanding detonation*, Appl. Phys. Lett. **52** (1988), pp. 1921–1923.
6. I. -L. Chern, J. Glimm, O. McBryan, B. Plohr, and S. Yaniv, *Front tracking for gas dynamics*, J. Comput. Phys. **62** (1986), 83–110.
7. A. J. Chorin, *Random choice solutions of hyperbolic systems*, J. Comput. Phys. **22** (1976), 517–533.
8. P. Colella and P. Woodward, *The piecewise parabolic method (PPM) for gas-dynamics*, J. Comput. Phys. **54** (1984), 174–201.
9. P. Colella, A. Majda, and V. Roytburd, *Theoretical and numerical structure for reacting shock waves*, SIAM J. Sci. Statist. Comput. **7** (1986), 1059–1080.
10. W. Doring, *Über den Detonationsvorgang in gasen*, Ann. Physik (5) **43(435)** (1943), 421–426.
11. C. Gardner, J. Glimm, O. McBryan, R. Menikoff, D. H. Sharp, and Q. Zhang, *The dynamics of bubble growth for Rayleigh-Taylor unstable interfaces*, Phys. Fluids **31** (1988), 447–465.
12. J. Glimm, *Solutions in the large for nonlinear hyperbolic systems of equations*, Comm. Pure Appl. Math. **18** (1965), 697–715.
13. J. Glimm, D. Marchesin, and O. McBryan, *Subgrid resolution of fluid discontinuities*. II, J. Comput. Phys. **37** (1980), 336–354.
14. J. Glimm, E. Isaacson, D. Marchesin, and O. McBryan, *Front tracking for hyperbolic systems*, Adv. in Appl. Math. **2** (1981), 91–119.
15. J. Glimm, C. Klingenberg, O. McBryan, B. Plohr, D. Sharp, and S. Yaniv, *Front tracking and two dimensional Riemann problems*, Adv. in Appl. Math. **6** (1985), 259–290.
16. J. Glimm, B. Lindquist, O. McBryan, and G. Tryggvason, *Sharp and diffuse fronts in oil reservoirs: Front tracking and capillarity*, in Mathematical and Computational Methods in Seismic Exploration and Reservoir Modeling (William E. Fitzgibbon, ed.), SIAM, Philadelphia, 1987, pp. 68–84.
17. J. Glimm, J. Grove, and X.-L. Li, *Three remarks on the front tracking method*, in Proceedings of Taormina Conference, Oct. 1987.

18. J. Glimm and D. H. Sharp, *Numerical analysis and the scientific method*, IBM J. Res. Develop. **31** (1987), 169–177.
19. J. Glimm, J. Grove, B. Lindquist, O. A. McBryan, and G. Tryggvason, *The bifurcation of tracked scalar waves*, SIAM J. Sci. Statist. Comput. **9** (1988), 61–79.
20. J. Glimm and X.-L. Li, *On the validation of the Sharp-Wheeler bubble merger model from experimental and computational data*, Phys. Fluids **31** (1988), 2077–2085.
21. J. Glimm, *The interactions of nonlinear hyperbolic waves*, Comm. Pure Appl. Math. **41** (1988), 569–590.
22. ——, *The continuous structure of discontinuities*, Lecture Notes in Physics, vol. 344, Eds: Rascle, Serre, and Slemrod, Springer-Verlag, New York, 1989, pp. 177–186.
23. ——, *Nonuniqueness of solutions for Riemann Problems*, Proc. 2nd Internat. Conf. on Hyperbolic Problems. Notes on Numerical Fluid Dynamics, vol. 24, Eds: Ballman and Jeltsch, pp. 169–178.
24. J. Grove, *The interaction of shock waves with fluid interfaces*, Adv. Appl. Math. (to appear).
25. J. W. Grove, *Anomalous waves in shock wave-fluid interface interactions*, Proc. Sixth Army Conf. on Applied Mathematics and Computing, U.S. Army Research Office, Research Triangle Park, 1989.
26. Eli Isaacson, D. Marchesin, and Brad Plohr, *The construction of nonlinear waves for conservation laws* (to appear).
27. Jim Jones, *The spherical detonation* (to appear).
28. J. Jones, *Asymptotic analysis of an expanding detonation*, New York University, Ph.D. Thesis, 1987.
29. P. D. Lax and B. Wendroff, *Systems of conservation laws*, Comm. Pure Appl. Math. **13** (1960), 217–237.
30 P. D. Lax, *Nonlinear partial differential equations and computing*, SIAM Rev. **11** (1969), 7–19.
31. B. van Leer, *Toward the ultimate difference scheme*. II, J. Comput. Phys. **14** (1974), 361–370.
32. ——, *Toward the ultimate difference scheme*. III, IV, J. Comput. Phys. **23** (1977), 263–275, 276–299.
33. ——, *Toward the ultimate difference scheme*. V, J. Comput. Phys. **32** (1979), 101–136.
34. C. Mader, *Numerical modeling of detonations*, Univ. of California Press, Berkeley, 1979.
35. C. Moler and J. Smoller, *Elementary interactions in quasilinear hyperbolic systems*, Arch. Rational Mech. Anal. **37** (1970), 309–322.
36. J. von Neumann, *Theory of detonation waves*, Progr. Report No. 238 and OSRD Report No. 549, 1942.
37. J. von Neumann and R. D. Richtmyer, *A method for the numerical calculation of hydrodynamic shocks*, J. Appl. Phys., **21** (1950), 232–237.
38. D. H. Sharp and J. A. Wheeler, *Late stage of Rayleigh-Taylor instability*, Institute for Defense Analyses, 1961.
39. D. H. Sharp, *Overview of Rayleigh-Taylor instability*, Physica **D12** (1984), 3–17.
40. G. A. Sod, *A survey of several finite difference methods for systems of nonlinear hyperbolic conservation laws*, J. Comput. Phys. **27** (1978), 1–31.
41. S. Stewart and J. B. Bdzil, *The shock dynamics of stable multidimensional detonation*, Combustion and Flame, **72** (1988), 311–323.
42. Y. B. Zeldovich, *On the theory of the propagation of detonation in gaseous systems*, Zh. Èksper. Teoret. Fiz. **10** (1940), 542–568. (Russian); English transl., Tech. Memos Nat. Advisory Committee to Aeronautics No. 1261 (1950).
43. Q. Zhang, *A model for the motion of single bubble and spike in Rayleigh-Taylor instability* (to appear).
44. Juan A. Zufiria, *Bubble competition in Rayleigh-Taylor instability*, Phys. Fluids **31** (1988), 440–446.

New York University

Proceedings of Symposia in Pure Mathematics
Volume **50** (1990)

Intermittency of Turbulence

ZHEN-SU SHE, ERIC JACKSON, AND STEVEN A. ORSZAG

1. Introduction. Turbulence, as one of nature's most complex phenomena, has long resisted analysis by most standard mathematical techniques. The traditional statistical mechanical approach is to regard turbulence as a random field, and to describe it probabilistically. However, it has long been recognized that turbulence, even in the fully developed regime, contains features that cannot be captured by a low-order statistical description. One such feature often mentioned is the phenomenon of intermittency. An intermittent signal is highly non-Gaussian, often containing more than one characteristic state, for instance, high amplitude fluctuations embedded in an essentially smoothly varying background. Experiments performed in various turbulent flows during the last forty years provide strong evidence that turbulence is intermittent. To investigate this intermittency, many phenomenological models have been constructed, most of which are based on the idea of a fractal concentration of turbulent fluctuations of activity.

In reality, turbulence is a continuous variation of statistical states, rather than jumps between clearly distinguishable states, as usually assumed in fractal models of intermittency. Detailed investigations of the properties of the distribution function of turbulent fields, which will be reported below, indicate that turbulence at moderately large Reynolds number is very different from a random Gaussian-distributed field. Normal (Gaussian) distributions appear naturally in systems with large numbers of degrees of freedom because of the classical law of large numbers. For this reason, normal distributions are the leading order assumption of most turbulence theories. However, the most important dynamical properties of turbulence, such as its energy cascade, depend critically on deviations from Gaussianity. Therefore, clarifying

---

1980 *Mathematics Subject Classification* (1985 *Revision*). Primary 76F05; Secondary 76-08, 76005, 35Q10.

Key words and phrases. Turbulence, intermittency, non-Gaussian statistics, probability distributions, numerical simulation, Navier-Stokes equations.

The final version of this paper will be submitted for publication elsewhere.

the nature of the non-Gaussianity of turbulence is of fundamental importance, and may result in deeper understanding of these flows.

In the present work we have conducted several direct numerical simulations of isotropic homogeneous turbulence by integrating the randomly forced three-dimensional incompressible Navier-Stokes equations:

$$\partial_t \mathbf{u} + (\mathbf{u} \cdot \nabla)\mathbf{u} = -\nabla p + \nabla^2 \mathbf{u} + \mathbf{f}, \tag{1a}$$

$$\nabla \cdot \mathbf{v} = 0. \tag{1b}$$

In the first set of runs reported below, the Gaussian force $\mathbf{f}$ is defined in terms of its correlation function

$$\langle f_i(\mathbf{k}, \omega) f_j(\mathbf{k}', \omega') \rangle = 2D_0 k^{-3} \delta(\mathbf{k} + \mathbf{k}') \delta(\omega + \omega') P_{ij}(\mathbf{k}), \tag{2}$$

following the "correspondence principle" postulated by Yakhot and Orszag **[16]**. Here, 2π-periodic boundary conditions in all three directions are assumed. Using a spectral code with resolution 96^3, the system, after an initial transient, reaches a statistically stationary state with a Taylor microscale Reynolds number $R_\lambda \approx 100$. In these runs, the force is applied to wavenumbers $1 \leq k \leq 15$ with no force applied for wavenumbers $k > 15$. The stationary state is characterized by a power law energy spectrum over a decade of wavenumbers, followed by a range with more rapid decrease of energy (see Figure 1). The exponent of the energy spectrum in the power law range is nearly $-5/3$, in agreement with the prediction of dimensional analysis and of the renormalization group (RNG) description (Yakhot, She, and Orszag **[17]** and references therein). The derivative flatness value is about 4.2, in agreement with results of other simulations without wide-band forcing in wavenumber space **[4]**. It is fair to conclude that by using the force (2) we have achieved a fairly good sample of isotropic homogeneous turbulence. Other features of these forced flows have been studied by Panda et. al. **[14]**, and show good agreement with measured turbulence properties.

In order to exclude any spurious effects of random forcing, especially on the properties of large scales inside the power law range, we have performed a second set of experiments by simulating decaying turbulence beginning from a field at the statistically stationary state obtained as described above. After about one large eddy-turnover time, the measured derivative flatness remains essentially unchanged while the skewness increases from 0.35 to 0.44. Since the derivative skewness measures the efficiency of enstrophy production and the energy cascade, this result suggests that the random force (2) does inhibit a part of energy transfer by the nonlinear interaction. However, the gross picture of intermittency of the field measured in terms of flatness is not much different from the first case. It is essential for our subsequent study to assume that this non-Gaussian character of the turbulent system we are simulating is fairly representative of typical real life turbulence at very small scales.

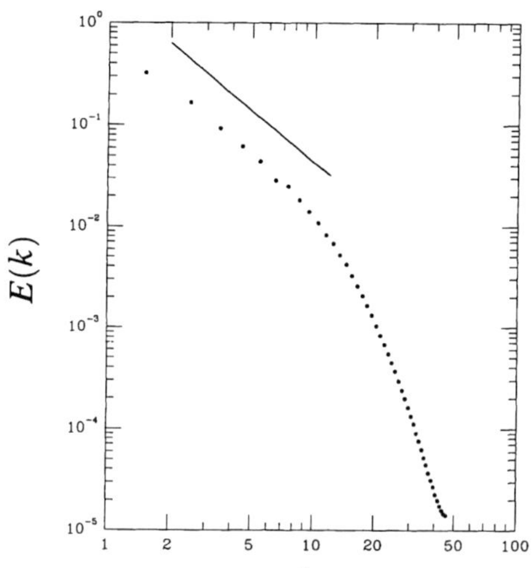

FIGURE 1. Energy spectrum of a turbulent field from direct numerical simulation with resolution 96^3. Note that the "inertial" range extends to $k \approx 10$.

2. Non-Gaussian statistics of the velocity and velocity derivative fields

2.1. *Univariate statistics of velocity.* It has long been known that the probability distribution of the full velocity field is basically Gaussian. It has also been known since the early work of Batchelor and Townsend [**1**] that the distribution of the velocity derivatives deviates significantly from Gaussian and, indeed, that this deviation seems to increase with increasing order of derivatives. It seems likely that such deviations in the derivative fields imply the existence of some deviation in the velocity itself. In order to investigate this hypothesis, we have attempted to isolate the contribution of different scales of the flow to the overall distribution by considering the physical space univariate distributions resulting from Fourier-band filterings of the velocity field. We observe that the physical space distribution corresponding to a single wavenumber shell in Fourier space has a Gaussian distribution for all wavenumbers. This implies that no correlations exist between modes within a single shell, so that the central limit theorem is applicable. When the width of the shell in Fourier space is increased we observe a gradual increase of the deviation of the distribution from a Gaussian one. This deviation has been measured quantitatively using the flatness of the resulting velocity field, $F_0 = \langle v^4 \rangle / \langle v^2 \rangle^2$.

In Figure 2, we show F_0 vs. k_2 for various k_1, where k_1 and k_2 are respectively the low and high wavenumber cutoff of the band in Fourier space. Interestingly, for k_1 greater than one, we observe an initial increase of the flatness as high wavenumber shells are added, which eventually saturates at a constant value. Moreover, the rate of increase rises with increasing k_1;

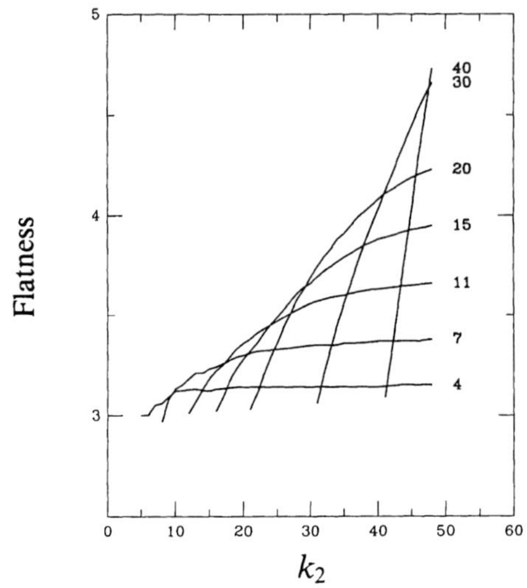

FIGURE 2. The flatness values for band-filtered velocity fields vs. high cutoff wavenumbers for various fixed low cutoff wavenumbers.

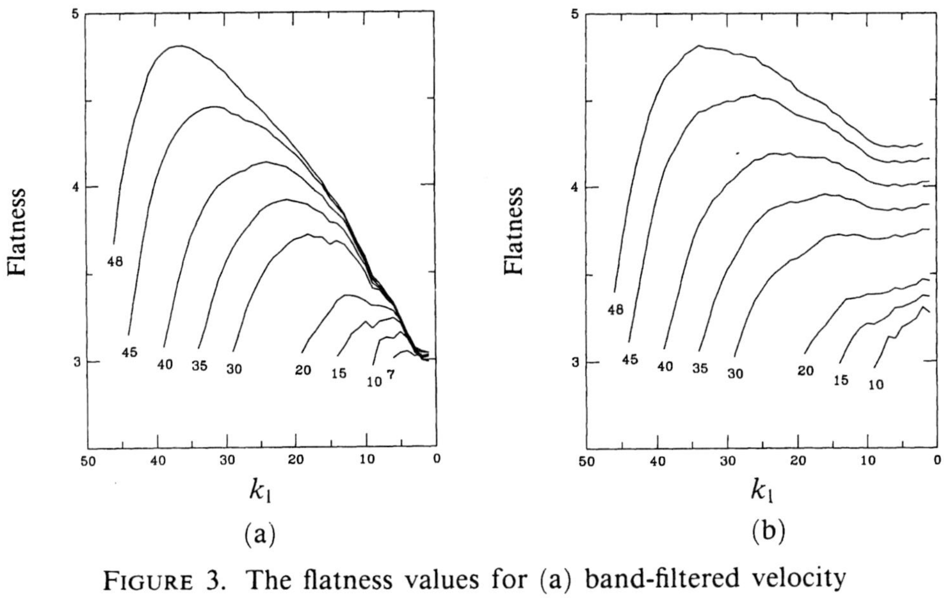

FIGURE 3. The flatness values for (a) band-filtered velocity fields and (b) longitudinal velocity derivative fields vs. low cutoff wavenumbers. Different curves are for various fixed high cutoff wavenumbers.

in other words, the deviation from Gaussianity becomes significantly more pronounced at small scales.

In Figure 3(a), we plot F_0 vs. k_1 for various fixed k_2, the high wavenumber cutoff. It is clear that as more low wavenumber shells of wavenumber $k_1 \ll k_2$ are added, the flatness value decreases, in contrast to the case when

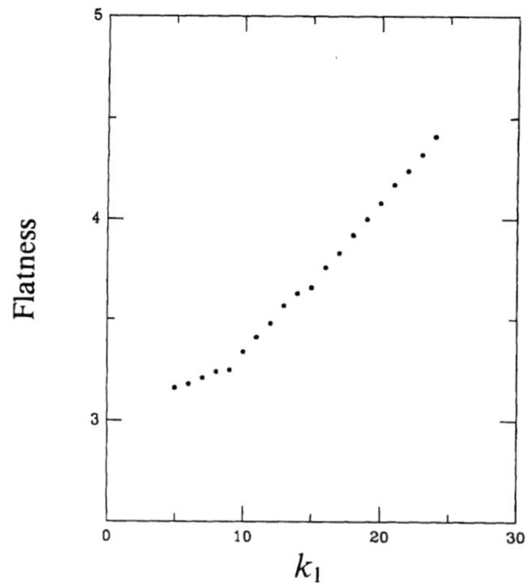

FIGURE 4. The flatness values for band-filtered velocity fields vs. low cutoff wavenumbers k_l with a band width $\Delta k = k_l$.

high wavenumber shells are added. In other words, low wavenumber shells tend to increase Gaussianity. When eventually all low wavenumber shells are added in, we get a nearly perfect Gaussian field. Fixing $k_2 = 47$ (the highest resolved wavenumber) and varying k_l, we find that the maximum flatness value appears at wavenumber $k_l = 37$ for the velocity and at $k_l = 34$ for its derivative (see Figure 3(b)). It is interesting to note that the maximum flatness value is about the same for both fields. The flatness value decreases quite rapidly with k_l (from $k_l = 37$) for the velocity field, while the decrease is much slower (from $k_l = 34$) for the derivative field (cf. Figure 3). This result may help to explain why higher order derivative fields appear to be more non-Gaussian. The wavenumber range which dominates the variance of first order derivatives is approximately $10 \leq k \leq 30$, which is not the most intermittent range according to the above result. Higher order derivatives will depend more critically on wavenumbers closer to the most intermittent range ($k \approx 37$ in the present case). The increase of flatness with the order of derivatives simply reflects the more pronounced non-Gaussian character of smaller scales in the near dissipation range. It should be noted that, at the present resolution, it is difficult to resolve completely the dissipation range scales; therefore, the most intermittent range may actually be at even higher wavenumbers than what we determine here; there may also be an underestimation of the maximum flatness value.

Figure 4 shows flatness values for wavenumber bands extending from k to $2k$. This bandwidth ensures that structures of length scale $1/k$ in the physical space velocity field are well resolved. The figure not only confirms the consistently greater flatness at small scales, it also shows a significantly higher rate of increase in the smallest, dissipation range scales, indicating that

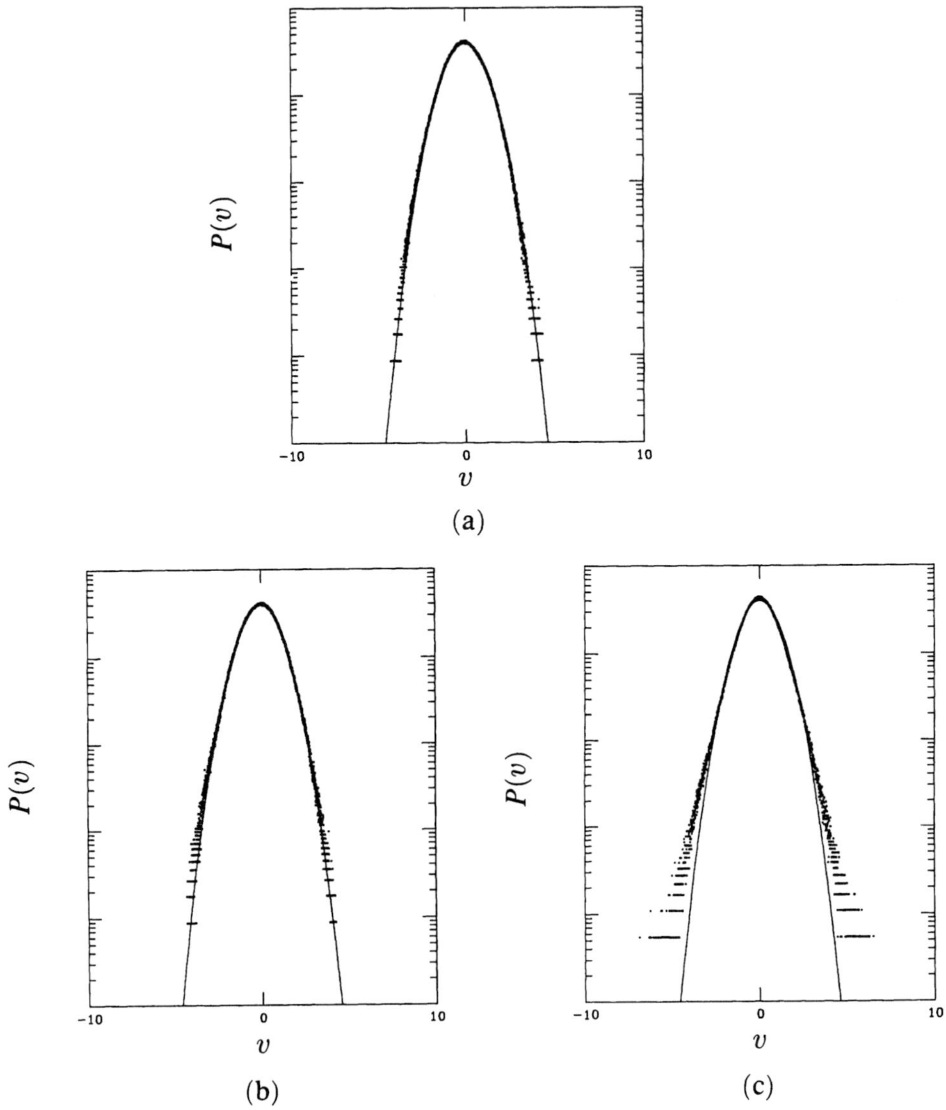

FIGURE 5. Linear-log plot of the PDF of velocity fields.
(a) the full field,
(b) an inertial range band-filtered field ($3 < k < 11$) and
(c) a dissipation range band-filtered field ($10 < k < 48$).

the mechanisms which lead to intermittency may be different in the inertial and dissipation ranges (see §4).

The precise character of the probability density function (PDF) is of interest. In Figure 5, various PDFs are shown for bands at different wavenumbers; they clearly show that the "inertial" range band has a nearly Gaussian distribution, while the "dissipation" range bands deviate significantly from it by the appearance of a near-exponential tail (at high amplitudes of velocity). It is clear that the appearance of this tail is consistently related to the increase of flatness. The low amplitude portion of the PDF appears to be basically Gaussian. This observation has several consequences. First, it gives an explanation of the saturation of flatness described above. Indeed, since the variance of the physical space velocity corresponding to shells at very high wavenumbers falls off rapidly, its contribution tends to be buried in the Gaussian core, and thus does not affect the value of the flatness.

In addition, the formation of a near-exponential tail suggests that the correlation between nearby shells in Fourier space is such that the corresponding physical space fields are correlated mainly in these high amplitude regions. Such nearly overlapping high amplitude regions contribute to establish a higher probability of high amplitudes. This kind of correlation is most efficiently established for shells in the near dissipation range, which contributes to the non-Gaussianity of the derivative fields.

2.2. *Two-point statistics.* Two-point statistical properties are usually investigated in terms of the velocity difference between two points of separation distance r. It seems clear that the velocity differences in physical space over a typical distance r are related to the Fourier amplitudes in the shell $k \sim \pi/r$. Assuming such a one-to-one correspondence holds, the investigation of velocity-difference statistics turns out to be easily accomplished by studying directly the statistics of the Fourier modes in a given shell.

Detailed study of the PDFs of Fourier amplitudes in wavenumber shells shows a consistent deviation from a Gaussian distribution for all k. Nevertheless, high wavenumber shells (i.e., small r) do tend to show more deviation than low wavenumber shells. There is a relatively higher probability for both near-zero and high amplitude values, characteristic of a truly intermittent distribution. The high amplitude behavior is similar to that observed in the small-scale physical-space velocity field. The manifestation of the deviation for near-zero values is also quite interesting—instead of a Gaussian core we observe a cusp-like form at the center of the distribution. These features may be seen in Figure 6, where the distribution function for amplitudes of Fourier modes residing in a single shell are plotted for a typical wavenumber. It would be interesting to investigate the relationship between the correlations indicated by the velocity-difference statistics and coherent physical space flow structures, and further, to determine how these may be related to the non-Gaussian form of the univariate distribution function. This topic should be explored in future studies.

The results described so far provide a consistent, unifying framework for the experimentally observed deviation from Gaussianity in turbulence. The near-Gaussian character of the velocity field reflects the Gaussianity of larger scales. In Fourier space, differentiation is simply multiplication by k, which changes the relative weight of different shells. More specifically,

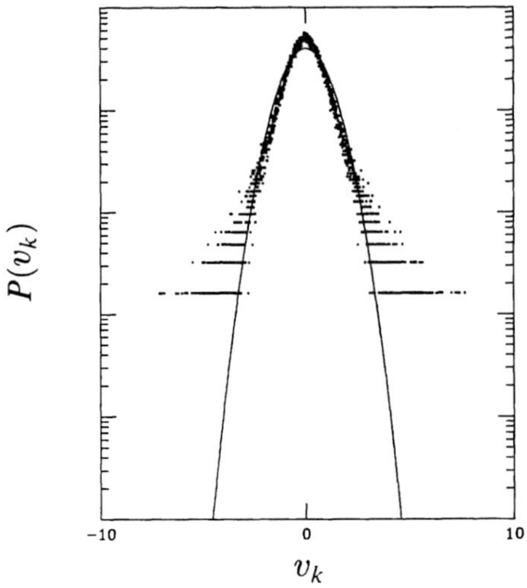

FIGURE 6. Linear-log plot of the PDF of Fourier amplitudes for a typical wavenumber shell ($k = 46$).

derivatives emphasize shells of wavenumbers near the Kolmogorov dissipation wavenumber. In turbulence of moderately large Reynolds number, it is precisely these scales which are intermittent and contribute to the deviation from Gaussianity. Therefore, the intermittency measured here and in experiments in terms of the flatness values is basically a dissipation range property. The inertial range appears not to contribute to deviations from Gaussianity for the derivative fields; rather it acts primarily to establish a Gaussian distribution for the total velocity field. In the inertial range, the variance of the physical space field resulting from different wavenumber shells scales as $k^{1/3}$ for derivative fields and as $k^{-5/3}$ for velocity fields. Therefore, very low wavenumber shells will not affect significantly the tail of the distribution function of the derivative field, so that by adding more inertial range scales we do not see a significant change of flatness.

2.3. *Strongly intermittent features of lateral derivative fields.* The picture developed above motivates the following question. Longitudinal differentiation changes the relative weights of different modes within a shell differently than lateral differentiation. If intermittency is basically an isotropic phenomenon, we may expect that measurements of flatness based on lateral derivatives will be consistent with those discussed above. In fact, our measurements show lateral derivative flatness values higher than longitudinal values, consistent with experimental measurements (Champagne [**2**]) and other numerical measurements (Kida and Murakami [**5**]). This difference suggests that there is a relatively higher probability for large velocity-difference events in transverse directions than in the longitudinal direction. To specifically check this, we have measured the PDF of Fourier modes in **k**-space for the y-component of velocity. The Fourier modes have been renormalized before

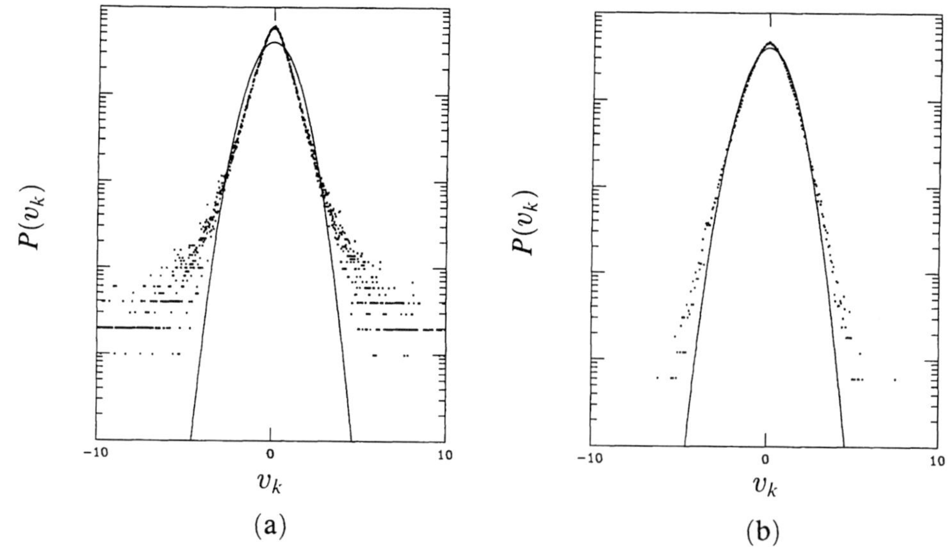

FIGURE 7. Linear-log plot of the PDF of Fourier amplitudes of the y-component of velocity for the ranges (a) $|k_y| < 15$ and (b) $30 < |k_y| < 48$.

the measurement so that each wavenumber shell has the same variance. Here **k**-space is divided into three regions according to the value of k_y, namely, $|k_y| < 15$, $16 < |k_y| < 30$, and $30 < |k_y| < 48$. The sample space is then defined by these bands for all k_x and k_z values. In Figure 6 we plot the PDFs with variance normalized to one for the first and third regions. It can be clearly seen that the "equator" region, $|k_y| < 15$, has many more large amplitude events, leading to a tail of the PDF which decays more slowly than exponentially.

In physical space the difference between the longitudinal and lateral derivative deviations from Gaussianity may be graphically seen in the appearance of a tail in the PDF of the lateral derivative distribution which clearly falls off more slowly than exponentially. We plot various PDFs for the longitudinal derivative fields with different k-band filterings in Figure 8, and for the lateral derivative fields in Figure 9. These results indicate that the exponential tail on a Gaussian core may not be a universal characteristic of non-Gaussianity in turbulence. There seems to be an anisotropic distribution of large amplitude Fourier modes on the surface of a given wavenumber shell for a given velocity component, i.e., a concentration of high amplitude modes near the "equator". The most plausible mechanism is that pressure, which keeps the field incompressible, increases transverse amplitudes. It is not clear, however, just how the pressure eliminates the high amplitude events near the "pole" region. These observations lead to the conjecture that, in physical space, there is much more high *shear* fluctuation than longitudinal fluctuation; in other words, there is less intermittent structure along the streamlines of velocity than across them.

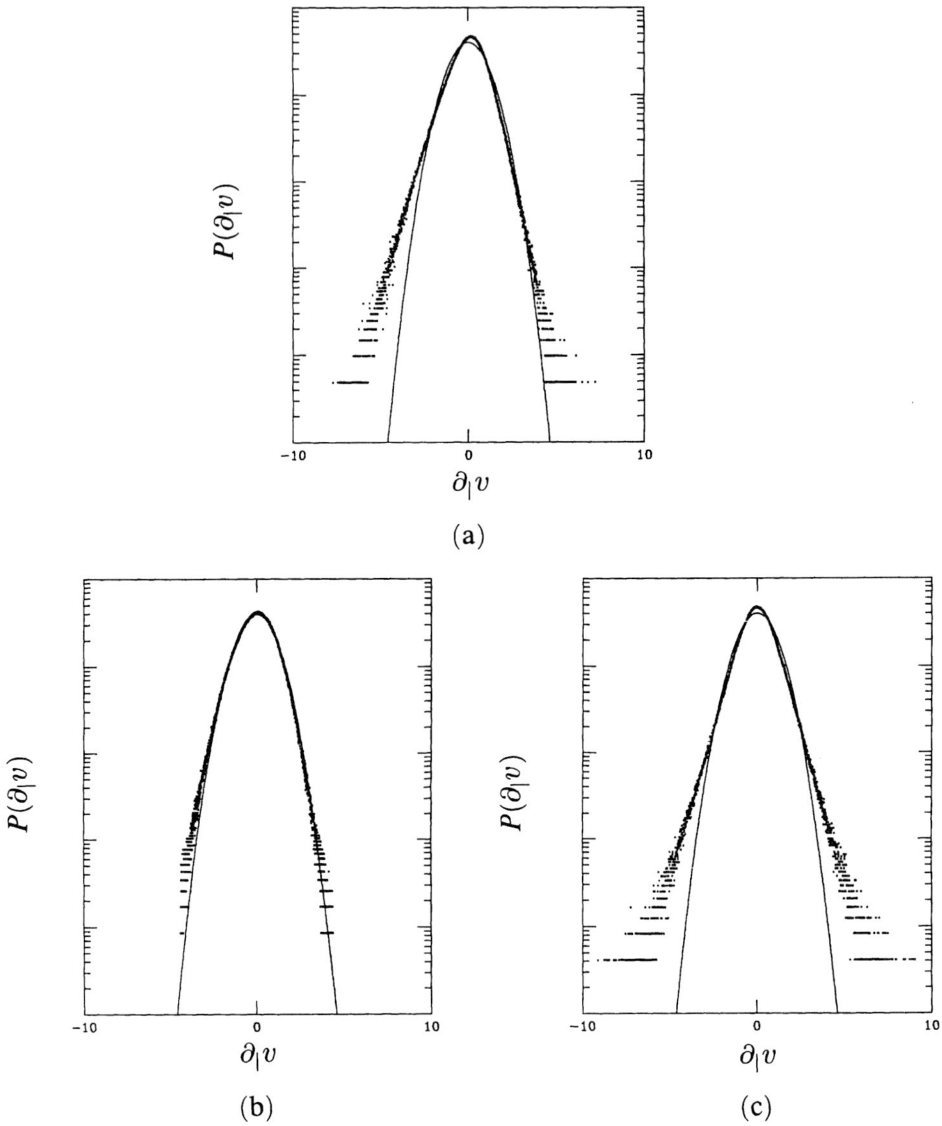

FIGURE 8. Linear-log plot of the PDF of longitudinal derivative fields: (a) the full field (b) an inertial range band-filtered field ($3 < k < 11$) and (c) a dissipation range band-filtered field ($30 < k < 48$).

It should be noted that the trends we have observed are insufficient to draw any major conclusion about what will happen when an extensive inertial range is involved at much higher Reynolds number. There are two possibilities: either inertial range scales are much less correlated so that their influence will simply maintain the Gaussian core of the PDF, or their summed correlations are finite so as to affect the lower-amplitude part of the tail and to increase the flatness. In the former case, the flatness will be independent of Reynolds

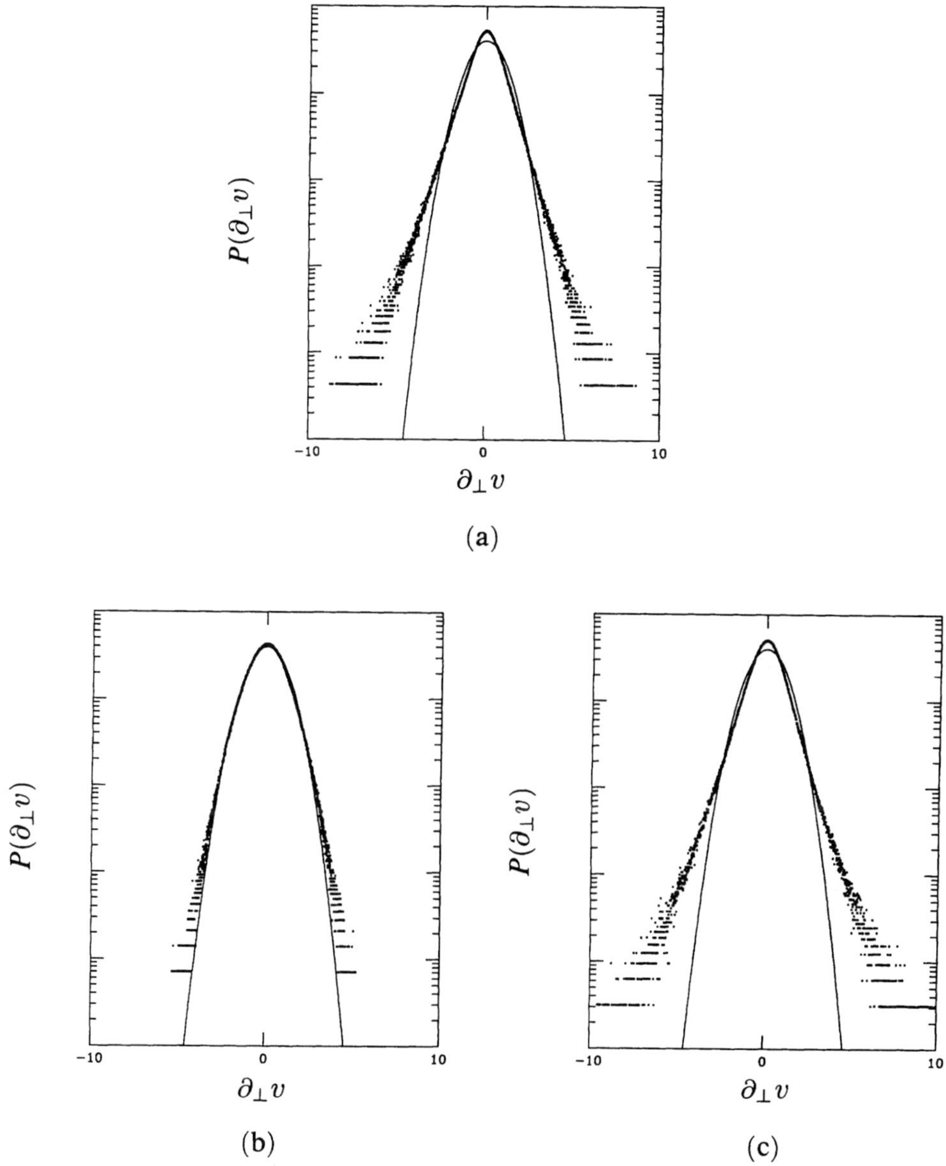

FIGURE 9. Linear-log plot of the PDF of lateral derivative fields. (a) the full field, (b) an inertial range band-filtered field ($3 < k < 11$) and (c) a dissipation range band-filtered field ($30 < k < 48$).

number and the deviation from Gaussianity will be restricted purely to the dissipation range. If this is so, the important conclusion is that the Kolmogorov [**6**] theory may remain valid. In this case, it is possible for the inertial range spectrum to be $k^{-5/3}$ while there are strong intermittency effects at dissipation scales. In the second case, the flatness will increase more rapidly with Reynolds number; at high enough Reynolds number intermittency effects will become a purely inertial range property. We cannot choose between these two scenarios on the basis of our present data.

3. Univariate statistics of dissipation. The dissipation rate

$$\varepsilon_{ij} = \frac{1}{2}\nu\left(\frac{\partial u_i}{\partial x_j} + \frac{\partial u_j}{\partial x_i}\right)^2$$

has played a special role in the study of intermittency effects in turbulence because its spatial fluctuations made Kolmogorov's original 1941 theory qualitatively suspect. On the other hand, the fundamental difference between turbulence and systems in statistical equilibrium is its dissipation, so that such issues as the nature of typical dissipation structures in turbulence are of central interest. Since the work of Kolmogorov [**7**] and Obukhov [**13**], various statistical models of spatial dissipation have been proposed (Mandelbrot [**10**], Novikov [**12**], Yaglom [**15**] and Frisch et. al. [**3**], among others). Despite the obvious fact that dissipation is a dissipation-range quantity, most of these models are based on assumptions about the inertial-range cascade process. In particular, a statistically independent step-by-step local energy cascade (Yaglom [**15**]) is consistent with a log-normal distribution of dissipation (Kolmogorov [**7**]; Obuhkov [**13**]). However, it was pointed out by Kraichnan [**8**] and Mandelbrot [**10**] that a log-normal distribution is unlikely to be an exact description, although it may be a good approximation based either on analysis of the nonlinear interaction or on a purely probabilistic basis. On the other hand, it has been shown experimentally that, in many situations, the distribution of the rate of strain $(\partial u_i/\partial x_j)^2$ has a nearly log-normal tail (see Monin and Yaglom [**11**]).

As shown in previous sections, small scale components of the turbulent velocity fields studied here have a near-exponential tail in the univariate PDF. This tail is manifested in the PDF of the derivative fields. In this section we will investigate its consequences for the distribution of the dissipation rate ε.

In Figure 10, we give a lin-log plot of the PDF of ε: $P(\varepsilon)$ vs. $\sqrt{\varepsilon}$. In this plot, there appears to be a straight line over an extensive range in the tail. It follows that the probability of having high ε drops off as $e^{-\sqrt{\varepsilon}}$. This behavior can be easily understood if we recall that ε is proportional to ω^2 which is proportional to $(\partial u/\partial x)^2$ and the PDF for $(\partial u/\partial x)^2$ has an exponential tail. Indeed, we have checked that the PDF for ω^2 has the same behavior as the PDF for ε, even though ε contains a mixture of longitudinal and lateral derivatives while ω^2 has only lateral derivative contributions. This result suggests that the univariate PDF of the dissipation rate may be derived in a straightforward way from that of spatial derivatives, and hence from that of the small scale components of velocity field. This remark is important in the sense that previous theoretical studies of intermittency of turbulence have always been restricted to independently modelling the spatial fluctuations of ε, in most cases artifically. One such case is the log-normal model, which

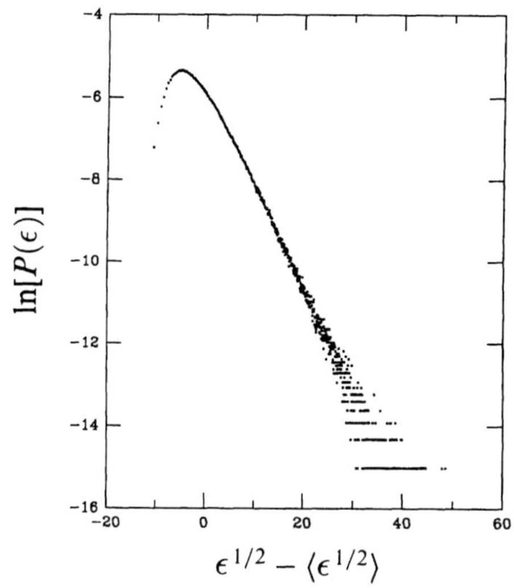

FIGURE 10. Linear-log plot of the PDF of the dissipation rate ε vs. $\sqrt{\varepsilon} - \langle\sqrt{\varepsilon}\rangle$.

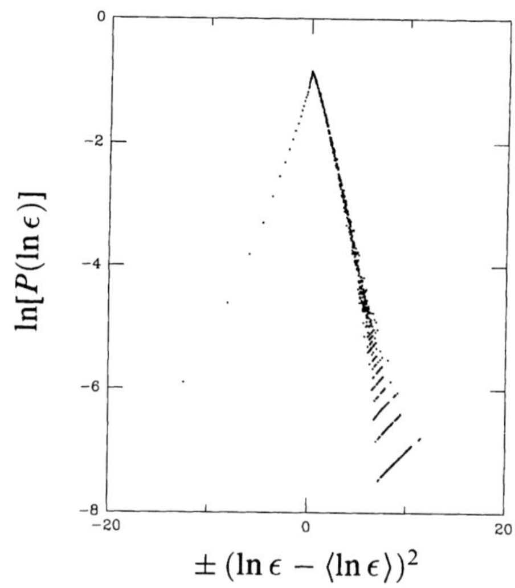

FIGURE 11. Linear-log plot of the PDF of the dissipation rate ε vs. $\pm(\ln\varepsilon - \langle\ln\varepsilon\rangle)^2$, where the sign is positive when $\ln\varepsilon - \langle\ln\varepsilon\rangle > 0$ and negative otherwise.

has seem to give a reasonable approximation for ε much higher than $\bar{\varepsilon}$ in some experiments (Monin and Yaglom [**11**, p. 627]). In Figure 11, we plot $\log(P(\varepsilon))$ vs. $(\log(\varepsilon) - \langle\log(\varepsilon)\rangle)^2$ for our computational results, in which we can also see a straight line region at high ε. However, we cannot conclude that a log-normal distribution is a good model because of the asymmetry which is required if $\log(\varepsilon)$ has a Gaussian distribution. Note that while we are not considering very high Reynolds number here, we are treating a case where the nonlinear interaction dominates the dynamics of the system and strong spatial fluctuations of ε are manifested. In the limit of infinite Reynolds number with the presence of very extensive inertial range, the present results may be qualitatively changed.

4. Discussion. In the present study, we have raised several questions:

(1) Why do spatial derivatives change the characteristics of the PDF (from nearly Gaussian to strongly non-Gaussian)?
(2) How should the non-Gaussianity of turbulence be characterized?
(3) How should the intermittent structures of turbulence be defined?
(4) What should be explained by a theory of intermittency in turbulence?

The first question is related to one of the most important properties of turbulence: the continuous excitation of all scales. The velocity field has such rapidly decreasing excitation with decreasing scales that small scale motions contribute little to the energy budget. In other words, the PDF of the velocity field is mostly determined by the large scale components of the motion. When derivatives of the field are taken, small scale motions become increasingly emphasized. It follows that large departures from Gaussianity for higher order derivative fields reflect the increasing non-Gaussianity of small scale motions.

The clear picture emerging here is that turbulence is basically non-Gaussian because the small scale motions intrinsic to energy dissipation are strongly non-Gaussian, while the inertial range eddies are nearly Gaussian. The non-Gaussian behavior of dissipation scales seems to be the basic intermittent feature of turbulence; however, it appears that high amplitude events at these scales are embedded in a nearly Gaussian background, in contrast to the assumptions of simple phenomenological models. It is interesting to ask whether the exponential tail observed in the PDFs of the near dissipation scales has some degree of universality.

Since high amplitude events make the primary contribution to the non-Gaussianity of turbulence, it would be natural to define intermittent structures as those flow regions occupied by high amplitude (of velocity, or vorticity, or of dissipation). In the past, this criterion has been adopted in flow visualization, in which cigar-like, worm-like, or sheet-like structures are noted. On the other hand, these objects do not seem to contribute much to the total energy budget, nor to the total dissipation. We may simply describe them as *intermittent structures*. They are those fine structures which depend on more subtle aspects of the nonlinear dynamics, distinguished from those more robust features of turbulence which may be obtained by simple dimensional analysis.

Developed turbulence contains structures of different correlation length. If one assumes that the departure from Gaussian statistics measures the degree of coherence, by the results we described in §2 we may reach the conclusion that structures of inertial range scales are less correlated than those of smaller dissipation range scales. We may speculate on how this may occur. The mechanism for the establishment of such coherence is viscous dissipation. Structures of typical length scale in the dissipation range are due to molecular dissipation, whereas those in the inertial range are due to eddy-viscosity effects. We would expect stronger coherence in the dissipation range since molecular viscosity acts in a deterministic way (due to the large separation between the scales of fluids motion and the mean free path of the molecular motion). The decreased coherence of eddies in the inertial range simply means that they are just barely distinguishable from the Gaussian background.

Intermittent flow structures also appear to be anisotropic. More specifically, these structures tend to be aligned with streamlines of velocity, so that they appear more clearly through lateral derivatives. This result is not simple to combine with the idea that turbulence is universally homogeneous and isotropic at small scales. To do this, we need a more subtle theoretical approach for the description of the dissipation scales of turbulence. For example, even the recently developed RNG theory of turbulence addresses only the structure of inertial range scales. A theory for the dissipation range may require fundamental new ideas on universality at these scales.

Acknowledgments. We would like to express our gratitude to Dr. V. Yakhot for many interesting and productive discussions. This work was supported by The Air Force Office of Scientific Research under Contract F49620-87-C-0036, The Office of Naval Research under Contract N0014-82-C-0451, and DARPA under Contract N00014-86-K-0759. Computations were performed on the ETA10 supercomputer at the John von Neumann National Supercomputer Center which is supported by the National Science Foundation.

References

1. G. K. Batchelor and A. A. Townsend, *The nature of turbulent motion at large wavenumbers*, Proc. Roy. Soc. London Ser. A **199** (1949), 238–255.
2. F. H. Champagne, *The Fine-scale structure of the turbulent velocity field*, J. Fluid Mech. **86** (1978), 67–108.
3. U. Frisch, P.-L. Sulem and M. Nelkin, *A simple dynamical model of intermittent fully developed turbulence*, J. Fluid Mech. **87** (1978), 719–736.
4. R. M. Kerr, *Higher-order derivative correlations and the alignment of small-scale structures in isotropic numerical turbulence*, J. Fluid Mech. **153** (1985), 31–58.
5. S. Kida and Y. Murakami, *Statistics of velocity gradients in turbulence at moderate Reynolds numbers*, Fluid Dynamics Res. **4** (1989), 347–370.
6. A. N. Kolmogorov, *The local structure of turbulence in incompressible viscous fluid for very large Reynolds numbers*, C. R. Acad. Sci. USSR **30** (1941), 301–305; *On degeneration of isotropic turbulence in an incompressible viscous liquid*, C. R. Acad. Sci. USSR **31** (1941), 538–540.
7. ——, *A refinement of previous hypotheses concerning the local structure of turbulence in a viscous incompressible fluid at high Reynolds number*, J. Fluid Mech. **12** (1962), 82–85.
8. R. H. Kraichnan, *On Kolmogorov's intertial-range theories*, J. Fluid Mech. **62** (1974), 305–330.
9. A. Y.-S. Kuo and S. Corrsin, *Experiments on internal intermittency and fine-structure distribution functions in fully turbulent fluid*, J. Fluid Mech. **50** (1971), 285–319.
10. B. Mandelbrot, *Intermittent turbulence in self-similar cascades*, J. Fluid Mech. **62** (1974), 331–358.
11. A. S. Monin and A. M. Yaglom, *Statistical fluid mechanics: Mechanics of turbulence*, Vol. 2, MIT Press, Cambridge, 1975.
12. E. A. Novikov and R. W. Steward, *Intermittency of turbulence and the energy dissipation-fluctuation spectrum*, Izv. Akad. Nauk SSSR Ser. Geophy. **3** (1964), 408–413.
13. A. M. Obukhov, *Some specific features of atmospheric turbulence*, J. Fluid Mech. **12** (1962), 77–81 [also J. Geophys. Res. **67**, 3011–3014].
14. R. Panda, V. Sonnad, E. Clementi, S. A. Orszag and V. Yakhot, *Turbulence in a randomly stirred fluid*, Phys. Fluids **A1** (1989), 1045–1053.
15. A. M. Yaglom, *The influence of fluctuations in energy dissipation on the shape of turbulence characteristics in the inertial interval*, Dokl. Akad. Nauk SSSR **166** (1966), 49–52.
16. V. Yakhot and S. A. Orszag, *Renormalization group analysis of turbulence*. I. *Basic theory*, J. Sci. Comput. **1** (1986), 3–51.
17. V. Yakhot, Z.-S. She and S. A. Orszag, *Space-time correlations in turbulence: Kinematical versus dynamical effects*, Phys. Fluids A1 (1989), 184–186.

Princeton University

Proceedings of Symposia in Pure Mathematics
Volume **50** (1990)

Supercomputer Simulations of the Interaction of Biomolecules in Solution

ENRICO CLEMENTI, STEVEN CHIN, GINA CORONGIU,
JAMES GIVEN, GEORGE C. LIE,
MICHELE MIGLIORE, AND PIERO PROCACCI

Abstract. In the first section we define the global simulation approach to complex systems and we point out the relationships between global simulation methodology, the supercomputer's evolution, and productivity of information. As an example we select the study of a liquid, specifically water. We start by building the molecules of the liquid from nuclei and electrons using quantum mechanics. Next we obtain the interaction potentials (two-, three-, and four-body) again by quantum mechanics. Then we use Monte Carlo and molecular dynamics to study the motions of a water molecule within its Onsager sphere, and its collective properties; subsequently, we overlap fluid dynamics by considering a flow along a channel with or without obstacles. Finally, we extend further and report a preliminary simulation of a Benard problem using Newton's equations. These simulations are performed on a parallel supercomputer which we have recently assembled. We then consider aspects of the solvations of DNA, myoglobin, and gramicidin *A*. All these simulations require availability of pair potentials, which are discussed with particular reference to a preliminary attempt to obtain by computations the tertiary structure of a protein.

1. Global simulations and productivity. A basic goal in science is to achieve universal understanding, thereby providing quantitative models for any event, in specific as well as for all events in general. Engineering attempts to put to work this understanding in practical applications and realizations of ever-increasing complexity and boldness. These goals and attempts have been with mankind since the beginning of our social evolution and today seem to be more and more at hand. A turning point on the learning curve is the ability to perform "global simulations" which, as we shall show, can be accomplished only if supercomputers are *extensively* available.

In the last forty years we have witnessed the birth and evolution of computers, minicomputers, personal computers, and supercomputers. However,

1980 *Mathematics Subject Classification* (1985 *Revision*). Primary 92A06, 92A08, 82A15, 76Zxx.

The final version of this paper has been submitted for publication elsewhere.

all the available indications point to the expectation that we are only at the beginning of a new socio-economic era where computing machines and signal processing machines play a determinant role. Ages are often classified in terms of some characteristic of its production tools, thus the *stone*, the *bronze*, and the *iron* ages. As we know, "tools" enhance the dimension and power of man; in the past the *physical* dimension was enhanced but today it is the *intellectual* and *mental* dimensions. We are clearly in the "computer" age.

There are many ways to define a supercomputer; for example, in terms of speed, cost, I/O, etc. However, as a "tool", we prefer to define it as *the necessary tool to perform global simulations.* Let us now once more define what we mean by "global simulations"; beforehand, however, let us note that obviously computers with one to a few gigaflops peak performance can be used to carry out computations requiring megaflop (or even lower) performance and a price/performance analysis might actually suggest a move in exactly this direction. Yet it would be a pity to use such beautiful machines for such "low" complexity tasks! Indeed, supercomputers should be used essentially for "super computations" and for the exploration of *new avenues.* Achieving the ability to simulate complex laboratory experiments *realistically* is one such goal.

We have been advocating for some time the "global simulation" approach to computer simulations in science and engineering **[1]**. The main idea is borrowed from history: from the events leading England to the first industrial revolution when "assembly lines" were introduced, thus enormously increasing productivity per worker unit. In an assembly line one starts with simple raw material and, step by step without interruption, transforms the raw material into more and more finished and complex products. In the "global simulation model" we start with the most "simple" assumption, namely that molecules are built up of point charge nuclei and electrons. With *quantum mechanics* we can then "assemble" simple molecules or larger ones; generally this is done in a timeless and temperatureless environment. With *statistical mechanics* we can consider many molecules at a given pressure and temperature, pass from an enthalpic description to a free energy description, and consider trajectories involving time. Finally, for even larger systems we can use *fluid dynamics* where viscosity, transport coefficient, convection, turbulence, etc., can be analyzed. Of course, all this is well known at the theoretical level, but the novel aspect we have been stressing is *how to move from one mechanic to the next one* in such a way that the entire process is self-consistent for computer simulations at the *operational* level. Another novel proposal is to use molecules, Newton equations, and boundary conditions to *solve* problems previously considered at the *fluid dynamical* level only.

Briefly stated, in the global simulation model we decompose a given problem into n subproblems, each one corresponding to a submodel $1, \ldots, i, \ldots, n$. The operational rules are that the input needed in the submodel "i" is fully

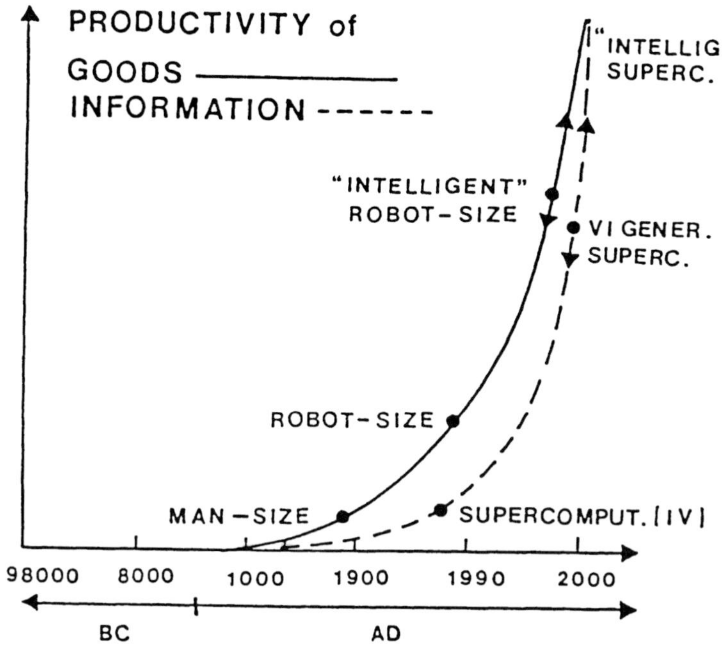

FIGURE 1. Sketch of some aspects of productivity's evolution. "Landmark's dates" for productivity of goods *and* information are drawn onto the time axis.

obtained as the output from submodel "*i*-1" and that the input for submodel "1" must be "very simple" in a computational sense. Thus, the global simulation approach is an "assembly line" designed to increase our productivity in generating *information*.

Let us now consider in some detail the relationships between the "global simulation modeling" and productivity; as it is known, the latter correlates both to *goods* and *information*. The main evolution of productivity in human history is sketched in Figure 1.

The productivity of information trails the main phases of the productivity of goods; it is therefore not surprising that even though production assembly lines revolutionized industrial output since the early 1800s, the "global simulation" method is *only now* being introduced. We recall that two conditions (market demand and a "tool"—the automated loom) were essential for initiating a "revolution" in the textile industry. Equivalently, the need and demand for interdisciplinarity of information, essential to avoid expensive errors in our highly technical and thus risk-prone society, had to wait for the appropriate tool—the supercomputer—before we could set up a *production assembly line for information*, namely the "global simulation" approach.

Advances in material sciences, brain research, networking, signal processing, cybernetics, and artificial intelligence are sufficiently reliable pointers for confidently predicting *intelligent robots* for productivity of goods, and *intelligent supercomputers* for productivity of information (see Figure 1). An "intelligent" supercomputer, in this context, is an engine of one to a few gigaflops sustained performance, thus with much higher peak performance (up to about 100 gigaflops) and with the attributes to hear, talk, and—within the limit of expert systems—understand some extremely rudimental language (like FORTRAN), and to see, display, and "understand" a few two-dimensional patterns like digits and logical-mathematical symbols. We recall that pattern and voice recognition and artificial intelligence are today capable of performing these tasks and could notably increase the user friendliness of computer systems and thus enhance the user's productivity. The raw speed we have forecasted for the "intelligent" supercomputer of the next decade is conservative. Indeed, an *l*CAP/3090 with 30 nodes (five clusters of IBM-3090/600) exhibits a peak performance of over three gigaflops; if we adopt the same architecture and software, but with a cycle time of two nanoseconds, then we would obtain a performance increase of about a factor of 10 for a total of about 30 gigaflops. In the 12 years between now and the end of this century we can expect three "new" generations with a speed up to a factor of three each time; thus, one would expect a performance not far from a teraflop. The forecast of an "intelligent" supercomputer within this century is most reasonable.

It is a particular privilege to stress that much of the above accomplishments have their root in John von Neumann's original papers. In this work we stress computations in chemistry and engineering. Let me recall that John von Neumann was a chemical engineer, who was enchanted by quantum theory (at the time of its beginning), the new base for understanding matter. From this developed his keen interest in attempting to solve the basic quantum theory equations and thus the need for computers. We feel that we are walking in his steps!

The "tool" to perform global simulations must be a supercomputer with super scalar, vector, and parallel characteristics. Let us recall then when we *operate on numbers* we can select between either the scalar (S), vector (V), or the parallel (P) modes of operation. Let us consider S, V, and P like vectors of a three-dimensional space. Any computation is then associated to a trajectory within the SVP space, as shown in Figure 2.

For complex algorithms, the corresponding trajectories will often occur within a preferential volume in the SVP space, thus defining an *envelope*, characteristic of the given algorithm. The ideal computer for an *algorithm* is the one which has *high performance for the algorithm envelope* in the SVP space. For "global simulations" the ideal computer must have super-performance along each one of the three dimensions, namely S, V, P, *being most unlikely that any simpler choice will suffice.* Alternatively stated, the envelope for global simulations extends over most of the SVP space. In the early 1980s when we extended our simulations to molecular dynamics and fluid dynam-

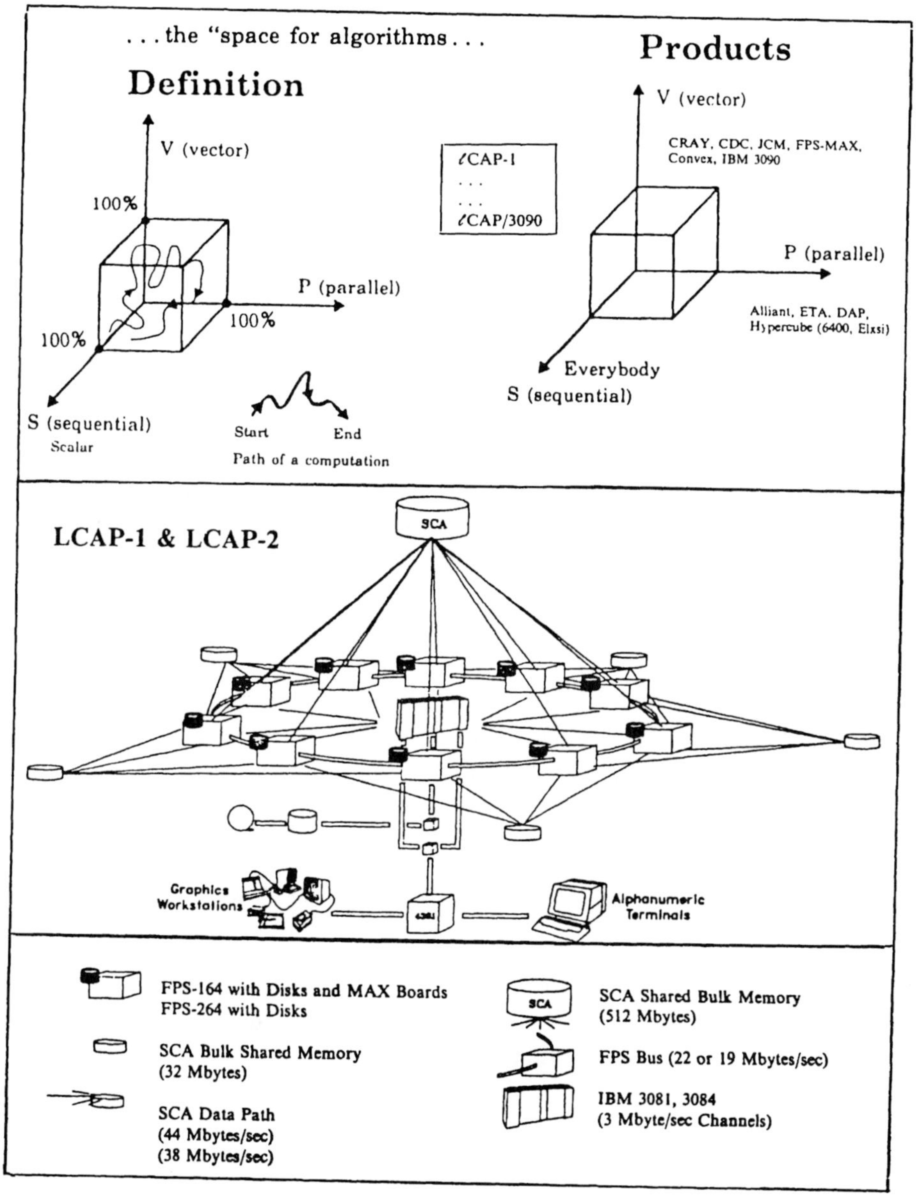

FIGURE 2. *l*CAP-1 and *l*CAP-2 supercomputers (see text) and the SVP space.

ics, there was no such supercomputer; the need therefore existed for us to build one.

Our first attempt resulted in the so-called Loosely Coupled Array of Processors #1 (*l*CAP-1, for short) consisting of ten FPS-164 array processors, each one with local disk packs and eight megabytes of random memory; two vector accelerators (Max-Board) brought the FPS-164 peak performance from 11 to 55 megaflops. These ten nodes are individually connected to an IBM mainframe via standard IBM channels. In addition, five 32-megabyte memories connect (with a high transmission rate of 44 megabytes/sec) four successive nodes, forming a double ring connectivity; further, a large memory of 512 megabytes connects each one of the ten nodes. Our Computer has about 550 megaflops peak performance. A second system, *l*CAP-2, was constructed with FPS-264s replacing the FPS-164s; this second system has about 380 megaflops peak performance. The two supercomputers were connected for experimentation in "cluster to cluster" parallel computing and the resulting system, with about one gigaflop performance, is called *l*CAP-3 and is reported in Figure 3.

The power of *l*CAP-3 can be exemplified by recalling the colossal *ab initio* SCF computation for gramicidin and a K^+ ion which required only ~ 11 hours of elapsed time. We note that (1) this system has 276 atoms with 1010 electrons, (2) we used 2736 primitive gaussian functions, (3) the SCF threshold was set at 10^{-5} atomic units, and (4) only integrals smaller than 10^{-5} atomic units were neglected. All other integrals, a total of 433,646,989, were computed in double precision and stored on disk-packs (working in parallel). For additional details on the *l*CAP system and performance relative to other commercial supercomputers, we refer to **[2–4]**.

Presently we are assembling one more supercomputer: *l*CAP-3090. In this system each cluster is an IBM-3090 system that can have up to six nodes (IBM-3090/600); we are considering up to five clusters.

In Figure 3 we have shown an *l*CAP-3090 built up by IBM-3090/400 clusters. The advantages of *l*CAP-3090 over *l*CAP-3 are: (1) one single vendor for all hardware and software, (2) very high hardware and system software reliability, (3) portability of application programs relative to *l*CAP-1 and *l*CAP-2, (4) peak performance up to three gigaflops, (5) growth opportunity starting from a single node, and (6) vector hardware, system software, and mathematical library availability.

In Figure 4 we report on the efficiency in parallelization for several (12) application areas. The optimum line (straight line) states that if we use N nodes we expect at best an N-fold speed-up relative to $N = 1$. Of course, one would expect some degradation due to overhead at the system software, hardware, and application program levels. Still, for the many applications reported the *ideal goal is reached* to a very considerable extent. We expect that the *l*CAP-3090 will cut down the overhead even more, thus yielding even better parallelization. For more details on the 12 examples reported in Figure 4, see **[5–9]**. Notice that most of *computational chemistry* and very important areas in *physics* and *engineering* are essentially represented in the examples of Figure 4.

The 12 examples reported in Figure 4 have few characteristics in common. To start with, the codes were written originally for sequential computers

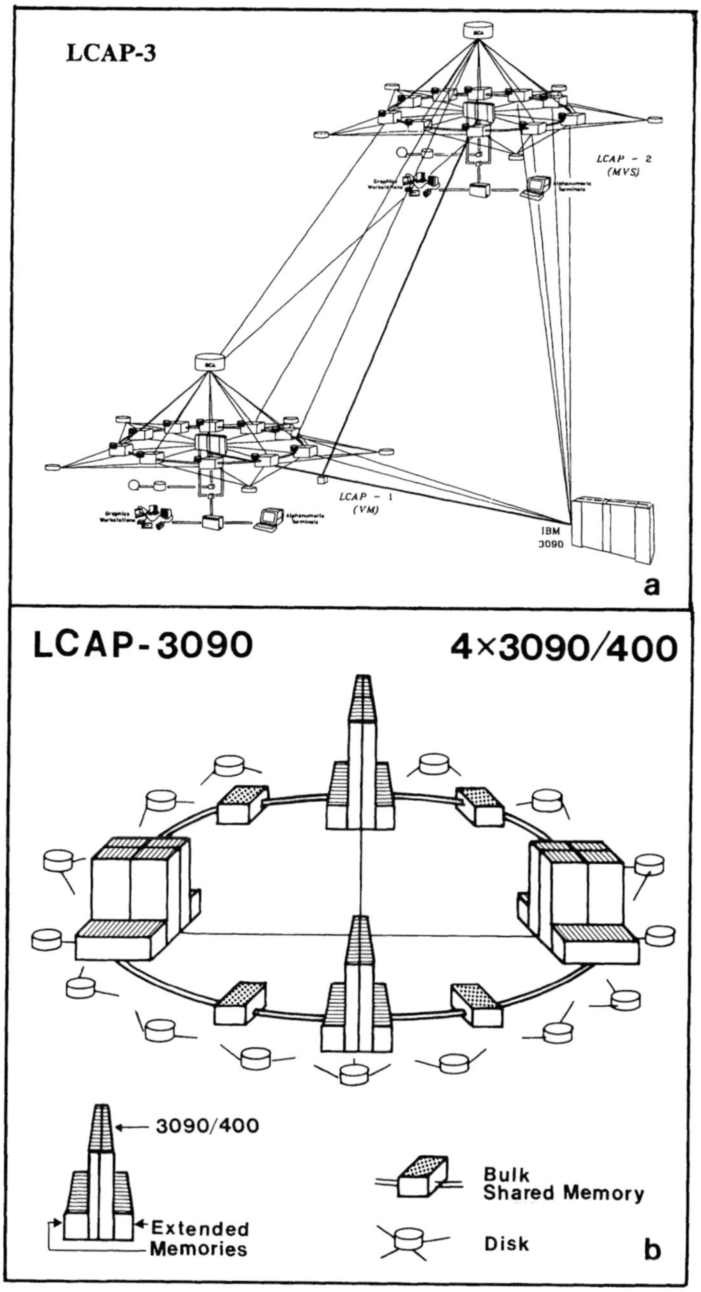

FIGURE 3. *l*CAP-3 (top) and *l*CAP-3090 (bottom): the ten nodes in each one of the two clusters of *l*CAP-3 are explicitly drawn; for *l*CAP-3090 the nodes of each cluster are compacted into one box.

and the parallelization was obtained using our "first-generation" precompilers (1983 to 1985 vintage). The reported "speed-up" (namely the ratio T_n/T_1,

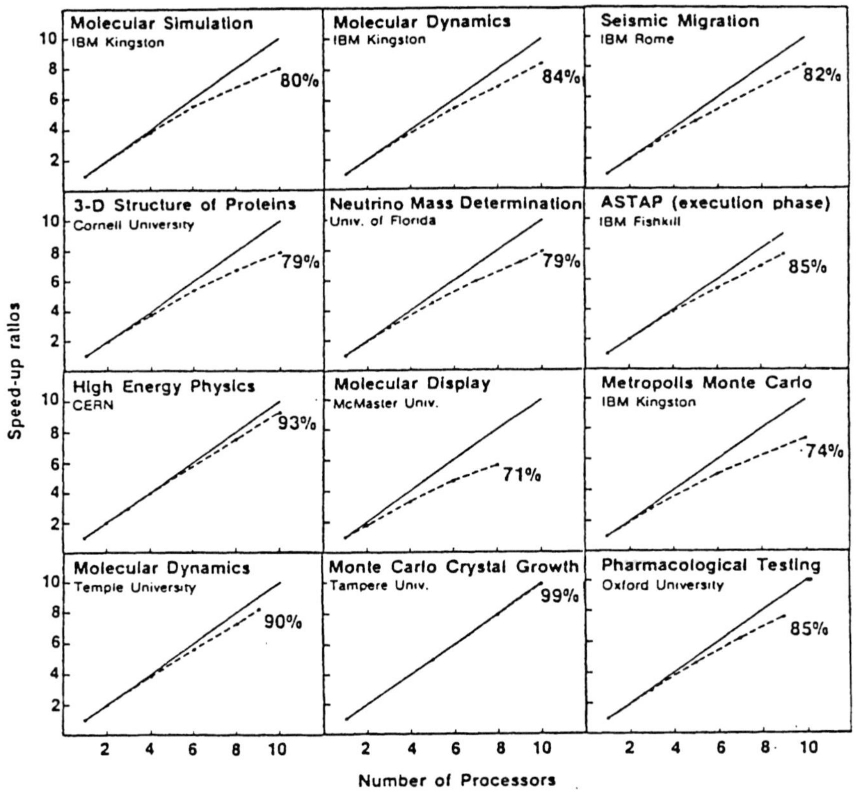

FIGURE 4. Percent parallelization in simulations for science and engineering.

where T_1 is the elapsed time for execution on one node or array processor (AP), and T_n is the elapsed time for the same job executed in parallel on n nodes) are obtained considering the full job, not only kernels; thus, input/output timing is included. Alternatively stated, the examples correspond to computations executed to elicit interest in a scientist and/or an engineer rather than in a computer-scientist or a computer-hardware architect. In general, we have learned that "grosso modo" 8 to 12 FPS-164, arranged as in *l*CAP-1, exhibit the performance of one CPU of a Cray XMP machine with a solid state disk. Since in the reported examples (as in most of our work) we have made no use of the FPS matrix accelerators (Max Boards), the measured *l*CAP-2 performance is three to four times better than the one above measured with *l*CAP-1. We recall that the FPS-264 performance is three to four times better than the FPS-164; thus, our previous statement is an expected one for simulations with high degrees of parallelization.

The 12 examples cited did require a little code modification for the translation from sequential to parallel and *no* algorithm modification. Indeed, the examples correspond to what we refer to as "natural parallelism". Very roughly, a "naturally" parallel code is one which mainly consists either of a set of nested DO loops (the larger the set, the more "naturally parallel") or of only one DO loop but with intensive computations at each loop step (i.e., case of large granularity); in addition the grains are relatively independent from each other. Notice that in the case of the set of nested DO loops, the innermost loop is often vectorizable and in this case we have optimal computational conditions, namely easy vectorization *and* easy parallelization. Quantum and classical many-body dynamics and also quantum or classical stochastic problems are most often "naturally" parallelizable. Other "natural" parallel codes are those which simulate computations easily *partitionable* in subtasks. An interesting example of the latter is *graphics*, where clearly the rotation and hidden line computing of a complex image can just as well be obtained by operating with many parcels of the image all at the same time.

"Natural" parallelism is often found in complex solutions since it tailors well the "scientific method". Indeed, in the case of a complex problem we most often "decompose" complexity by solving first for only some of the variables (the most relevant ones) and then coupling the obtained partial solutions either with perturbation or relaxation techniques or appropriate many-body expansion.

However, codes and algorithms are not always "naturally" parallel! In such cases it is *basic* that each node communicates efficiently with other (either a few or many) nodes. This can only be poorly accomplished with today's slow channels, but rather efficiently with "busses" and "bulk shared memories". For this reason we have devoted much attention in the 1985 to 1987 period to these "communication" devices, which change our "*l*CAP" into "*t*CAP", i.e., we go from "loose" to "tight" coupling.

The FPS-BUS software is based on a message-passing scheme; the user has to specify from which end to which processor data is sent and received. If the same data has to be sent from one processor to all the others, then the broadcast mode can be used.

The facilities provided by the precompiler for using the *bulk memories* offer a choice of two modes: the shared and the message-passing mode. In the shared mode the data to be shared is stored in the memory at locations known to all the processors, and it is passed between processors by specifying a *write* of data from one processor and by a *read* of data by another one. In the message-passing mode, messages in the form of contingous strips of data are passed between processors via a network of paths transparent to the user, who specifies in this code the "send" and "receive" transactions along with the targeted processors.

Performance measurements have been done using either the FPS-164 or the FPS-264 as processors, exploiting the three communications schemes described above, and choosing three patterns of data transmission. Pattern A has been called STAR: in this configuration one processor (considered the master) sends data to the others (the slaves); pattern B has been called RING (RNG for short): in this case each processor sends and receives data from its

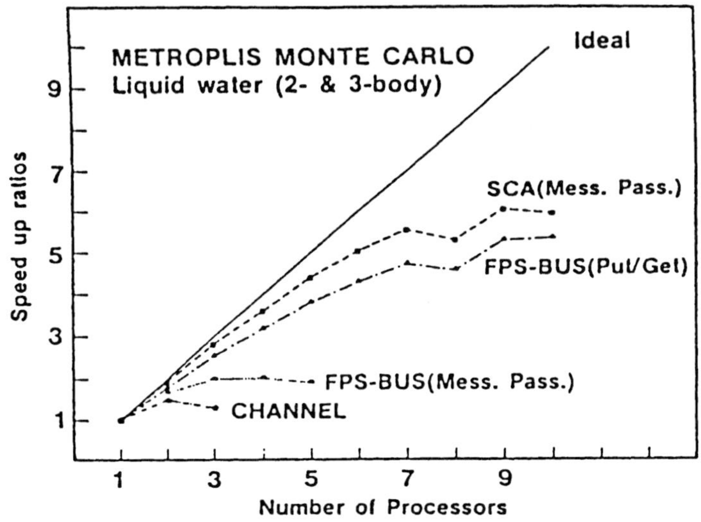

FIGURE 5. Parallelization with either channels of FPS-BUS or SCA bulk shared memories (from [**10**]).

two neighboring processors; pattern C has been called LINE (LN for short): this configuration can be achieved by cutting the RING configuration so that two end processors have only one neighbor. For details of performance, see [**4–10**].

An interesting example of the use of these communications devices reports on the "speed-up" for the Metropolis Monte Carlo study of water using two- and three-body potentials for a box containing 512 particles [**10**]. The net conclusion of this study is summarized in Figure 5; loose parallelism is insufficient for this case and, all in all, the bulk shared memory is preferable to the BUS. We stress, however, that these timing examples should not be extrapolated to machines and software much different than the one used. Indeed, for example, the latency time for transmission of data depends very much on the specific system software and hardware one has [**11**]. We conclude this section by providing a comparison of timing performance on *l*CAP-3090 (simulated data) relative to that obtained on *l*CAP-1, *l*CAP-2, and one CPU of a Cray XMP (with solid state disk).

2. An example of global simulation. Let us consider, as an example, the simulation of some chemical in the liquid phase and, to be more specific, let us consider water, a rather complicated but most interesting liquid. To start with (the "input" for the first submodel), all one needs to know is that one molecule of water has the stoichiometric composition of two hydrogen atoms and one oxygen atom which, to fit our submodel, we translate into the notion that our molecule is made up of ten electrons and has two one-charge point-like nuclei plus one with eight positive charges. Standard quantum chemistry yields a stable molecule; within the Hartree-Fock approximation

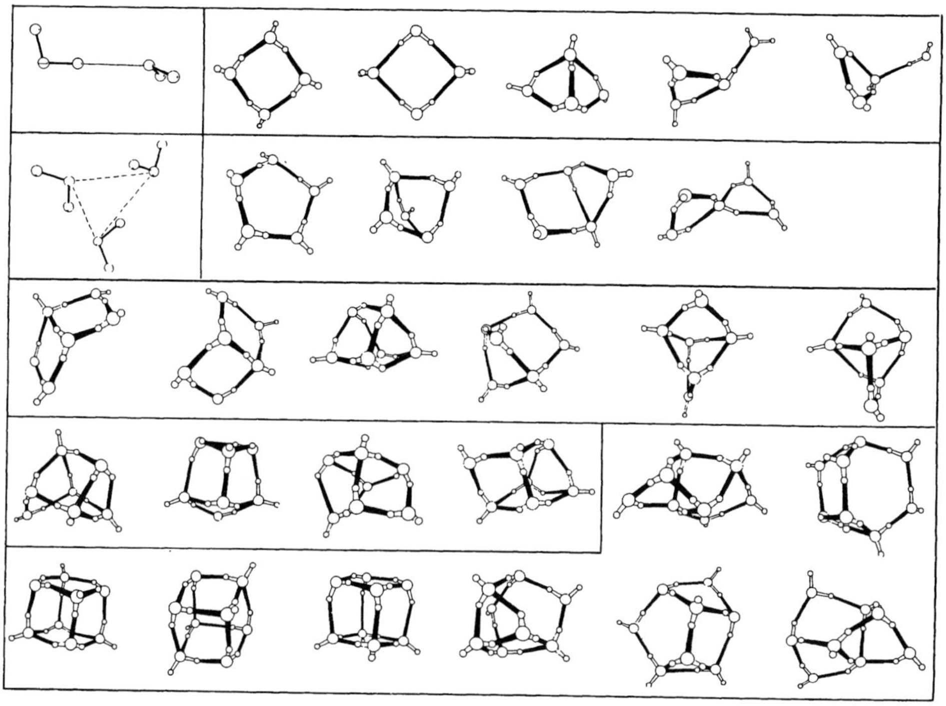

FIGURE 6. Clusters of water molecules $(H_2O)_n$. For $N \geq 4$ the most stable configuration is given at the left, the less stable at the right. For details see [**25**].

[**12**] the simulated water molecule has an HOH angle slightly too wide relative to experimental data [**13**]. The correct (within experimental accuracy) value is easily obtained by correlating the electrons' motions [**14**]. Thus we now know the geometry and the energetic of the water molecule from simulations; it is not difficult to also obtain the vibrational motions by mapping the intra-atomic potential surface in the vicinity of the equilibrium configuration [**14**].

Let us move from one molecule of water to two, three,..., several [**15**] and finally many [**16**]. Computations of the interaction between two molecules even at the near to Hartree-Fock limit [**16**], yield the well-known hydrogen-bond dimer (see Figure 6). This picture can be improved by using correlated wave functions and energies [**17**] (e.g., configuration interaction approach). The computed energy points, sampling the potential energy surface, can be fitted by some analytical expression, within the "guidelines" of some simple physical model. This way we obtained the two-rigid-waters intermolecular potential (two-body) known as MCY [**17**], or, by relaxing the water "rigidity" and allowing vibrational flexibility, the MCYL two-flexible body potential [**18**]. In the same way one could proceed and consider trimers, tetramers, etc. A study of small water clusters [**15**] remind us that even for a few water molecules the cluster can have many stable geometries, all very near in energy, namely that we must face a difficult minimization problem. It should be noted that the cluster study provides us also with very interesting geometrical arrangements of hydrogen bonded structures, leading to "clathrate-type" complexes. Obviously one should not build up the cluster size indefinitely, not only because it would become computationally impractical and eventually impossible, but mainly because it would be the logically incorrect approach to a problem whose answers are to be found in the understanding of statistical fluctuations.

It is time to move to the second submodel, namely equilibrium statistical mechanics, limiting ourselves to two main techniques: Metropolis Monte Carlo **[19]** and molecular dynamics **[20]**. In the former, N molecules are placed in a volume V at a given temperature T and many (millions) geometrical configurations are randomly generated, their energy computed with the interaction potentials obtained from quantum mechanics. Properties are statistically averaged in the Boltzmann sense. Periodic boundary conditions are imposed on the N particles. In molecular dynamics again we place N molecules in a given volume V and let it interact according to the quantum-mechanically derived potential; for each particle the Newton's equations of motion are solved, namely we compute molecular trajectories, i.e., each molecule will modify its translational and rotational motion reacting to the field exerted by the $N-1$ surrounding molecules. Only motions corresponding to a very small temperature range (ideally *one* temperature value) are considered. The motions are discretized in time steps generally of 0.5×10^{-15} seconds for rigid molecules and 0.125×10^{-15} seconds for vibrating molecules.

Figure 7 reports some of the Monte Carlo results **[18, 21–23]**. At the top of the figure we compare the computed and the experimental scattering intensity for liquid water at room temperature using either X-ray (left) or neutron beam (right), comparing simulations at the two-body level with those which include three- and four-body corrections. As we can see, the latter brings about a marginal improvement to an already good match. However, for the enthalpy the many-body corrections are very important. Thus, static and structural properties are expected to be well described at the two-body approximation (and even better agreement can be obtained using well calibrated *effective* potentials) but this might not be the case for dynamical properties.

The latter type of properties can be subdivided into two groups, namely those relating the motions (translations and rotations) of one water molecule within its solvation cell **[24]** and those describing the behavior of many water molecules **[25]** within a box of size L and with wavenumber $K = 2\pi/L$.

From Figure 8 we learn that *rotational motions* (angular correlation functions, ACF) are nearly five times faster than the *translational motions* (velocity correlation functions, VCF) for rigid molecules **[24]**. These motions and the corresponding orientational correlation functions are at the origin of the infrared and Raman spectra for liquid water. The computed spectra

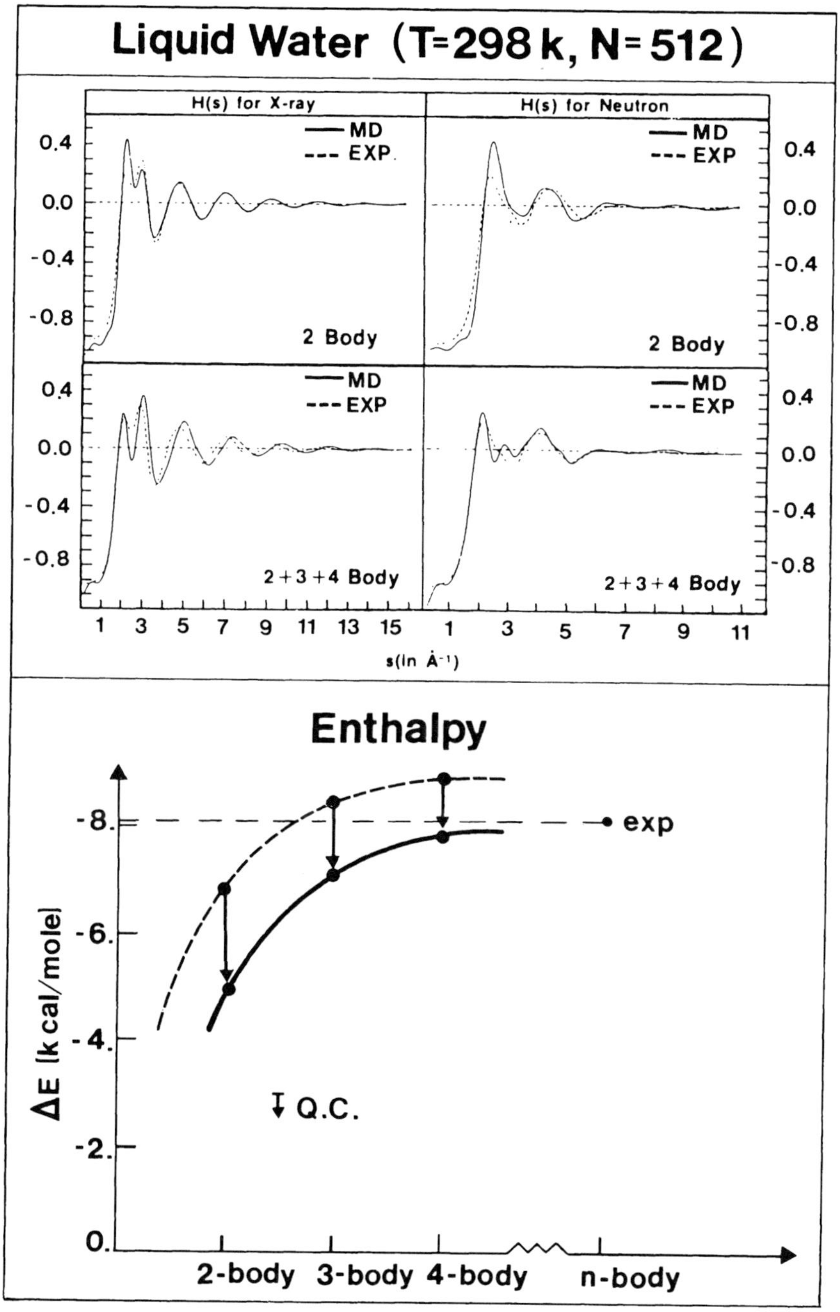

FIGURE 7. Top: X-ray and neutron-beam spectra of liquid water at room temperature either from experiment or from simulations with or without three- and four-body corrections. Bottom: Monte Carlo enthalpy of liquid water from two-, three-, and four-body potentials.

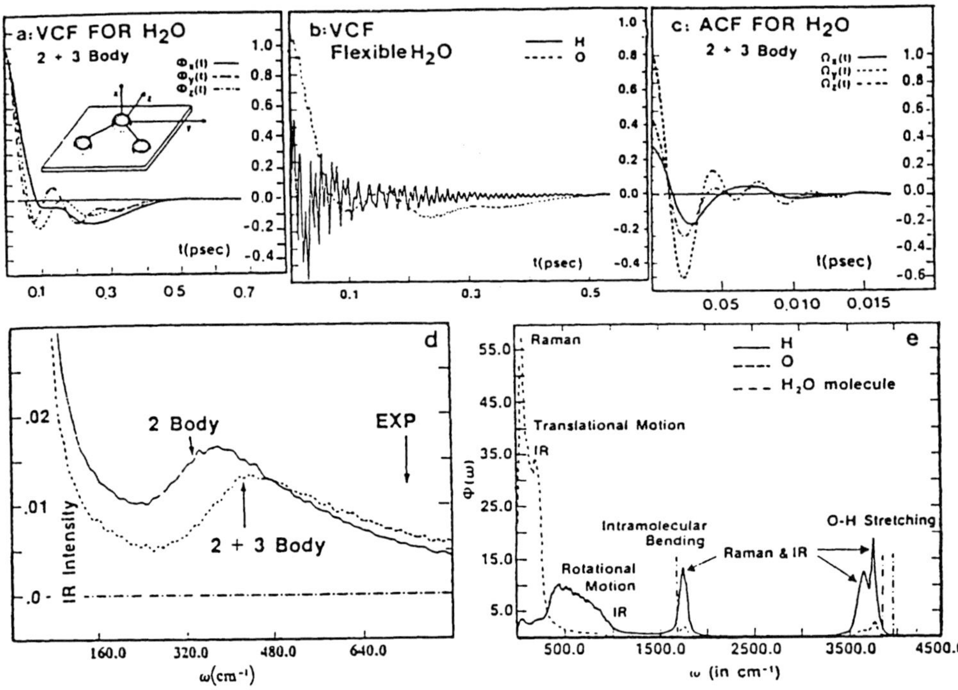

FIGURE 8. Top: angular, ACF, and translational, VCF, velocity correlation functions for liquid water at room temperature from molecular dynamics simulations with three-body corrections and with flexible water (middle). Bottom: computed IR and Raman spectra.

are reported at the bottom of Figure 8; note that the vibrational IR spectra is incorrectly shifted to low values (bottom left insert) while the intramolecular motions (bottom right insert) are computed in remarkable agreement with experiments [**18**]. For the intramolecular motions we have also reported equivalent spectra for the gas phase. The spectral shifts, comparing gas to the liquid, are very accurately reproduced in our simulation. In Figure 9 we report the dynamical structure factor, a "collective" property, computed from molecular dynamics and from hydrodynamics (namely, where the value of K approaches zero). Note how the shape of the structure factor for $K = 0.319$ is not too dissimilar from the one predicted from hydrodynamics [**25**].

The simulated results presented in Figures 7, 8, and 9 show that within the "global simulation spirit" we can indeed progress from nuclei and electrons to a few water molecules, to the hydrodynamic limit. Let us now progress into the area hitherto exclusively reserved for fluid dynamics, i.e., for solutions of Navier-Stokes equations. We ask the question: can we trust the Navier-Stokes equations for *very small systems* of a few hundred *Å* size and for events

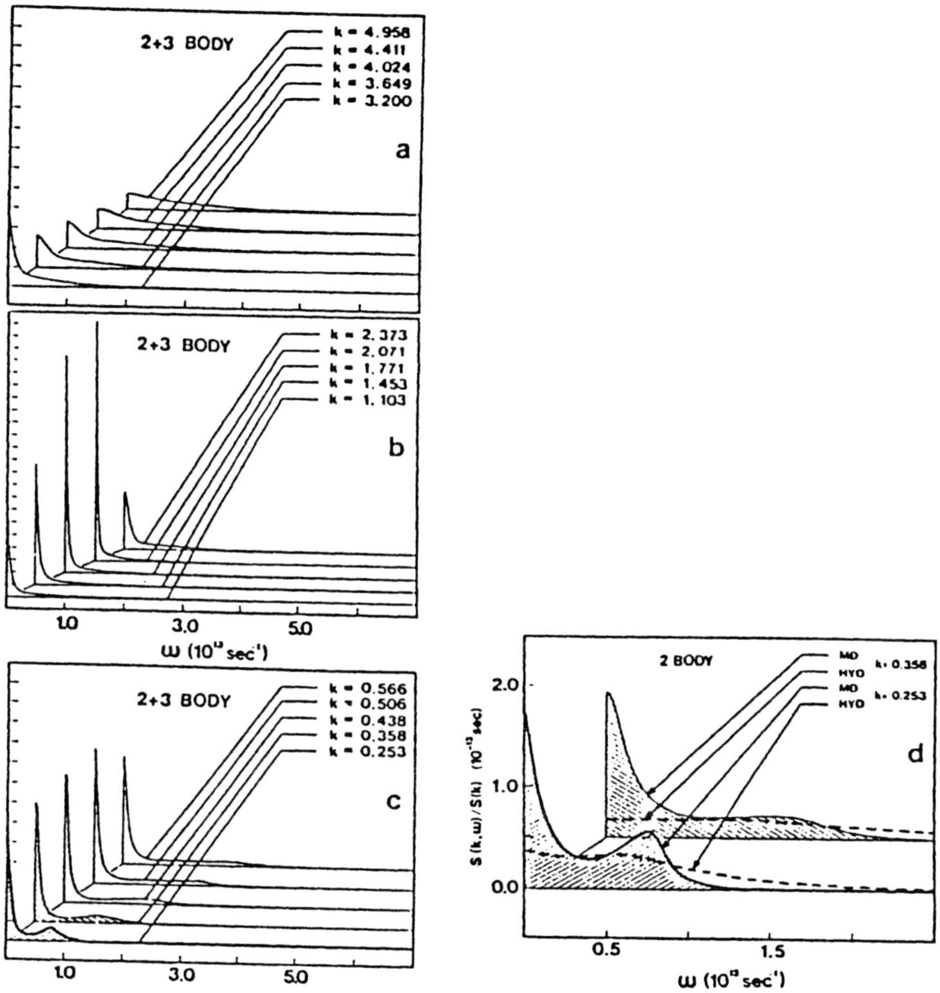

FIGURE 9. Dynamical structure factor from molecular dynamic simulations (left three inserts) and from hydrodynamics (right).

having a lifetime in the *pico-* or *nanosecond* range? Alternatively stated, the question concerns the description of systems generally larger than those analyzed by molecular dynamics but much smaller relative to fluid dynamics. We answered these basic questions by performing molecular dynamic-type computations with molecules or atoms interacting via intermolecular potentials, but in an open (nonequilibrium) system and with those boundary conditions which are typical in fluid dynamics. For example, in Figure 10 we consider a flow past a plate [**26**] (upper insert) or past a cylinder [**27–29**] (bottom insert). Our simulations were performed on $\sim 10^5$ particles for events in the picosecond to nanosecond range over a few hundred $\AA$ scale. We have reported that

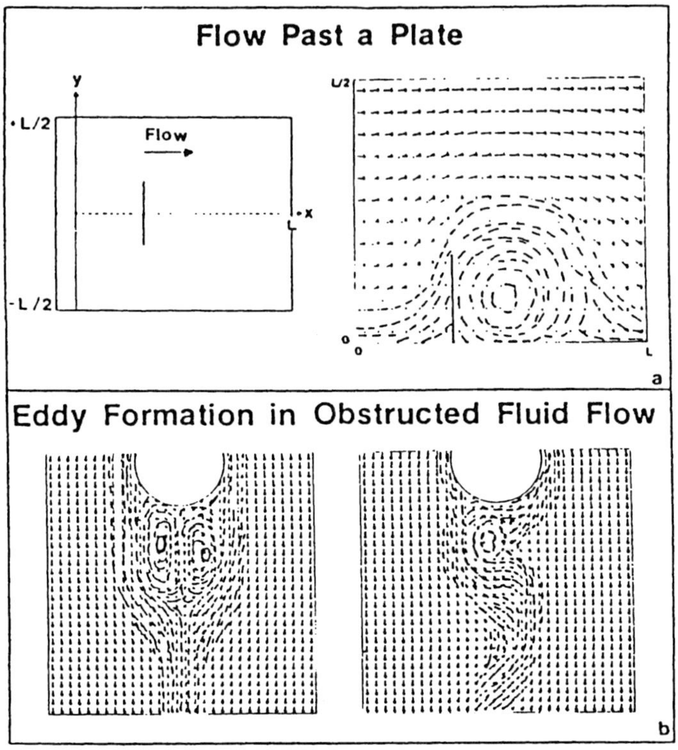

FIGURE 10. Top: schematic of a flow past a plate (left) and visualization of velocity field for steady state. Arrow lengths are scaled as the square root of the ratio of individual cell velocity to maximum cell velocity. Bottom: velocity field for flow past a cylinder.

such systems amazingly well and very accurately scale from results and laboratory experiments of systems at the cm^2 size and with a lifetime of seconds! For example, the Raynolds number at which we observe wakes past the cylinder is about 25 ± 10, i.e., essentially equal to 35, the Raynolds number for which laboratory data observed the same behavior [**27**].

Equally gratifying are the results of Figure 11 where we report on Benard type flow obtained by considering 200,000 argon atoms [**30**]. This rather astonishing accomplishment, even if preliminary since we have not obtained true Benard's patterns, can be appreciated by comparing our simulation with Benard patterns observed in laboratories at a trillionth larger scale. We returned to the problem focusing on the vertical boundaries since we noticed that the particles, when reflected, entered the fluid with excess energy, namely in a "destructive" way, and thus prevented the formation of convective flows. In Figure 12 we report such flows, which exhibit the correct sinusoidal (rather than linear) temperature profile. For this and other details, see [**31**].

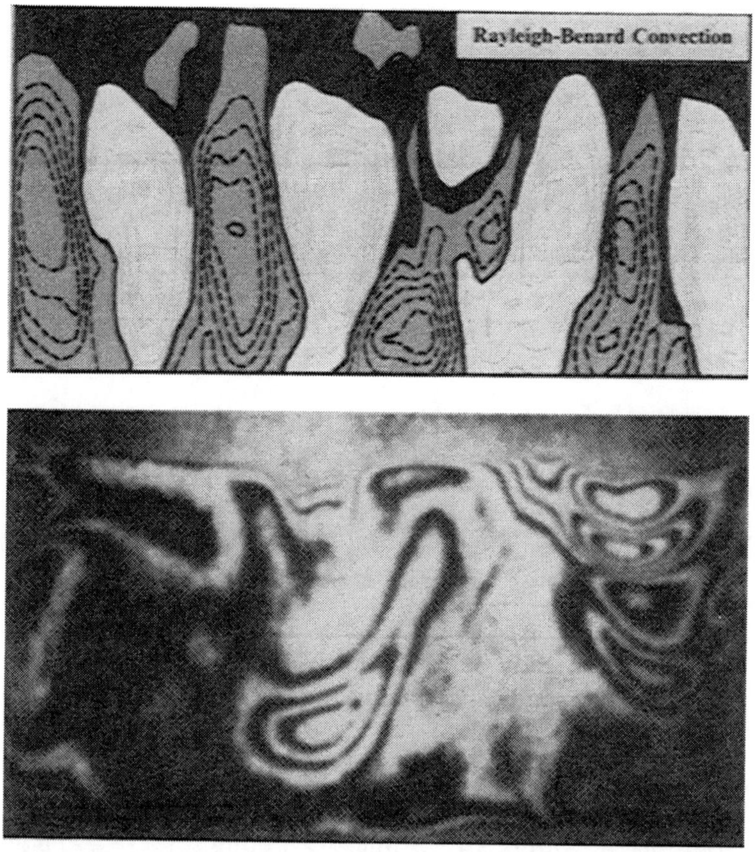

FIGURE 11. Top: simulated Benard-flow obtained from 200,000 neon atoms. Bottom: laboratory Benard-flow taken from a film with the kind permission of P. Bergé and M. Dubois, C. E. N. Saclay, Gif-sur-Yvette, France.

In closing we note that the flow and the "pipes" in our simulations are about 1500 Å large and are thus of definitive biochemical interest. These results are exciting but, in the context of this paper, the most relevant aspect is that indeed our "global method" is *verified*, starting from systems with a few nuclei and electrons up to vastly larger systems previously described only with fluid dynamics.

3. Water as a solvent: Simple ions, DNA, proteins and gramicidin as "solute". The above study on a liquid is cast into the form of an example of the "global simulation" approach; to be sure we did not work in our laboratory on water so as to have a show test case for the global simulation methodology. Our interest in water relates to our work in biophysics and biochemistry: as is well known, water is the most abundant solvent on earth and *the* medium to carry life.

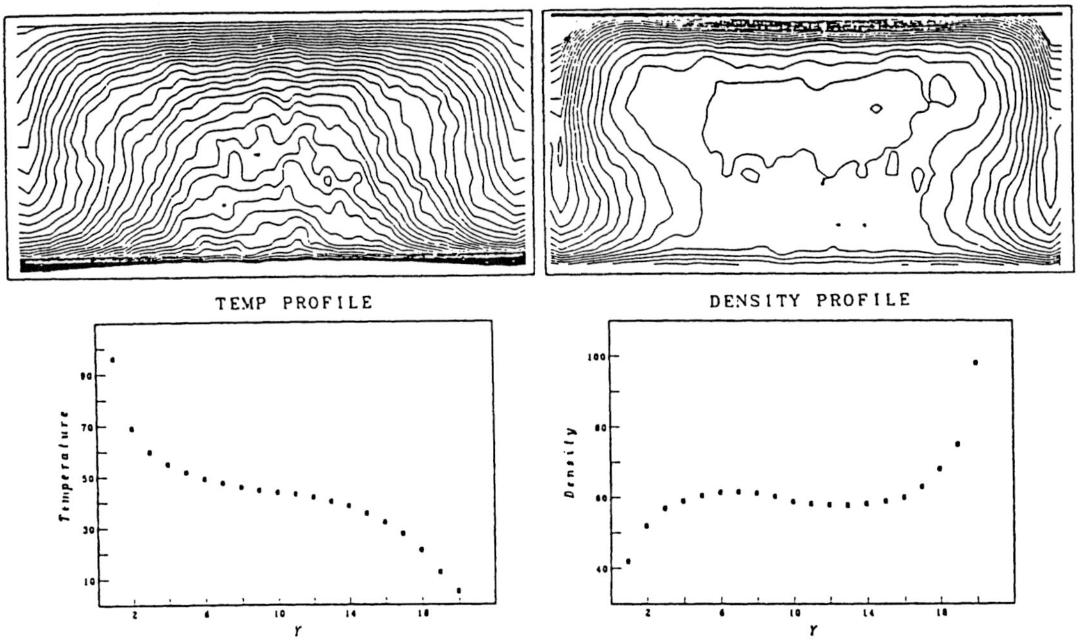

FIGURE 12. Steady state Rayleigh-Benard convective flow from J. Givens and E. Clementi.

Water, as a solvent, is a network of hydrogen bonded molecules in which rearrangement cost is low in terms of internal energy but not in terms of entropy variations. Figure 6 is revealing; for a long time free energy of hydrations was a difficult task, but recently this has been changing because of a perturbational technique with a coupling parameter [**32**] aimed at determining the free energy difference between the solutes. In our laboratory we have extended these simulations to large energy differences [**33**], like those between ions like Li^+, Na^+, K^+, F^-, and Cl^-. As shown in Table I, the computed energies are in relatively good agreement with experimental values (given in parentheses) and the relative trends of positive versus negative ions correlate excellently [**34**]. The values reported in Table I have been obtained using two-body potentials. Introduction of three-body corrections bring the simulated free energy even closer to the experimental values. A simulation performed for Li^+ yielded a value of 114 ± 8 kcal/mol.

In Figure 13 we condense aspects of our simulations of water solvating biomolecules. Specific iterations of water molecules and local hydrogen bond rearrangements at the active sites are a concomitant aspect not here discussed (for papaine see [**35**], for carbonic anhydrase see [**36**], and for lysozyme see [**37, 38**]). The top three insets (left to right) show the water network in the *B*-DNA grooves as first reported by our Monte Carlo simulations [**39–41**], later confirmed by Dickersen et al. [**42**], and more recently reconfirmed by our molecular dynamics study where the interest was on the time-dependency of

TABLE 1

Free energy (in Kcal/mol) of hydration from Monte Carlo [33] and laboratory data.

Ion:			
Li^+	140.9 ± 0.06^1	118.1^2	123.5^3
Na^+	109.6 ± 1.8^1	90.6^2	98.6^3
K^+	76.7 ± 2.7^1	73.1^2	80.8^3
F^-	99.8 ± 2.1^1	110.7^2	103.8^3
Cl^-	56.7 ± 0.6^1	81.5^2	75.8^3

[1]See [**33**]. [2]See [**34a**]. [3]See [**34b**].

the water motions [**43**, **44**]. Indeed, our simulation of the dipole correlation functions as computed is in good agreement with recent relaxation experiments by Lindsay et al. [**45**]. The middle and right insets report on our computed positions for the counter-ions in *B*- and *Z*-DNA, respectively, at a concentration of one counter-ion for phosphate group. We have stressed [**40**, **41**] that a study of DNA solvation without explicit account of counter-ions (and vice-versa) is meaningless; in addition, we should note that the exact location of the counter-ions around DNA is dependent on the counter-ions concentration (or excess).

In the second row of Figure 13 we report a view of myoglobin, a cross-section of myoglobin unit cells with those water molecules identified in X-ray diffractions and the much larger population of water molecules located by preliminary Monte Carlo simulations. This work initiated in collaboration with Prof. Parak is an attempt at: (1) refining the position of about 165 water molecules suspected from the X-ray diffractions, and (2) determining the distribution of about 300 water molecules not likely accounted for presently. Work is still at a very early stage [**46**].

A related simulation is the determination of the number and positions of the water molecules hydrating a crystal of proflavine deoxydinucleotide phosphate complex [**47**, **48**]. The oxygen-atoms diffraction pattern (left inset) obtained experimentally [**49**] agrees well if we simulate with Monte Carlo 61 or 66 water molecules per unit cell (middle inset), but not 50 ± 2 water molecules as assumed in the laboratory work [**49**].

Finally, in the bottom three insets of Figure 13 we consider the transport of K^+-ions through a transmembrane channel, gramicidin *A*, using molecular dynamics. At the left we report on the K^+ gramicidin interaction potential obtained by *ab initio* computations of K^+ with the entire gramicidin. With our *l*CAP-3 computer, each interaction energy costs about four hours of computer time (however, the initial one cost 11 hours); we recall that the system contains 256 atoms and 1050 electrons and is thus the largest system studied, up to now, by *ab initio* quantum-mechanical methods [**50**, **51**]. The middle

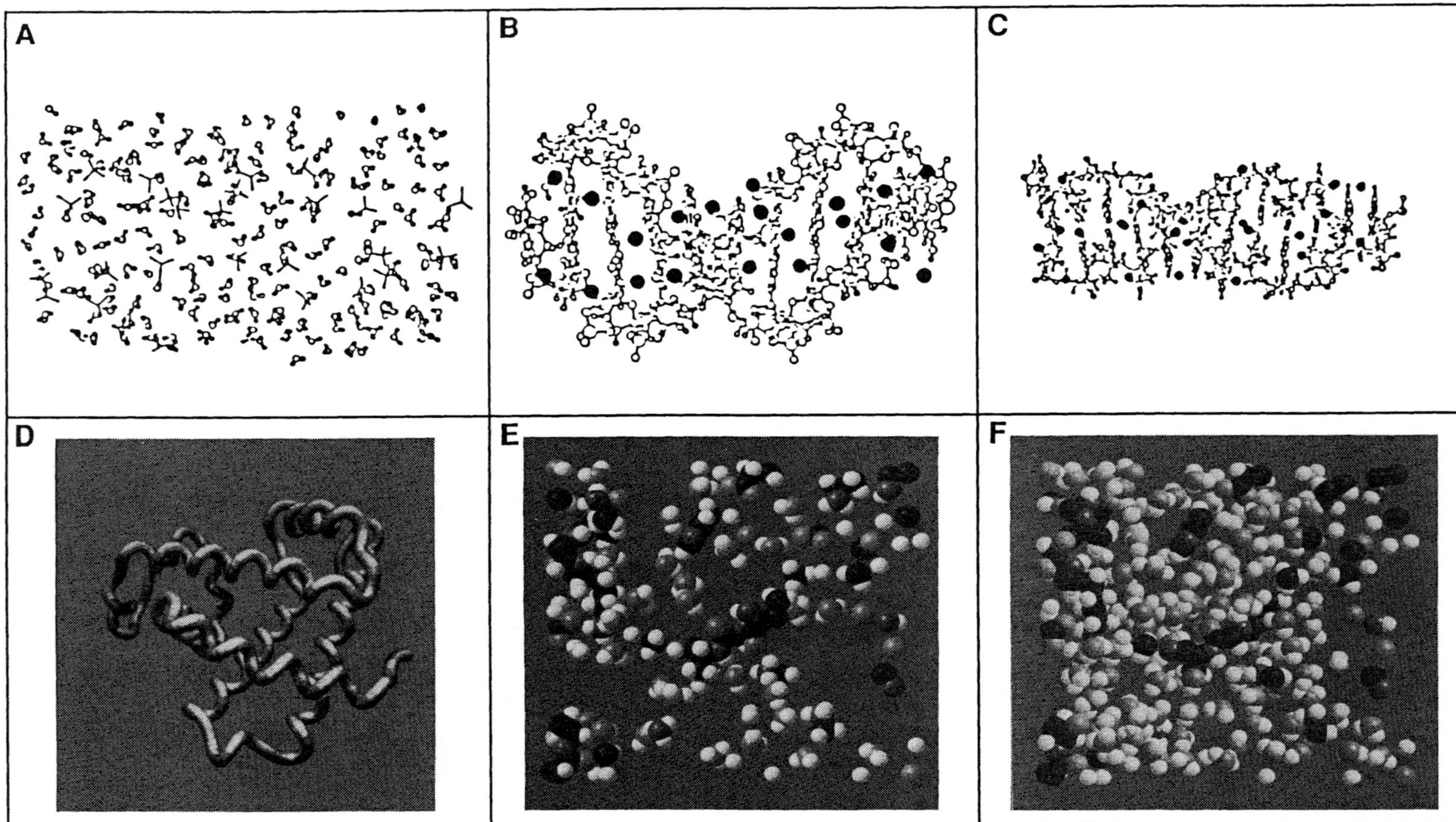
A
B
C
D
E
F

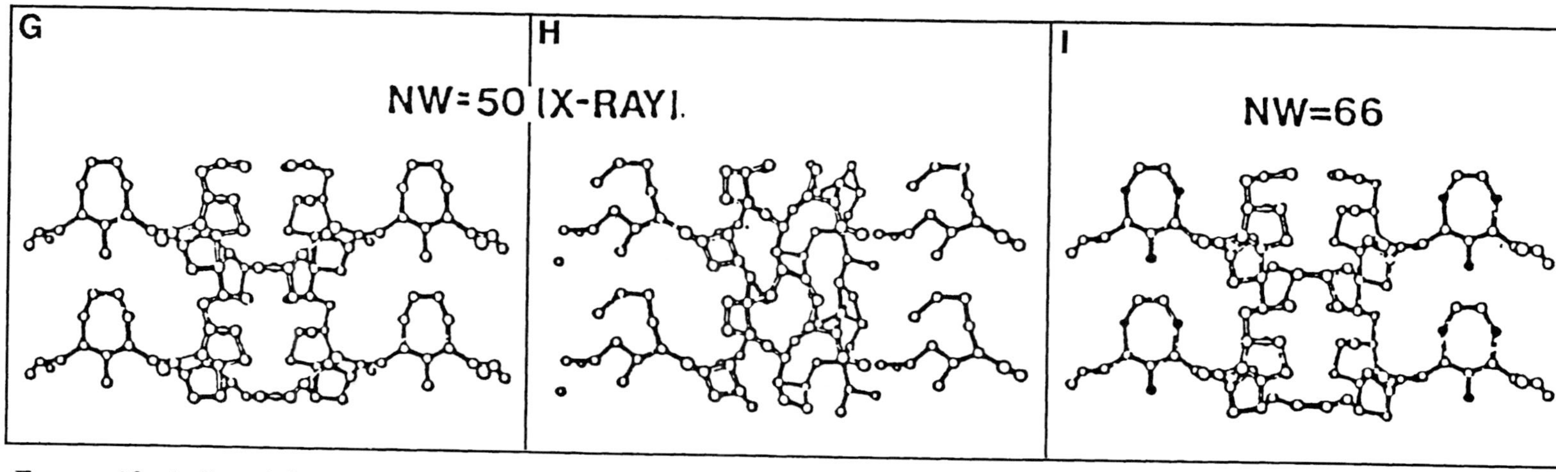

FIGURE 13. A, B and C: Water network in *B*-DNA grooves, k^+-ions positions in *B*-DNA and in *Z*-DNA. D, E and F: Ribbon model of X-ray crystal structure for myoglobin, and view of myoglobin crystallization molecules from X-ray diffraction and from Monte Carlo simulations. G, H and I: Oxygen-atom from crystallization water from X-ray diffraction data, assuming 66 water molecules (simulated) and 50 water molecules (laboratory data).

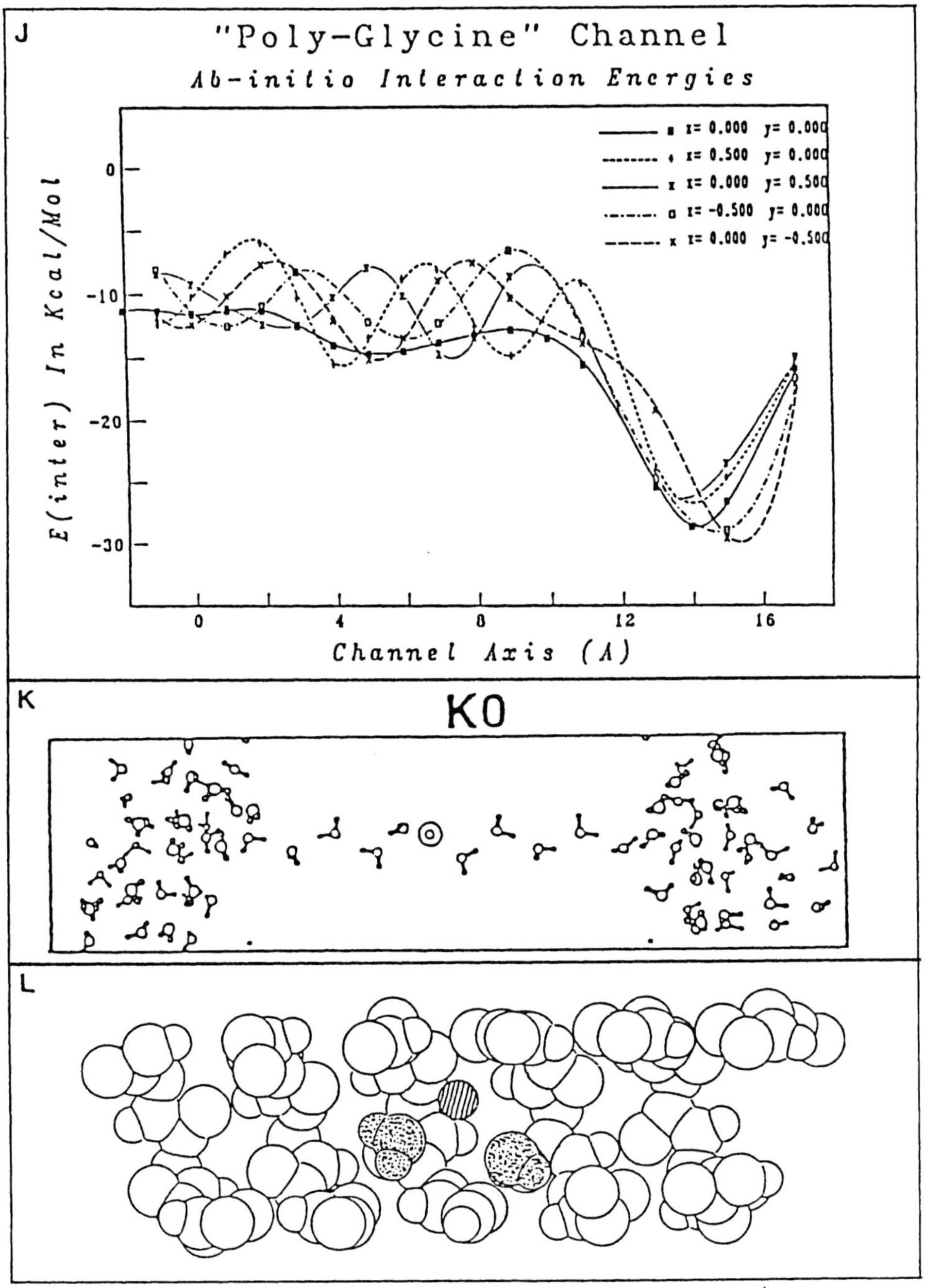

FIGURE 13 (continued). J, K and L: *ab initio* interaction for K^+ with the entire gramicidin channel, Monte Carlo water molecules around K^+ inside the channel, and two water molecules sandwiching K^+ inside the channel from molecular dynamics simulations (frame from computer animated video [**54**]).

inset reports the organization of water molecules outside the channel and inside when an ion is in the middle of the channel [**52**, **53**]. In the right inset we present one frame from the computer animated video [**54**]. It is very interesting to recall that upon application of a voltage differential across the channel a K^+ does not go through but is trapped at the mouth-end unless it is sandwiched in between two water molecules. The latter undergoes a continuous series of vibrations or even rotations, changing the electrostatic interaction with K^+ and thus pushing and pulling it through the channel.

4. Conclusions. We have not mentioned the very obvious point that *l*CAP needs specialized system software. "Master-slave" and "fork-joint" concepts are very familiar and now new, but still much skill is required to provide for an operational system. This aspect of our department's research is discussed in a recently reviewed chapter [**11**].

The "global simulation" methodology is expected to become the standard approach to complex engineering problems, especially to those in which materials are subjected to stress, rapid temperature differentials, and unusual environments.

Because they are related to *l*CAP and to our work, two areas are worth mentioning: biophysics (with biochemistry and bioengineering) and modern organic syntheses of material for future computers. In Figure 14 we report a matrix taken from a rather old work [**55**] where we summarize the pair-potentials we have been and are working on. The assembly of these pair-potentials is, in itself, a supercomputer task. The *use* of the pair-potential will impact profoundly either specific problems (like understanding liquid water) or entire fields (electrochemistry, for example) and again this will require supercomputers, possibly of *l*CAP-type. Presently we have completed the pair-potential for the matrix elements of the first row and for the first and last elements of the second row. Concerning the third row, it is our pleasure to report that the amino-acid, amino-acid potentials have been very recently obtained and about 1200 interaction energies have been computed for the interaction of amino acids, DNA. The amino-acid, amino-acid potential required about 5000 *ab initio* computations; the amino acids-DNA will require about two to three times as much. The amino-acid, amino-acid pair-potential library is complemented by the equivalent library for amino acid-water [**56–58**].

As is known, today the potentials needed in Monte Carlo and in molecular dynamics to simulate protein's motions and active site reactivity and to refine or even determine the protein structure, are fitted from experimental data, particularly X-ray data. The number of parameters needed are about three per pair-potential. Even if most atoms can be grouped under a given type of pair-potential (what we call "class", see [**56–58**]), still we have to determine hundreds of parameters. The problem is that there are not sufficient numbers of mutually consistent experiments to obtain a reliable fit. Even for the

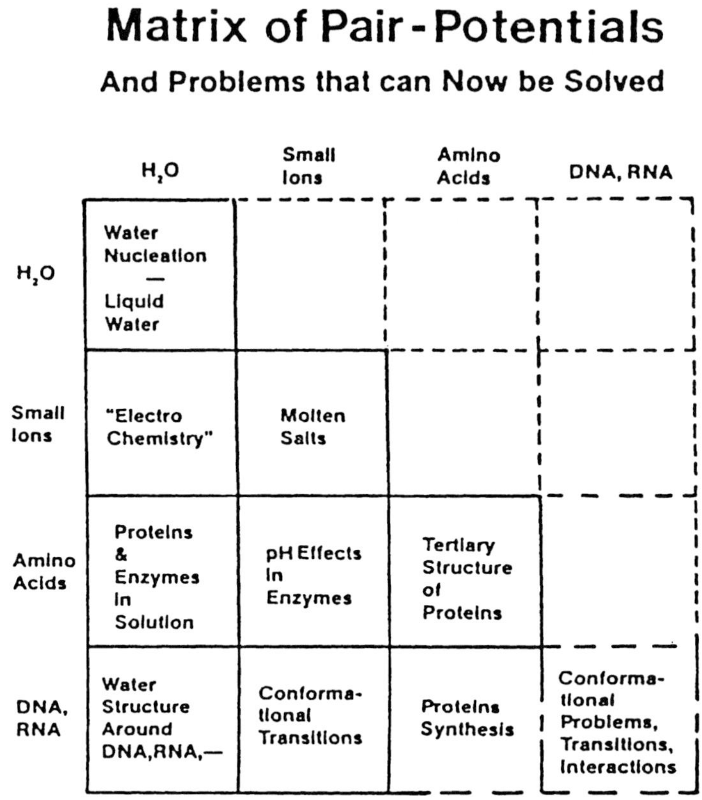

FIGURE 14. Matrix of pair-potentials

water-water interactions, despite the hundreds of experiments, ambiguities have plagued the literature with quantatively unreliable pair potentials. This problem only gets more serious if we consider the pair-potentials for the interaction of proteins with DNA.

Let us discuss, as an example, attempts to obtain protein structure from computer simulations. For the past several years we have been working in collaboration with Prof. Scheraga's groups to predict the three-dimensional structures of proteins, and in particular, the structure of interferon using the technique of direct energy minimization. Human leukocyte interferons (IFN-α) are antiviral proteins of 155 or 156 amino acid residues, which are synthesized and secreted by leukocytes in response to several types of viral or bacterial infection [**60**]. At least 15 different IFN-α's have been identified and sequenced [**61**, **62**]. Attempts have been made to determine the regular backbone structure with the general agreement that the protein is rich in α-helix with a small amount of β-sheet [**63**].

In our work on interferon we have concentrated on building the residues towards the C-terminus end of the protein. A previous publication [**64**] had reported the building of three polypeptide fragments of 27, 9, and 8 residues

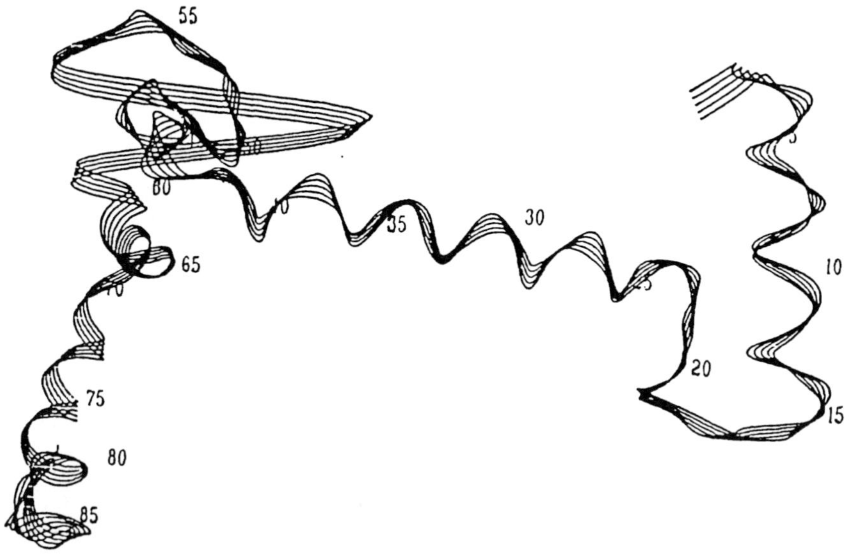

FIGURE 15. Ribbon representation for the computed lowest energy conformation of the 98 residue fragment of interferon. The labels are not actual residue numbers, but are relative to the N-terminus end of the peptide fragment.

in length. We have continued this work, eventually minimizing a 98 residue fragment of the interferon, encompassing residues numbered from 52–149.

This 98 residue fragment was built by combining two fragments, one of 47 residues and the second of 51 residues. Both fragments exhibited a good deal of α-helix content, the N-terminus fragment being somewhat of a straight α-helix, and the C-terminus fragment containing in part a β-sheet surrounded by part of a globular packing followed by an α-helix.

A ribbon model of the lowest energy minimum for the 98 residue fragment is given in Figure 15. The labels in this figure correspond to relative residue numbers beginning from the N-terminus residue. The picture clearly depicts the presence of local secondary structures such as α-helix and β-sheet. The figure also clearly shows the fact that the two fragments have not folded, but instead have been simply joined together. Upon closer interpretation of the structure, it also becomes obvious that certain hydrophilic residues such as aspartic acid, glutamic acid, and threonine, labeled in Figure 15 as residue numbers 43, 45, 55–57, and 62–64, are buried in the inside of the protein, away from the surface. This is contrary to chemical and physical expectations of protein structure, obviously pointing out limitations in the build-up procedure for predicting structures of globular proteins.

Limitations can be categorized into three areas. In the first category is the effect of the solvent. In this study we have chosen to include the solvent effect based on a shell model by Gibson and Scheraga [**65a**], and modified slightly for our work. However, the model suffers from the limitation that few reliable experimental data exist on prototype molecules in order to properly parametrize the model. Therefore, the Gibson and Scheraga caveat, "... present version of the algorithm must be regarded with caution" [**65b**] is in order, as is also obvious from our results.

The second area in which the above procedure fails is in the poor representation of charged amino acids, which were assumed essentially as neutral. We know that the charged amino acids play an important role in determining the interactions with the solvent as well as contribute to the overall structure of the molecule.

The third area is the neglect of long-range interactions. As mentioned previously, the build-up procedure assumes that the three-dimensional structure of protein is dominated by short-range interactions. From our results this statement appears to be true if one is interested in *secondary* structures. However, for predictions of full *three-dimensional* structures, the effect of long-range interactions should be important. Future efforts to alleviate these shortcomings are discussed below. Here, however, we stress that this type of experimentation is truly important. Notice that simulations on small polypeptides would likely tell us little about the last two types of problems.

Molecular Dynamics (MD) might provide an alternative method for studying the problem of determining protein structure. MD techniques can be used to simulate protein structure in two ways: first, to study detailed properties of particular conformations and, second, to study conformational changes in proteins.

Conformational changes in a protein can be simulated with molecular dynamics techniques by solving the equations of motion, incorporating a coupling to a constant temperature bath, and using relatively large time steps. The procedure is briefly as follows. After equilibration, the temperature of the simulation can be raised by coupling to a temperature bath. After equilibration at the raised temperature, the temperature can again be lowered, and the simulation equilibrated. This process may be repeated several times. Conformational changes can be identified by studying the fluctuations in the kinetic, potential, and total energy. A conspicuous change in the total energy possibly signifies a conformational change.

The conformational changes induced by the temperature bath in the molecular dynamics simulations should serve to refine the results of the energy minimized structure. This refined structure, or structures, can then be subject to further molecular dynamics simulation to understand detailed properties of the protein molecule.

Solvent can be explicitly included into MD simulations, provided the potentials are already derived. The use of MD simulations with and without the solvent included provides a method to understand the influence of the solvent on the solute, and vice-versa.

Once the secondary structures are determined, either by energy minimization or MD, it is necessary to find the protein topology which represents trying to arrange together different local structures into the most favorable three-dimensional structure. Our strategy is to implement a kind of expert system to help in providing possible folding pathways to be analyzed by numerical techniques, following similar published work [**66**]. The idea is to take advantage of the huge amount of information that is available in the protein data banks, and to design a program which is capable of using statistical methods to decide upon the best alternative conformations in which to combine the local secondary structures into supersecondary structures.

These supersecondary structures should be good starting points for investigating final folding of the protein by additional simulation, either using energy minimization or molecular dynamics. Further simulation should result in low energy minima of three-dimensional structures.

It seems clear from the above that the problem simulating the three-dimensional structure of protein is still open, but perhaps we can see some light at the end of the tunnel. First, one will have to "clean up" the methodological problems mentioned above and then address the problem of the "reliability" of present-day empirical potentials [**67**, **72**] and, of course, improve on the *ab initio* one. Notice that for water-water the *ab initio* potentials are at a rather accurate level; for water-DNA we are now recomputing the previous potentials [**73**] at the Hartree-Fock level and even for the amino-acid, amino-acid potentials [**59**] we have included the superposition correction for the small basis set. We refer the reader interested in this problem to a recent paper by H. Scheraga [**74**]. Clearly, the many-dimensional space to be searched in order to find the minima corresponding to the stable tertiary structure can be substantially reduced if we impose experimental constraint as done, for example, in K. Wuthrich's group. The importance of this simplication—as a learning step—is stressed.

In Figure 16 we plot and extrapolate the expected miniaturization of *electronic components* (given in a number of atoms, 10^N, rather than in units of volume) and the molecular complexity one will be able to deal with quantum- and statistical-mechanical *simulations.* Each curve is drawn three times, since we wish to give lower and upper error bounds on our extrapolations. Notice that the lower bound for molecular simulations (the year 2000) is nearly met today; for example, in an unpublished study at our laboratory, we are expanding from $\sim 200,000$ atoms to $200,000 \times m$ atoms (with $15 < m < 25$) for the Benard problem. Therefore, Figure 16 is somewhat conservative, possibly because it is by now six to seven years old. The molecular simulations we are particularly interested in in Figure 16 are for *electronic devices at the molecular level.* We must recall and stress that modern organic chemistry more and more deals with the problem of synthesizing molecular structures for chemicals with *very specific functionality*, like for transport, conduction of electrons or protons, optical effects, pharmacological effects, etc. If we draw a simple and classical electric circuit with resistances, impedances, condensers, and wires, etc., we can ask the question, "which chemical system will repre-

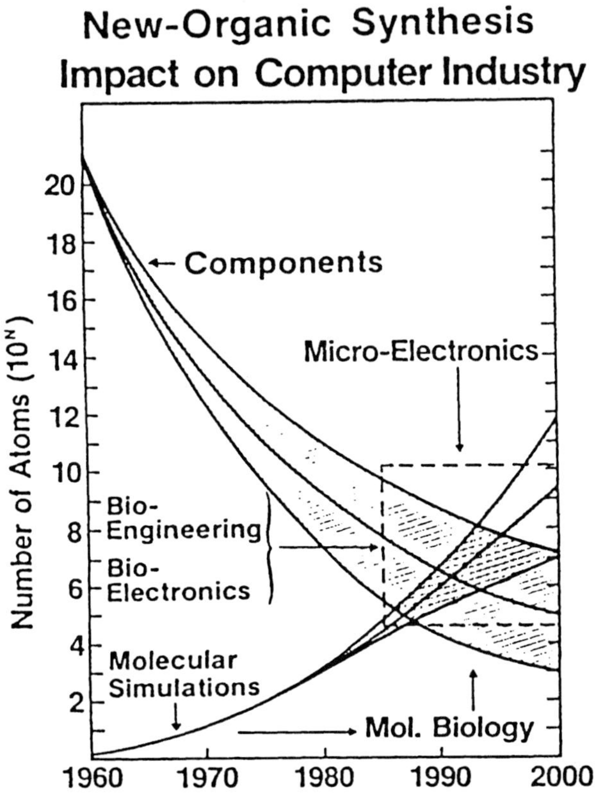

FIGURE 16. Expected miniaturization of electronic components.

sent the *resistance*, or the *condenser*, or the *wire*?" In time, today's obvious but naive answers (like *saturated hydrocarbons, dimers of planar molecules with π electron mobility and charge transfer*, and *unsaturated olephins with proper terminals*) will become more and more detailed, concrete, and subject to theoretical and laboratory verification. Slowly complicated circuits will have optimal "organic molecules analogs" [75]. Likely, the syntheses of these "exact" chemicals will be obtained by bioengineers using new engineered proteins and enzymes. One prediction is quite easy and almost certain: when vastly different fields merge, as indicated in Figure 16, then the chances for profound innovations are maximal.

Acknowledgments. The interaction with Professors H. Scheraga and K. D. Gibson (Cornell University) and of Professor F. G. Parak (University of Munster) is most gratefully acknowledged.

REFERENCES

1. E. Clementi, J. Phys. Chem. **89** (1985), 4426.
2. E. Clementi, S. Chin, and D. Logan, Israel J. Chem. **27** (1986), 127.
3. E. Clementi and D. Logan, *Special purpose computers* (B. Alder, ed.), Academic Press, New York, 1987.

4. D. Logan, J. Saarinen, and E. Clementi, *l*CAP/3090: *Genesis and evolution of a parallel processing system*, Proc. Third Internat. Conf. on Supercomputing, Vol. 1, Boston, 1988, p. 79.
5. H. L. Nguyen, H. Khanmohammadbaigi, and E. Clementi, J. Comput. Chem. **6** (1985), 634.
6. E. Clementi, G. Corongiu, and J. Detrich, Comput Phys. Comm. **37** (1985), 287.
7. L. Domingo and E. Clementi, IBM Technical Report KGN-17, 1985.
8. E. Clementi and G. Corongiu, *Advances in biophysics* (A. Wada, ed.), Elsevier, Limerick, 1985, p. 75.
9. E. Clementi, D. Logan, V. Sonnad, Z. Christidis, and A Capotondi, *Computers in mechanical engineering*, Vol. 5, No. 3, 1986.
10. G. Corongiu, J. Detrich, and E. Clementi, IBM Technical Report KGN-170, 1987.
11. E. Clementi and S. Chin, eds., *Biological and artificial intelligence systems*, ESCOM Science Publishers B. V., The Netherlands, 1988. (Review of the architecture of *l*CAP-1, *l*CAP-2, *l*CAP-3, and *l*CAP-3090 and of the system software and applications.)
12. E. Clementi and H. Popkie, J. Chem. Phys. **57** (1972), 1077.
13. W. S. Benedict, N. Gailer, and E. K. Plyler, J. Phys. Chem. **24** (1956), 1139.
14. R. J. Bartlett, I. Shavitt, and G. D. Purvis, J. Chem. Phys. **71** (1979), 281.
15. K. S. Kim, M. Dupuis, G. C. Lie, and E. Clementi, Chem Phys. Lett. **131** (1986), 451.
16. H. Popkie, H. Kistenmacher, and E. Clementi, J. Chem. Phys. **59** (1973), 1325.
17. O. Matsuoka, E. Clementi, and M. Yoshimine, J. Chem. Phys. **64** (1976), 1351.
18. G. C. Lie and E. Clementi, Phys. Rev. (4) **33** (1986), 2679.
19. N. Metropolis, A. W. Rosenbluth, M. N. Rosenbluth, A. H. Teller, and E. Teller, J. Chem. Phys. **21** (1953), 1087.
20. B. J. Alder, J. Chem Phys. **40** (1964), 2724.
21. G. C. Lie, E. Clementi, and M. Yoshimine, J. Chem. Phys. **64** (1976), 2314.
22. E. Clementi and G. Corongiu, Internat. J. Quant. Chem. Symp. **10** (1983), 31.
23. J. Detrich, G. Corongiu, and E. Clementi, Chem. Phys. Lett. **112** (1984), 426.
24. M. Wojcik and E. Clementi, J. Chem. Phys. **85** (1986), 3544.
25. ———, J. Chem. Phys. **85** (1986), 6085.
26. L. Hannon, G. C. Lie, and E. Clementi, J. Sci. Comput. **1** (1986), 145.
27. D. C. Rapaport and E. Clementi, Phys. Rev. Lett. **57** (1986), 695.
28. L. Hannon, G. C. Lie, and E. Clementi, Phys. Lett. A **119** (1986), 174.
29. L. Hannon, G. C. Lie, E. Clementi, and V. Yakhot (to appear). (See IBM Kingston Technical Report KGN-128, 1987.)
30. L. Hannon, G. C. Lie, and E. Clementi (to appear).
31. J. Givens and E. Clementi, IBM Technical Report KGN-119 (July 1988); J. Statist. Physics. (submitted).
32. T. P. Lybrand, I. Ghosh, and J. A. McCammon, J. Amer. Chem. Soc. **107** (1985), 7793; W. L. Jorgensen and E. Ravimohan, J. Chem. Phys. **83** (1985), 3050; P. A. Bash, U. C. Singh, R. Langrige, and P. A. Kollman, Science **236** (1987), 564..
33. M. Migliore, G. Corongiu, E. Clementi, and G. C. Lie, J. Chem. Phys. **88** (1988), 7766.
34a. R. Gomer and G. Tryson, J. Chem. Phys. **66** (1977), 4413.
34b. H. L. Friedman and C. V. Krishnan,, *Water: A comprehensive Treatise, Vol. 3*, (F. Franks, ed.), Plenum Press, New York, 1973.
35. G. Bolis, M. Ragazzi, D. Salvaderi, D. R. Ferro, and E. Clementi, Internat. J. Quant. Chem. **14** (1978), 815.
36. E. Clementi, G. Corongiu, B. Jonsson, and S. Romano, FEBS **100** (1979), 313.
37. E. Clementi, G. Ranghino, and R. Scordamaglia, Chem. Phys. Lett. **49** (1978), 218.
38. G. Ranghino and E. Clementi, Gazz. Chim. Ital. **108** (1978), 157.
39. G. Corongiu and E. Clementi, Biopolymers **20** (1981), 551.
40. ———, Biopolymers **20** (1981), 2427.
41. E.Clementi and G. Corongiu, J. Biol. Phys. **11** (1983), 33.
42. G. G. Price, U. Heinemann, S. Chandresegaran, Lou-Sing Ken, M. L. Kopka, and R. E. Dickerson, Science **238** (1987), 498.
43. K. N. Swamy and E. Clementi, Biopolymers **26** (1987), 1901.
44. ———, Biopolymers **28** (1989), 1497.

45. S. Lindsay, Biopolymers **26** (1987), 439.
46. For previous works see F. Parak, *Methods in enzymology* (L. Packer, ed.), Vol. 127, Academic Press, 1986, p. 196; H. Hartmann, W. Steigemann, R. Reuscher, and F. Parak, Eur. Biophys. J. (1986).
47. K. S. Kim and E. Clementi, J. Amer. Chem. Soc. **107** (1985), 227.
48. ———, J. Phys. Chem. **89** (1985), 3655.
49. H. S. Shieh, H. M. Berman, M. Dabrow, and S. Neidle, Nucleic Acids Res. 8, 85 (1980); S. Neidle, H. M. Berman, and H. S. Shieh, Nature **288** (1981), 129.
50. R. Gomperts and E. Clementi, IBM Research Report KGN-107 (February 1987).
51. ———, IBM Research Reports KGN-158 and KGN-159 (September 1987).
52. K. S. Kim and E. Clementi, J. Amer. Chem. Soc. **107** (1985), 5504.
53. K. S. Kim, H. L. Nguyen, P. K. Swaminathan, and E. Clementi, J. Phys. Chem. **89** (1985), 2870.
54. *Unity of science, engineering and supercomputing*, a 45-minute computer-animated video with audio commentary in English, French, Italian, Spanish or Chinese (1986); available from IBM-Kingston, Dept. 48B/Mail Station 428, Kingston, New York 12401.
55. E. Clementi, *Computational aspects for large chemical systems*, Lecture Notes in Chemistry, Vol 19, Springer-Verlag, New York, 1980.
56. E. Clementi, F. Cavallone, and R. Scordamaglia, J. Amer. Chem. Soc. **99** (1977), 5531.
57. M. Ragazzi, D. Ferro, and E. Clementi, J. Chem. Phys **70** 1979, 1040.
58. G. Ranghino, E. Clementi, and S. Romano, Biopolymers **22** (1983), 1449.
59. S. Chin, J. J. Perez, M. Re, and E. Clementi, Proc. Third Internat. Conf. on Supercomputing Vol. 1, p. 109, Boston (April 15–18, 1988).
60. P. Lengyel, Ann. Rev. Biochem. **51** (1952,), 251; S. Pestko and S. Baron, Meth. Enzymol. **78** (1981), 3.
61. W. P. Levy, M. Rubinstein, J. Shively, U. Del Valle, C.-Y. Lai, J. Moschera, L. Brink, L. Gerber, S. Stein, and S. Pestko, Proc. Nat. Acad. Sci. U.S.A. **78** (1981), 6186.
62. J. Collins, Symp. Soc. Gen. Microbiol. **35** (1983), 35.
63. D. L. Miller, H.-F. Kung, T. Staehelin, and S. Pestko, Meth. Enzymol. **79** (1981), 3.
64. S. Chin, K. D. Gibson, H. A. Scheraga, and E. Clementi: a library of the minima for the 400 dipeptides is available from the authors. See also K. D. Gibson, S. Chin, M. R. Pincus, E. Clementi, and H. A. Scheraga, *Supercomputer simulations in chemistry*, Lecture Notes in Chemistry (M. Dupuis, ed.), Vol 44, Springer-Verlag, New York, 1986, p. 198.
65a. K. D. Gibson and H. A. Scheraga, Proc. Nat. Acad. Sci. U.S.A. **58** (1967), 420; G. Némethy, Z. I. Hodes, and H. A. Scheraga, Proc. Nat. Acad. Sci. U.S.A. **75** (1978), 5760; Z. I. Hodes, G. Némethy, and H. A. Scheraga, Biopolymers **18** (1979), 1565.
65b. K. D. Gibson and H. A. Scheraga, J. Comput. Comput. Chem. **8** (1987), 826.
66. D. P. Dolata, A. R. Leach, and K. Prout, J. Comput.-Aided Mol. Design **1** (1987), 73; B. Robson, E. Platt, R. V. Fishleigh, A. Marsden, and P. Millard, J. Mol. Graphics **5** (1987), 8.
67. J. Hermans and D. Ferro, Biopolymers **10** (1971), 1121.
68. R. DeSantis, E. Giglio, A. M. Liquori, and A. Ripamonti, Nature **206** (1965), 456.
69. M. Goodman and N. S. Choi, *Peptides* 1968 (E. Bricas, ed.), Wiley, New York 1968, pp. 1–17.
70. S. Lifson, A. T. Hagler, and P. Dauber, J. Amer. Chem. Soc. **101** (1979), 5111.
71. P. Weiner and P. Kollman, J. Comput. Chem. **2** (1981), 287.
72. B. R. Brooks, R. E. Bruccoleri, B. D. Olafson, D. J. States, S. Swaminathan, and M. Karplus, J. Comput. Chem. **4** (1983), 187.
73. R. Scordamaglia, F. Cavallone, and E. Clementi, J. Amer. Chem. Soc. **99** (1977), 5545.
74. H. Scheraga, *Biological and Artificial Intelligence Systems* (E. Clementi and S. Chin, eds.), ESCOM Science Publishers B.V., The Netherlands, 1988, p. 1.
75. M. Dupuis, *Biological and Artificial Intelligence Systems* (E. Clementi and S. Chin, eds.), ESCOM Science Publishers B.V., The Netherlands, 1988, p. 185.

IBM Corporation

Proceedings of Symposia in Pure Mathematics
Volume **50** (1990)

Von Neumann and Neural Networks

JACK D. COWAN

0.1 Introduction. Among John von Neumann's many interests was how the brain works. Indeed, in his last years he was very much preoccupied with the topic. In part he was stimulated by his contacts with Norbert Wiener, and particularly with Warren McCulloch and Walter Pitts, who in 1943 formulated a theory of how networks of model nerve cells or neurons embody "ideas" in the form of logical propositions (McCulloch and Pitts 1943). Von Neumann was very much taken with the McCulloch-Pitts theory and used its formalism in his early work on digital computers, automata theory, and reliability (von Neumann 1956), and in the (incomplete) manuscript of his Silliman lectures on the brain-computer analogy (von Neumann 1958). What follows is a simple account of McCulloch-Pitts networks, a description of von Neumann's contributions, and a review of some of the subsequent developments in the theory of neural networks.

0.2 The McCulloch-Pitts neuron. Let $u_i(t)$ denote the state of the ith neuron at time t. Suppose $u_i = 1$ if the neuron is active, 0 otherwise. Let $\Theta[v]$ be the Heaviside step function, $= 1$ if $v \geq 0$, 0 if $v < 0$. Let time be measured in quantal units Δt, so that $u(t + \Delta t) = u(n\Delta t + \Delta t) = u(n + 1)$. Then the activation of a McCulloch-Pitts neuron can be expressed by the equation:

$$u_i(n+1) = \Theta\left[\sum_j w_{ij} u_j(n) - v_{TH}\right] \tag{1}$$

where w_{ij} is the strength or "weight" of the $(j \to i)$th connection, and where v_{TH} is the voltage threshold. Evidently activation occurs if and only if the total excitation $v = \Sigma_j w_{ij} u_j(n) - v_{TH}$ reaches or exceeds zero.

1980 *Mathematics Subject Classification* (1985 *Revision*). Primary 92A90; Secondary 68M15, 01A60.

Supported in part by Grants from the University of Chicago Brain Research Foundation, and the US Department of the Navy, Office of Naval Research (Grant # N 00014-89J-1099).

Completed at the Santa Fé Institute, Santa Fé, New Mexico.

This paper is in final form and no version of it will be submitted for publication elsewhere.

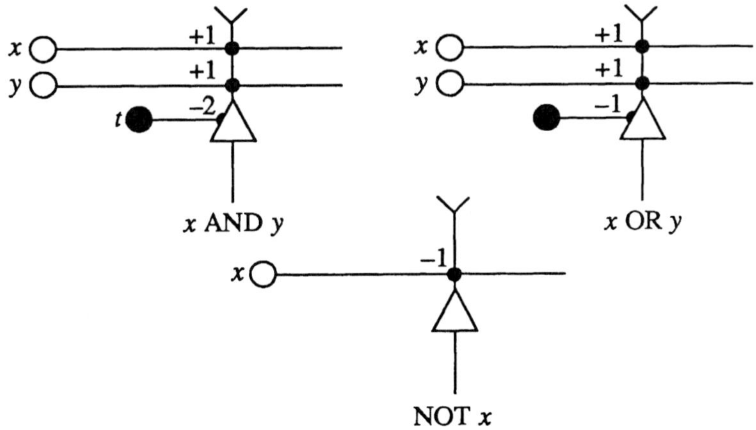

FIGURE 1. McCulloch-Pitts neurons. Each unit is activated if and only if its total excitation reaches or exceeds zero. For example, the first unit is activated if and only if both the units x and y are activated, for only then does the total excitation, $(+1)x + (+1)y$, balance the threshold bias of -2 set by the threshold unit, t, whenever both x and y equal $+1$ (activated). The t-unit is always active. The numbers (± 1) etc. shown above are called "weights." Positive weights denote *excitatory* synapses, negative weights *inhibitory* ones. Similarly, open circles denote excitatory neurons; filled circles, inhibitory ones.

McCulloch-Pitts neurons are synchronous binary elements or switches, which can be combined in a variety of circuits to implement logical functions such as $x(n)$ AND $y(n)$, $x(n)$ OR $y(n)$, NOT $x(n)$, $x(n)$ OR ELSE $y(n)$, NOT ($x(n)$ OR ELSE $y(n)$), etc. Figure 1 shows some simple examples, and Figure 2 a more complicated one.

The logical function NOT ($x(n)$ AND $y(n)$) shown in Figure 3, sometimes referred to as the NAND logic function, is particularly important in that it is a *universal* element: all other logical functions can be expressed by combining NAND logic elements and no others (Sheffer 1913). Even the logical quantifiers $(\exists n)$ "THERE EXISTS AN $n \ldots$ SUCH THAT" and $(\forall n)$ "FOR ALL n" can be represented by simple MP circuits, as seen in Figure 4. These circuits are of interest in providing simple examples of nets with an internal state. The second network in Figure 4 represents the logical function IF $y(1) = 1$ AND $y(n+1) = 1$ THEN $(\forall m)_{m \le n} x(m) = 1$; i.e., it implements the function $(\forall m)_{m \le n} x(m)$ only if y was active at $n = 1$ and all subsequent times. Similarly the first network represents the logical function IF $y(1) = 0$ AND $y(n+1) = 1$ THEN $(\exists m)_{m \le n} x(m) = 1$; i.e., it implements the function $(\exists m)_{m \le n} x(m)$ only if y was quiescent at $n = 1$ and changed its state to $y = 1$ at some later time. Thus specific initial states are required to implement such functions. These considerations led McCulloch and Pitts to introduce the

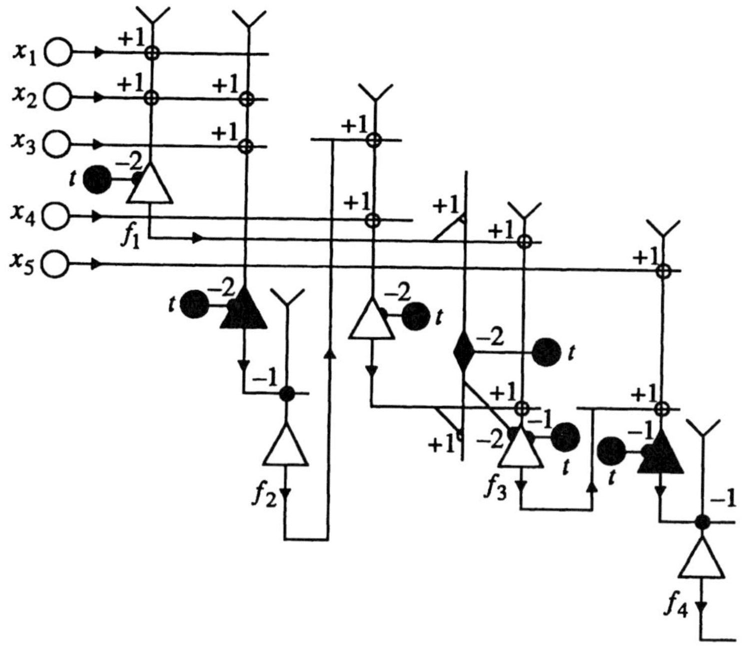

FIGURE 2. MP-network implementing the logical function NOT [[(x_1 AND x_2) OR ELSE (NOT (x_2 AND x_3) AND x_4)] OR x_5].

concept of a *finite state automaton*, as distinguished from a Turing machine (Turing 1936–1937). As McCulloch and Pitts asserted, it is easily shown that:

> every net, if furnished with a tape, scanners connected to efferents, and suitable efferents to perform the necessary motor operations, can compute only such numbers as can a Turing machine ... [and] that each of the latter numbers can be computed by such a net ... and that nets with circles [feedback loops] can compute, without scanners and a tape, some of the numbers the [Turing] machine can, but no others and not all of them.

Thus McCulloch-Pitts or MP nets, as they are sometimes called, are finite state automata in which the present state, i.e., the pattern of activation of all neurons in the network, together with the afferent stimulus, determines the next state. Such nets embody the logic of propositions, as McCulloch wished, and permit the framing of sharp hypotheses about the nature of brain mechanisms, in a form equivalent to computer programs. This was a remarkable achievement. It established once and for all the validity of making formal models of brain mechanisms, if not their veridicality. It also

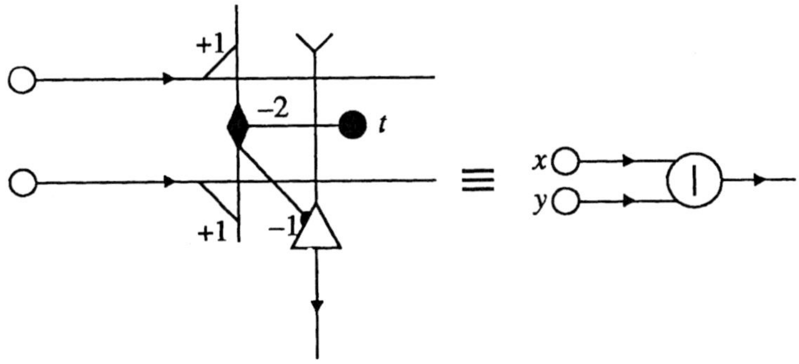

FIGURE 3. NAND logic function implemented by an MP network comprising two units.

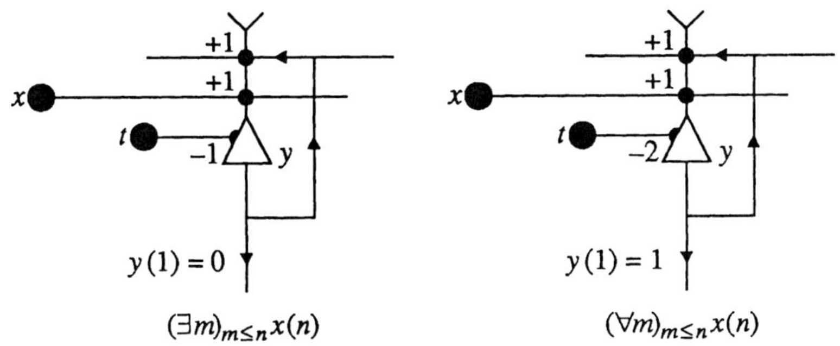

FIGURE 4. MP nets implementing the logical functions $(\exists m)_{m \leq n} x(m)$ and $(\forall m)_{m \leq n} x(m)$. In the first case y must be initially quiescent, in the second case it must be active.

established the possibility of a rigorous theory of mind, in that neural nets with feedback loops can exhibit purposive behavior.

0.3. Fault tolerance. In 1951–1952, following conversations with McCulloch and with K. Bruckner and M. Gell-Mann, von Neumann took up the problem of designing MP nets to function reliably despite malfunctions and failures of their component neurons, or misconnected wires, or damage to portion of the network. Von Neumann solved the reliability problem in two differing ways. His first solution was to make use of the error-correcting properties of *majority logic* elements. Such an element executes the logical function $m(a, b, c) = (a \text{ AND } b) \text{ OR } (b \text{ AND } c) \text{ OR } (c \text{ AND } a)$. Figure 5 shows a simple implementation of this function. The procedure is to *triplicate* each logical function to be executed; i.e., execute each logical function three times in *parallel*, and then fee the outputs through majority logic elements as shown in Figure 6.

Let ε be the probability of a majority element malfunctioning, and let η be the probability of error on any of its input lines. Suppose all such input

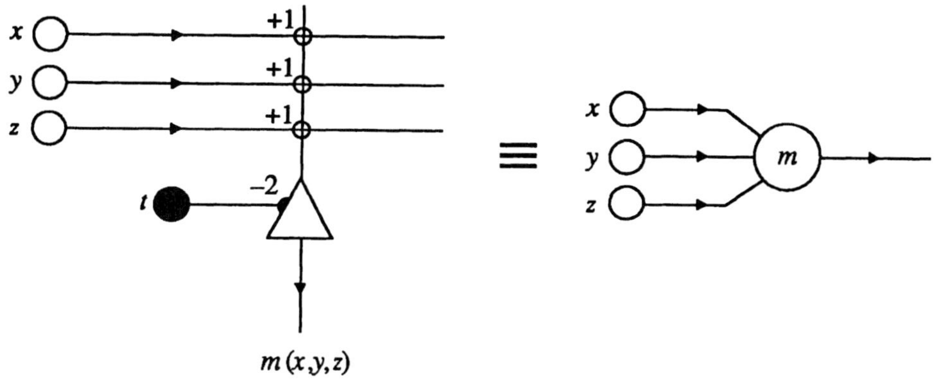

FIGURE 5. Majority logic function implemented by an MP neuron with threshold 2. (See text for details.)

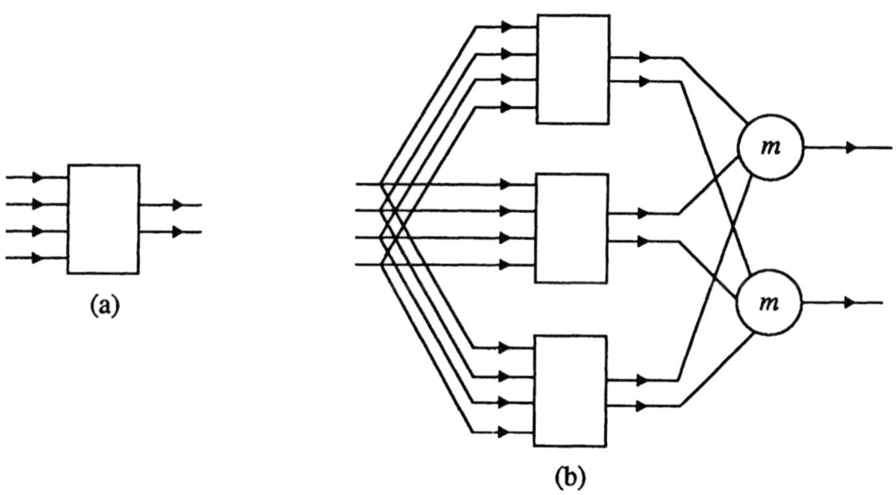

FIGURE 6. (a) Single MP network. (b) Triplicated version of the same network with majority logic error correction.

lines are either all activated or all quiescent. Then $\theta = 3\eta^2 - 2\eta^3$ is an upper bound for the probability that at least two input lines are in error, and $\varepsilon' = (1-\varepsilon)\theta + \varepsilon(1-\theta) = \varepsilon + (1-2\varepsilon)\theta = \varepsilon + (1-2\varepsilon)(3\eta^2 - 2\eta^3) \le \varepsilon + 3\eta^2 = \eta^*$ is an upper bound for the probability of error on the output line. This can lead to improved reliability since $\eta^* < \eta$ is possible. In fact von Neumann showed that the probability of error on the output lines of the triplicated network was bounded by η^*. Figure 7 shows a graph of $\eta^*(\eta)$ in the case when $\varepsilon < 1/6$.

Inspection of this graph indicates that η^* will converge to η_0 if $\eta < 1/2$, or to $1 - \eta_0$ if $\eta > 1/2$, provided the process is *iterated*; i.e., if the triplication-majority logic operation is repeated a number of times. In such a case the

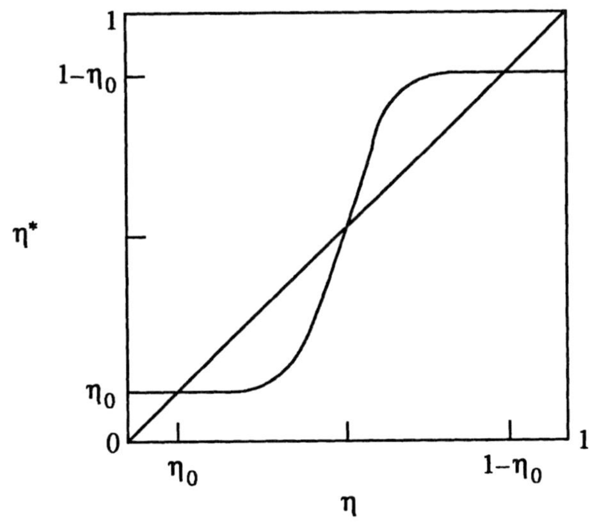

FIGURE 7. Graph of output error η^* against input error η for a triplicated MP network with majority logic error correction, in the case where $\varepsilon < 1/6$.

resulting *serial-parallel* MP network will have an output error of η_0 approximately equal to $\varepsilon + \varepsilon^2$, provided $\varepsilon < 1/6$. Thus if $\varepsilon = 0.08$, $\eta < 1/2$, $\eta^* = 0.1$. It is important to note that this result depends on the assumption that all input lines are either activated or quiescent. Von Neumann was able to extend the result to the general case, but only at a cost of requiring a further triplication and a tighter bound on ε. The general result is that provided $\varepsilon < 0.007$, an η^* equal to about 4ε can be achieved. However the redundancy is high. Let μ be the *logical depth* of the function to be computed; i.e., the longest serial chain of MP units in the original network. Then about 3^μ MP neurons are required to achieve outputs whose probability of error is about four times the probability of errors of the MP units. If $\mu = 3$ the requisite redundancy is 9:1, but if $\mu = 6$ it is about 700:1, and if $\mu = 10$ is is about 60,000:1. Von Neumann's second solution to the reliability problem was to *multiplex*; i.e., use many MP circuits to do the job of one. In such nets one bit of information (the choice between "1" and "0") is signaled not by the activation of one MP neuron, but instead by the synchronous activation of many MP neurons. Let Δ be a number between 0 and 1. Then "1" is signaled if ξ, the fraction of activated MP neurons involved in any job, exceeds Δ; otherwise "0" is signalled. Evidently a multiplexed MP network will function reliably only if ξ is close to either 0 or 1.

Von Neumann achieved this as follows. Consider nets made up entirely of NAND logic functions, as shown in Figures 3 and 8. Von Neumann called the network shown in Figure 8 an *executive organ*. The random permutations P are necessary to achieve random activation of lines in any bundle of N lines.

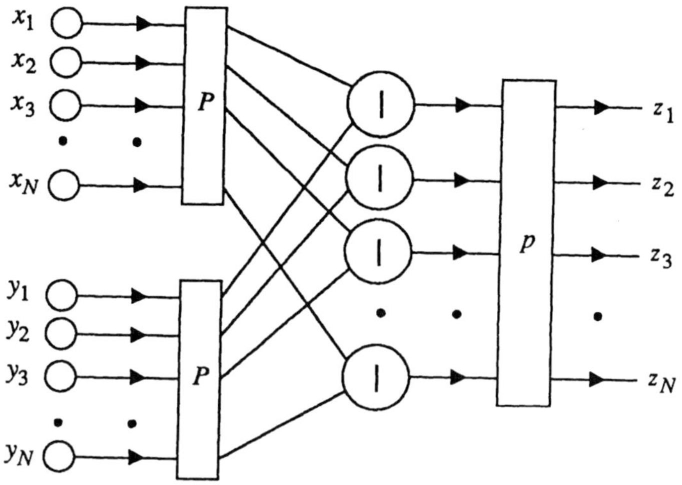

FIGURE 8. Multiplexed NAND logic network. P is a random permutation of each bundle of N lines.

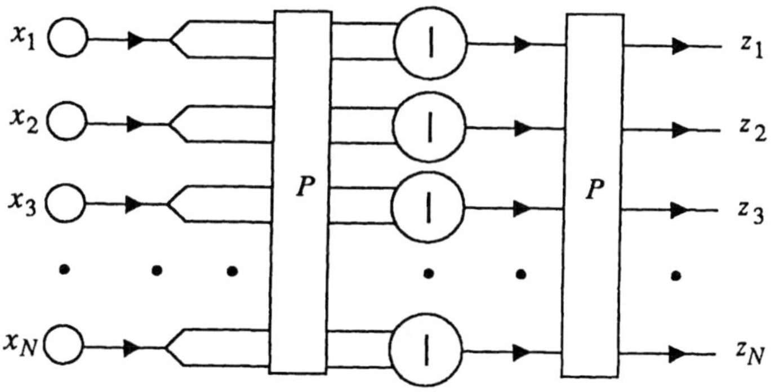

FIGURE 9. Von Neumann's restoring organ. (See text for details.)

Let ξ and η be the fractions of activated lines in the "x" and "y" bundles, respectively, ζ be the corresponding fraction in the "z" bundle and let ε be the probability of malfunction of a NAND logic element. Von Neumann proved that in the limit of large N, ζ is normally distributed with mean $1-\xi\eta+2\varepsilon(\xi\eta-1/2)$ and a standard deviation that is $\theta(1/\sqrt{N})$. To control error in such a network von Neumann introduced the *restoring organ* shown in Figure 9. This network is just a special executive organ in which $x = y$, so that ζ is again normally distributed with mean $1-\xi^2+2\varepsilon(\xi^2-1/2)$.
Von Neumann subsequently proved that if circuits are built from executive organs followed by two restoring organs in series, then for N large and ε sufficiently small, the fraction of activated lines in the output bundle is $1-(2\xi\eta-(\xi\eta)^2)^2$. It is easy to verify that this is $\theta(\Delta^2)$ if $\xi \geq 1-\Delta$, $\eta \geq 1-\Delta$; $1-\theta(\Delta^2)$ if $\xi \leq \Delta$, $\eta \geq 1-\Delta$; and $1-\theta(\Delta^4)$ if $\xi \in \Delta$, $\eta \leq \Delta$; so that if Δ is sufficiently small, such circuits implement the NAND logic function. It turns

out that with $\Delta = 0.07$ and $\varepsilon < 0.0107$, the probability of circuit malfunction η^* can be made to decrease with increasing N. With $\varepsilon = 0.005$, von Neumann showed that $\eta^* \sim aN^{-1/2}10^{-bN}$, where $a = 6.4$ and $b = 8.6 \cdot 10^{-4}$. It follows that η^* can be made less than ε provided $N \geq 2,000$. This is achieved with a redundancy of $3N$: 1 that is independent of the logical depth μ; thus for logical computations of large depth the method of multiplexing is superior to majority logic decoding.

This solution to the problem of reliable computing with unreliable elements is general since the NAND logic element is universal. Is it biologically plausible? The answer seems to be negative since real neurons have *thousands* of synaptic contacts, so that it is not necessary to concatenate many NAND elements in circuits of large logical depth to implement logical functions of many variables. This suggests that real neurons can implement logical functions of many variables with a probability of error ε not much greater than that of the NAND element. This observation motivated Winograd and Cowan to study the limiting case in which ε is independent of the logical complexity of the function to be implemented. In such a case it is possible to use *error-correcting codes* just as is done in noisy communication channels (Shannon 1948, Hamming 1950) in which a message comprising k symbols is transmitted via a signal comprising n signals, $n - k$ of which are used by the receiver for error detection and correction. In the computing case this implies that k *computations* be implemented by an MP network comprising n/k times as many elements as would be required in the error-free case. Figure 10 shows an example of the application of a $(5,2)$ Hamming code to the MP network shown in Figure 2. The first layer of elements combines the logical functions x_1 AND x_2 and NOT (x_2 AND x_3) with the encoding function of the $(5,2)$ Hamming code. This takes the following form. Let y_1 and y_2 be symbols to be transmitted (or functions to be computed). The $(5,2)$ encoding consists in also transmitting (or computing) $y_3 = y_1$ OR ELSE y_2, $y_4 = y_2$, $y_5 = y_1$. In the case shown above both y_1 and y_2 correspond to the parallel pair of logical functions x_1 AND x_2, and NOT (x_2 AND x_3). Therefore the first layer elements implement the logical functions $f_{11} = f_{15} = x_{11}$ AND x_{21}, $f_{12} = f_{14} = x_{12}$ AND x_{22}; $f_{21} = f_{25} =$ NOT (x_{21} AND x_{31}), $f_{22} = f_{24} =$ NOT (x_{22} AND x_{23}); $f_{13} = f_{11}$ OR ELSE f_{12}, $f_{23} = f_{21}$ OR ELSE f_{22}. The second layer elements are even more complicated since they implement the combination of two copies of the logical function z_1 OR ELSE (z_2 AND x_4), where $z_1 = x_1$ AND x_2 and $z_2 =$ NOT (x_2 AND x_3), with the decoding function of the $(5,2)$ Hamming code (needed to detect and correct errors occuring in the first layer), and the encoding function again (needed for transmission to the third layer). Finally the third layer elements combine the decoding function with two copies of the final logical function NOT (z_3 OR x_5), where $z_3 = z_1$ OR ELSE (z_2 AND x_4). The decoding function of the $(5,2)$ Hamming code is just a parity check sum, implemented via the logical functions (y_1 OR ELSE (y_1 OR ELSE y_5) AND (y_1 OR ELSE y_2 OR ELSE y_3) AND (y_2 OR

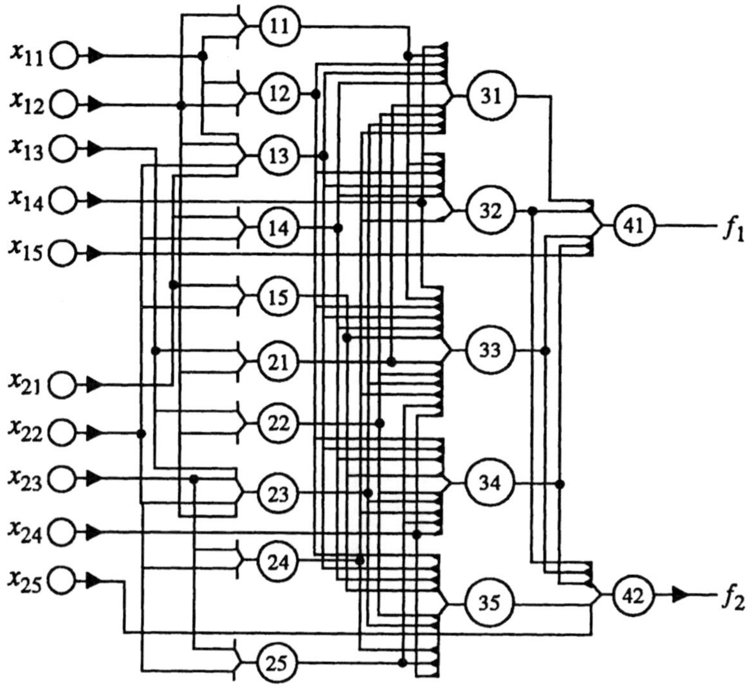

FIGURE 10. Redundant MP network in which two parallel implementations of the logical function NOT [[(x_1 AND x_2) OR ELSE (NOT(x_2 AND x_3) AND x_4)] OR x_5] are combined with a (5,2) Hamming code. (See text for details.)

ELSE y_2 OR ELSE y_4) AND (y_1 OR ELSE y_2 OR ELSE y_3). Thus the third layer elements implement the logical functions NOT ((f_{31} OR ELSE (f_{31} OR ELSE f_{35}) AND (f_{31} OR ELSE f_{32} OR ELSE f_{33})) OR x_{51} and NOT ((f_{32} OR ELSE (f_{32} OR ELSE f_{34}) AND (f_{31} OR ELSE f_{32} OR ELSE f_{33})) OR x_{52}, where f_{3i}, ($i = 1, 2, \ldots, 5$) are the second layer logical functions.

The overall effect of such an encoding scheme is to *distribute* the logical functions to be implemented over the entire MP network. Such a scheme works most sufficiently with large k and n; in effect in a parallel distributed architecture. Thus the Winograd-Cowan scheme is an early example of *Parallel Distributed Processing* or *PDP*. Of course it can be argued that the scheme is not realistic, in that all the extra coding machinery is assumed to be error free. As we have noted this may not be true for simple logical elements, but it may be more plausible for real neurons. In unpublished work Winograd and Cowan studied this issue in more detail and found a realistic optimal scheme involving both multiplexing and PDP (Cowan and Winograd, in preparation).

0.4. The "integrate and fire" neuron. The MP neuron is of course a very simplified representation of real neural properties. Since the 1940s it has been extended and elaborated in a number of ways. Perhaps the first significant extension was what is now called the "integrate and fire" neuron. This derives from the Eccles model of synaptic transmission (Eccles 1957, 1964). Eccles studied the current flows produced by synaptic activation of somato-dendritic membrane. Figures 11 and 12 show the relevant details. It is the synaptic current I_s generated by potential changes in the axon terminal which modulates the ionic conductances of somatic or dendritic membrane. Let $v(x,t)$ be the transmembrane potential at any point along the axon. Elementary circuit theory indicates that $I_m = -\frac{\partial I_a}{\partial x} = \frac{1}{R}\frac{\partial^2 v}{\partial x^2}$, where R is the axoplasmic resistance per unit length, and $\frac{\partial}{\partial x}$ is the usual symbol for partial differentiation with respect to the variable x. The membrane has both a capacitance C and numerous *ion channels*. Thus $I_m = C\frac{\partial v}{\partial t} + I_i$ where C is the total capacitance of the membrane covering one unit axon length, $\frac{\partial}{\partial t}$ denotes differentiation with respect to the variable t, and I_i is the total ionic current. It follows that the partial differential equation $\frac{1}{R}\frac{\partial^2 v}{\partial x^2} = C\frac{\partial v}{\partial t} + I_i$ governs the evolution of v. In many cases somato-dendritic membrane is passive, so that the ionic current flow may be written as $\frac{1}{R_m}(v - E_r) + I_s$ where R_m is the resistance at resting equilibrium of unit length of the membrane. The resulting space-clamped equation, which we shall refer to as the Eccles equation, takes the form:

$$(2) \qquad C\frac{\partial v}{\partial t} + \frac{1}{R_m}(v - E_r) + I_s = 0.$$

Eccles demonstrated that such effects can be described by adding currents of the form $I_e = \overline{g_e}u_e(t)(v - E_e)$ at an *excitatory* synapse and $I_i = \overline{g_i}u_i(t)(v - E_i)$ at an *inhibitory* synapse, where $\overline{g_s}(t)$ ($s = e$ or i) is a constant conductance, $u_s(t) = 1$ if the axon terminal is activated and 0 otherwise, and where E_s is the associated equilibrium potential for synaptic current flow. The equilibrium potentials E_e and E_i are respectively, 0 and $-80mV$. It follows that the activity at an excitatory synapse *depolarizes* the membranes (makes the inside less negative), whereas activity at an inhibitory synapse *hyperpolarizes* it (makes the inside more negative). Evidently inhibitory synaptic activity effectively raises the membrane voltage threshold. Figure 13 shows the time course of such potential changes, which are known as excitatory and inhibitory postsynaptic potentials, or EPSPs and IPSPs. The Eccles equation and its associated PSPs obtain in the case of synapses on a neuron soma, or locally on a dendrite.

Consider now multiple activations of the Eccles circuit. Let $\delta(t)$ be the Dirac delta function. Assume that the membrane response to incoming excitation $\Sigma_{jk}\delta_j(t - t_k)$ is of the form $C\frac{\partial v}{\partial t} + \frac{1}{R_m}(v - E_r) = \Sigma_{jk}\tau i_j \delta_j(t - t_k)$; i.e., the term $\overline{g_j}(v - E_j)$ in the Eccles equation is approximated by the constant current i_j. This is known as the "current source" approximation to synaptic

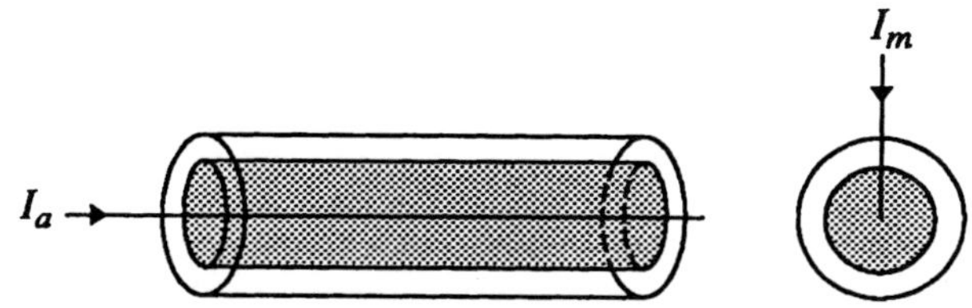

FIGURE 11. Axoplasmic current I_a and membrane current I_m flowing in a section of axon.

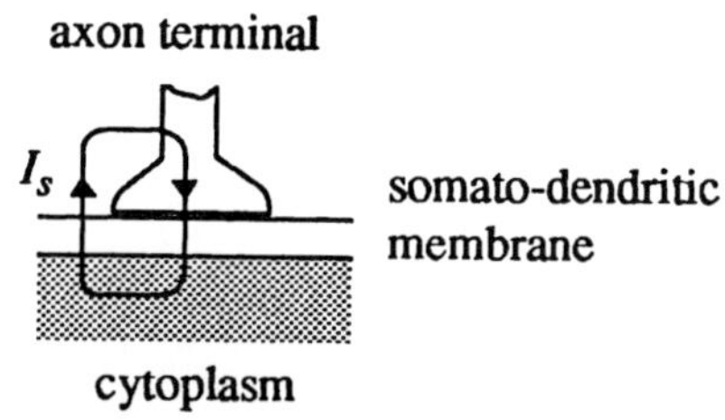

FIGURE 12. Current flows at an axo-somatic or axo-dendritic synapse. I_s is the synaptic current.

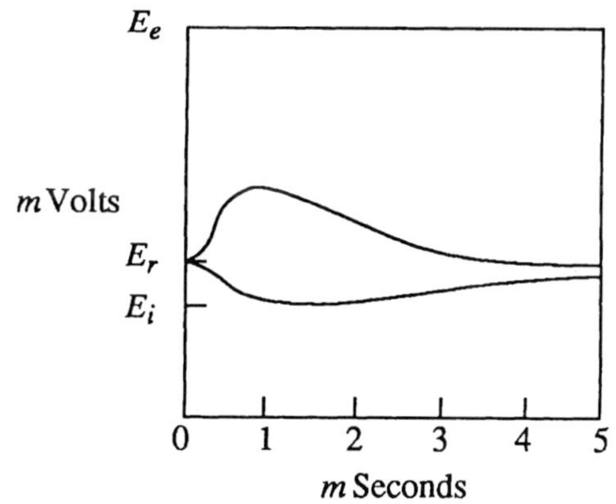

FIGURE 13. Typical EPSP and IPSP generated by presynaptic activity. (See text for details.)

conductance changes. So Equation (2) can be rewritten in the form:

$$\frac{\partial v_i}{\partial T} + v_i = \sum_{jk} w_{ij}\delta_j(T - T_k) \tag{3}$$

where $v_i = v - E_r$ is the voltage built up in the ith neural membrane (relative to the resting potential E_r), and where $w_{ij} = R_m i_j$ is the weight of the $j -> i$th synapse. The solution to this equation is easily obtained. it is $v_i(T) = \int^T dT' e^{-T'} \Sigma_{jk} w_{ij}\delta_j(T - T' - T_k)$, sometimes written as $\Sigma_{jk} w_{ij}\delta_j(T - T_k) \otimes u(T)e^{-T}$ where $\otimes$ represents temporal convolution. But $\delta_j(T - T_k) \otimes u(T)e^{-T} = u_j(T - T_k)e^{-(T-T_k)}$. It follows that the neural firing condition can now be written as:

$$u_i(T+\Delta) = \Theta\left[\sum_{jk} w_{ij}u_j(T-T_k)e^{-(T-T_k)} - v_{TH}\right]. \tag{4}$$

The reader should compare Equation (2) with the McCulloch-Pitts firing condition. Neural models with such a firing condition are known as "integrate-and-fire" models and show more complex and realistic threshold properties. Equations of this type were first introduced by E. R. Caianiello (1961) as a model for network interactions.

0.5. "Analog" neurons. The "integrate and fire" neuron, although more realistic than the MP neuron, is hard to analyze. A smoothed version was introduced by J. D. Cowan (1967) as a way to mitigate the difficulties generated by the discontinuous nature of equation (4). The resulting equation takes the form:

$$e_i(T+1) = \Phi\left[\tau\sum_j w_{ij}e_j(T) + h_i\right] \tag{5}$$

where $e_i(T)$ is the firing rate of the ith element (a time-averaged version of $u_i(T)$), $h_i(T)$ is an additional external stimulus, and $\Phi[x]$ is the logistic function $(\frac{1}{\Delta})[1+\exp(-\gamma(x-\theta))]^{-1}$, shown in Figure 14. The differential equation version of Equation (5) is simply:

$$\frac{de_i}{dT} = -e_i + \Phi\left[\tau\sum_j w_{ij}e_j(T) + h_i\right]. \tag{6}$$

These equations have proved to be very useful in the study of the dynamics of neural nets and also in connection with what McCulloch called the modelling of "mind": the programming of nets by way of the synaptic weights w_{ij} for a variety of information processing tasks.

0.6. Continuum neurodynamics. Before dealing with such problems however, we first discuss some of the dynamical properties of the nets described by Equation (6). In this connection it is interesting to compare Equation (6) with those equations obtained from a continuum approximation to neural activity (Beurle 1956, Wilson and Cowan 1972, 1973) in which the relevant observable is not e_i but $e(\underline{x},T)$, the proportion of neurons activated per unit time in a given volume element d^3x. The equation corresponding to (6): where $q(x,T) = 1 - \Delta_r e(\underline{x},T)$ is the proportion of neurons which have not been activated in the interval $(T-\Delta_r, T)$ and are therefore *sensitive.* {In the network or lattice case, neurons are either sensitive or activated, so that $q_i = 1$ or else $\Delta_r e_i = 1$.} Wilson and Cowan studied this equation and extensions incorporating separate inhibitory as well as excitatory neurons, in the case $w(\underline{x},\underline{x}') = (w|\underline{x}-\underline{x}'|)$, the translation invariant case. They found

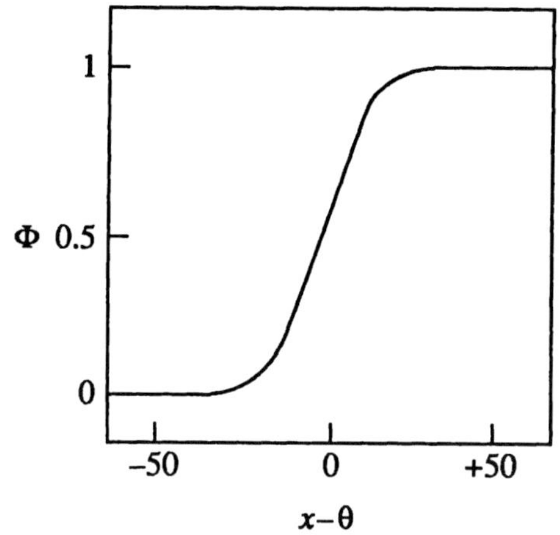

FIGURE 14. The logistic function $\Phi = [1+\exp(-\gamma(x-\theta))]^{-1}$. The parameter γ determines the slope of the function at the point $x = \theta$.

that such nets can support a variety of dynamical states: stationary equilibria, limit cycles, active transients, propagating steps and waves, etc. Further analytical studies by G. B. Ermentrout and J. D. Cowan (1979) showed that two-dimensional sheets of such a neural "medium" could support standing wave patterns in the form of blobs and stripes, similar to those generated by the Turing equations of morphogenesis (Turing 1952). Chaotic dynamics has also been demonstrated (Sbitnev 1982).

0.7. Statistical neuromechanics. There is another approach to the analysis of neural network dynamics, that based on statistical mechanics. Consider Equation (6) in the form:

$$\frac{dv_i}{dT} = -v_i + \sum_j w_{ij}x_j(T) + h_i, x_j = \Phi[v_j], T = \frac{t}{\tau} \tag{8}$$

It follows that if $w_{ij} = -w_{ji}$, $w_{ij} = 0$, and if τ is large, then this equation can be rewritten in the form (Cowan 1968, 1971):

$$\frac{dv_i}{dT} = \sum_j w_{ij}\frac{\partial G}{\partial v_j}, \tag{9}$$

where $G = \sum_i \{\log[1 + q_i \exp(v_i)] - q_i v_i\}$, $h_i + \sum_j w_{ij}q_j = 0$, is a "constant of motion" of the network. Since the matrix iW, $W = \{w_{ij}\}$, is Hermitian, one can form the block diagonal matrix W', made out of blocks of the form:

$$\begin{pmatrix} 0 & 1 \\ -1 & 0 \end{pmatrix}$$

by way of the congruence transformation $W' = \Lambda W \widetilde{\Lambda}$, where Λ is built up from the eigenvectors of W. Then if $y_i = \Lambda v_i$, we have the Hamiltonian equations $\frac{dy_i}{dT} = \frac{\partial G}{\partial y_j}$. The physical content of this result is that a network of neural-like elements with skew-symmetric coupling constants can generate neutrally stable oscillations in x_i in the range between 0 and 1. Moreover, because G is a constant of the motion, one can introduce a form of equilibrium statistical mechanics, in which the probability of being in the state $\{y_1, y_2, \ldots, y_N\}$ takes the form $Z^{-1}\exp[-\alpha G]$, where $Z = \sum_{\{y\}} \exp[-\beta G]$. One can then compute various statistical averages of the behavior of such nets. Of course the skew-symmetry of the coupling coefficients w_{ij} is rather artificial, as is neglect of the damping produced by the term $+v_i$ in Equation (8).

Some seven years ago, J. J. Hopfield (1982) made the important observation that if $w_{ij} = w_{ji}$ instead of $w_{ij} = -w_{ji}$, one can rewrite Equation (8) in the form $\frac{dv_i}{dT} = -\frac{\partial E}{\partial v_i}$ where:

$$E = -(1/2)\sum_{ij} w_{ij}x_i x_j - \sum_i h_i x_i + \sum_i \int^{x_i} \Phi^{-1}(x)\,dx. \tag{10}$$

The symmetry condition implies that E is a Liapunov function for the network (LaSalle and Lefschetz 1961); i.e., the network always settles down into a stable, nonoscillatory configuration $\{x_1, x_2, x_3, \ldots, x_N\}$ which minimizes E. If γ, the slope of the function $\Phi[x]$, is large, then each x_i tends to either 0 or 1 (just the opposite of the skew-symmetric case). Moreover it can be shown that in the limit of large γ, $\sum_i \int^{x_i} \Phi^{-1}(x)dx$ is small, and so E is approximately

$$-(1/2)\sum_{ij} w_{ij}x_i x_j - \sum_i h_i x_i = -(1/2)\sum_{ij} w_{ij}\Phi_i\Phi_j - \sum_i h_i\Phi_i. \tag{11}$$

This is recognizable as the Sherrington-Kirkpatrick (SK) spin-glass Hamiltonian (Sherrington and Kirkpatrick 1975), in which the spins Φ_i can take the values 0 or 1 (or else ± 1). There is therefore a striking analogy between neural nets with symmetric connections and magnetic materials, an observation which was anticipated by Cragg and Temperley (1954, 1955).

Since the publication of Hopfield's paper, many physicists have used the analogy between spin-glasses and neural nets to analyze what are now referred to as Hopfield nets. Again one can carry out calculations using the methods of equilibrium statistical mechanics via the function $P\{\Phi\} = Z^{-1}\exp(-\beta H)$. The partition function

$$Z = \sum_{\{\Phi\}} \exp[-\beta H] \text{ is usually written by physicists as}$$
$$\text{trace }\Phi[\exp(-\beta H(\Phi))] = \text{tr}[\exp(-\beta H(\Phi))],$$

since it is the sum of the diagonal elements of a matrix. Then $\langle\Phi_i\rangle$, the average value of Φ_i is obtained from $\text{tr}[\Phi_i P\{\Phi\}] = \text{tr}[\Phi_i Z^{-1}\exp(-\beta H(\Phi))] = \beta^{-1}\partial(\log Z)/\partial h_i$, and r_i, the mean firing rate of element i, equals $(1/2)\cdot(1+\langle\Phi_i\rangle)$. It is evident that calculating Z is the key. In many cases this can be done by using the properties of Gaussian integrals, and the method of *steepest descents*, which reduces the integral $\int dx \exp[-NH(x)]$ to $\exp[-NH(x_0)]\times$ (a Gaussian integral), when N is large and $H(x)$ is minimized at x_0.

0.8. Applications. Given the neural machinery described above, to what can it be applied? The dynamical studies we have described provide models for changes of gross brain states, such as those occuring in epileptic seizures and migraines, and in drug-induced hallucinatory states (Ermentrout and Cowan 1979a, b). They also provide a basis for studies of the development and regeneration of brain circuitry (Willshaw–von der Malsburg 1976, 1979, Whitelaw and Cowan 1981, Swindale 1980, 1982, Linsker 1986), and for a variety of problems in vision (Marr 1974, Marr and Poggio 1976), to name but a few studies. However, undoubtedly the most intriguing applications are to be found in the modelling of the neural basis of learning and memory. In what follows we give a short account of recent advances in some of these areas.

0.9. Pattern recognition and learning. Given a complicated logical function to be implemented with either discrete MP or analog neurons, there remains the formidable problem of constructing a circuit. An even more difficult problem is common: given some logical task—say, distinguishing the letter **T** from the letter **C**, how can one construct a circuit to implement this, given only an informal understanding of the circuit's architecture? In addition one usually requires the circuit to be immune to a certain amount of damage, and to distinguish a variety of **T**s from a variety of **C**s.

1.0. Perceptrons and adalines. A start on these and related problems was made in the late 1950s by F. Rosenblatt (1958), who almost solved the entire problem by demonstrating that McCulloch-Pitts nets with modifiable connections could be "trained" to classify certain sets of patterns as similar or distinct. Rosenblatt called such nets "*Perceptrons*", and we shall use this term in what follows. Figure 15 shows the architecture of a typical "elementary" Perceptron. It consists of a set of "sensory" units connected through a single layer of McCulloch-Pitts neurons (henceforth referred to as M-P units), to a set of "motor" units. Initially, the strengths or weights of all contacts or synapses in the network are set to arbitrary values, so that stimulation will generate an arbitrary response. To obtain a desired response from the network requires that all these synaptic weights be adjusted. Rosenblatt found a way to do this with the following *training procedure*: first note an M-P unit's response to a given stimulus. Some responses will be correct, i.e., correspond to desired ones; others will be incorrect. Therefore adjust the unit's weights as follows: make no adjustments if the unit's response is correct. However, if it is incorrect, increase the weights of all activated synapses if the unit should be activated and isn't, and decrease them if the opposite obtains. Do this for all possible desired stimulus-response patterns. It can be shown that after only a finite number of presentations of stimulus-response patterns, the

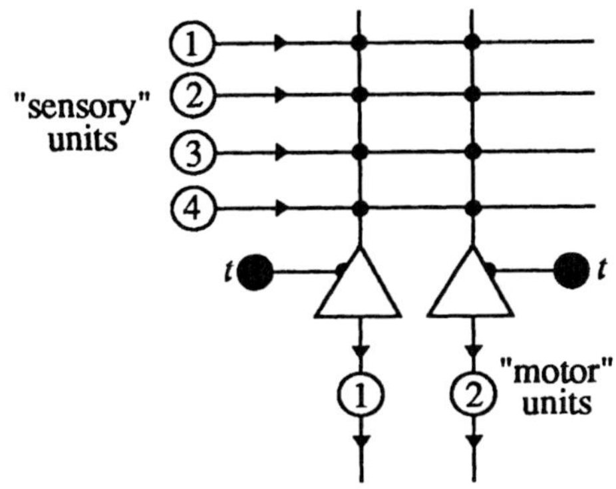

FIGURE 15. An "elementary" Perceptron, i.e., an adaptive McCulloch-Pitts neural network with modifiable synaptic weights which are changed if they generate incorrect responses. (See text for details.)

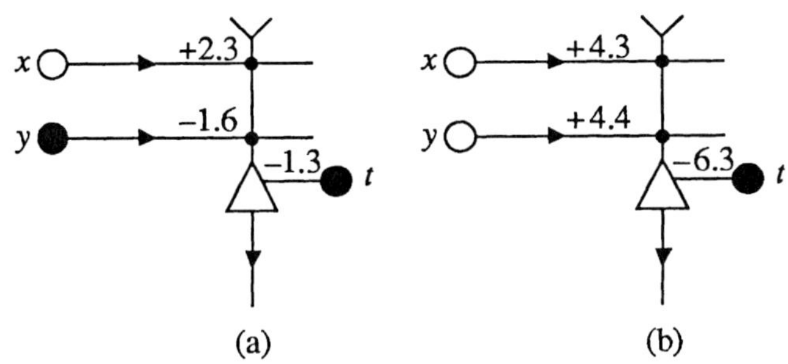

FIGURE 16. (a) A Perceptron (M-P unit) with randomly chosen synaptic weights. (b) The same unit after it has been "trained" to compute the logical function x AND y. (See text for details.)

weights converge to a set of values representing whatever computation or classification is embodied in these patterns (Novikoff 1963).

The following example should clarify the nature of the procedure. Consider the M-P unit with arbitrary weights shown in Figure 16(a). Its "truth table," i.e., its response to stimulus patterns, is shown in Table 1(a). It will be seen that this unit computes the function x AND NOT y. Suppose, however, that the desired computation is x AND y, whose truth table is shown in Table 1(b). Activate the x, y units. (The t unit is always active). The net excitation received by the unit (See Figure 1's legend) is then (see Figure 16): $(2.3) \cdot 1 + (-1.6) \cdot 1 + (-1.3) \cdot 1 = -0.6$, which is negative. The M-P unit will not be activated, as is evident from Table 1(a). But Table 1(b) indicates that the unit should be activated: therefore add $+1$ to all the weights, which then become 3.3, −0.6, and −0.3 respectively. Now activate the x unit, but not the y unit. The net excitation is then $(3.3) \cdot 1 + (-0.6) \cdot 0 + (-0.3) \cdot 1 = 3.0$, which is positive. The M-P unit will be activated. However this is not the desired response. Therefore subtract +1 from the x and t weights, giving 2.3, −0.6, and −1.3 respectively. Continue the process by activating the y unit, but not the x unit, and so on. After 57 iterations, using all possible activation patterns, a set of weights is reached, namely 4.3, 4.4, −6.3, which generates only correct responses. The unit then implements the desired logical function x AND y, for only when both x and y are activated is the net excitation $(4.3).x + (4.4).y$ greater than the threshold value of 6.3.

x	y	x AND NOT y	x	y	x AND y
1	1	0	1	1	1
1	0	1	1	0	0
0	1	0	0	1	0
0	0	0	0	0	0
		(a)			(b)

TABLE 1. (a) "Truth Table" corresponding to the random weights shown in Figure 6. (b) "Truth Table" of desired computation. (See text for details.)

Shortly after Rosenblatt's first publications there appeared a closely related variant of the Perceptron, invented by B. Widrow and M. E. Hoff (1960), called *Adaline* (for adaptive linear neuron). The only difference between Perceptrons and Adalines lies in the training procedure. Let w_{ij} be the weight of the ijth contact, $e(s)$ the excitation delivered by a given stimulus pattern s to motor units, and $a(s)$ the required activation pattern to be triggered by this stimulus. Let $\varepsilon(s) = a(s) - e(s)$. Adaline operates so as to *minimize* $(1/2)\varepsilon^2(s)$. It can be shown that this is achieved by adjusting the weights via the rule $\Delta w_{ij} = s_i\varepsilon(s)$, where Δw_{ij} denotes the change in w_{ij}, and s_i the state of the ith sensory unit (equal to ± 1). It follows from this rule that w_{ij} increases if s_i is active (equals +1) and $e(s)$ is less than $a(s)$, and decreases if s_i is active and $e(s)$ is greater than $a(s)$; and conversely if s_i is quiescent (equals -1). It will be seen that this rule corresponds closely to the Perceptron's, for if an M-P unit is not activated by a given sensory unit when it should be, the weight of the relevant synapse will increase, and conversely. Adalines matched closely the performance of Perceptrons in a variety of tasks.

Consider again the example described above. Initially the weights are 2.3, −1.6, and −1.3. Following the first stimulation pattern $(+1, +1)$, $\varepsilon = 1 - (-0.6) = 1.6$, so all the weights are increased by an amount proportional to ε, say $\varepsilon/8 = 0.2$, to the values 2.5, −1.4, and −1.1 respectively. After the next pattern $(+1, -1)$, $\varepsilon = -1 - 2.8 = -3.8$, so the weights become 2.0, −0.9, and −1.6 respectively, and so on. After only 27 iterations, the weights become 0.5, 0.6, and −0.7, and the excitation $(0.5).x + (0.6).y$ exceeds the threshold value of 0.7 if and only if both x and y are activated.

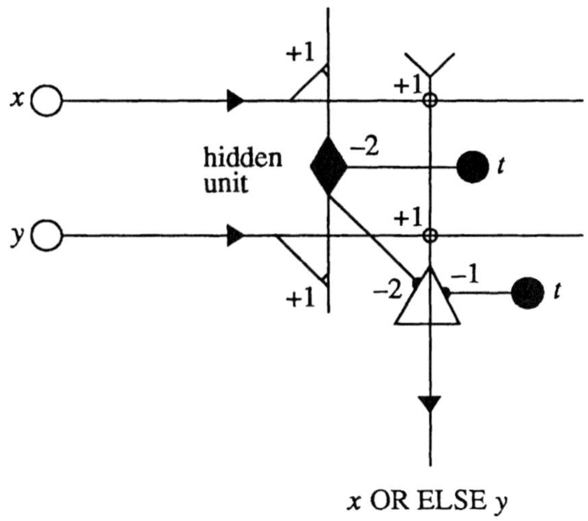

FIGURE 17. M-P network that implements the logical function x OR ELSE y. The network comprises two M-P units, one of which is a "hidden" unit, and two threshold units. The weights of the hidden unit cannot be correctly modified using the procedures described earlier.

There are, however, limits to the performance of elementary Perceptrons and Adalines. S. A. Papert and M. L. Minsky (1969) proved that elementary Perceptrons cannot distinguish between such simple patterns as **T** and **C**. The difficulty lies in the nature of the M-P neurons. As we described earlier, single units of this type can compute only such simple logical functions as "x AND y", "x OR y", "NOT x", "x AND NOT y", and so on. However the function "x OR ELSE y" and its negation, "NOT (x OR ELSE y)," each require several M-P unit. The reason for this is simply that "x OR ELSE y" is the same as "(x AND NOT y) OR (y AND NOT x)". This is unfortunate because the function "NOT (x OR ELSE y)" is computationally *universal* (Turing, 1936): every other function can be expressed as a string of "NOT (x OR ELSE y)"s. Figure 17 shows the architecture of the simplest M-P network that implements the function "x OR ELSE y". It comprises two M-P units and two t-units. One of the M-P units lies entirely within the network: its output does not drive a motor unit, but only the other M-P unit. In current terminology the interior is said to be "hidden"; and what Papert and Minsky proved is that an elementary Perceptron or Adaline, which consists of only a single layer of M-P units, is not computationally universal, even with modifiable connections. In addition, they conjectured that hidden units in multilayer Perceptrons cannot be trained, or in other words, that the problem of *assigning credit* to hidden units is unsolvable. It turns out that the limitations of simple Perceptrons and Adalines can be overcome with a

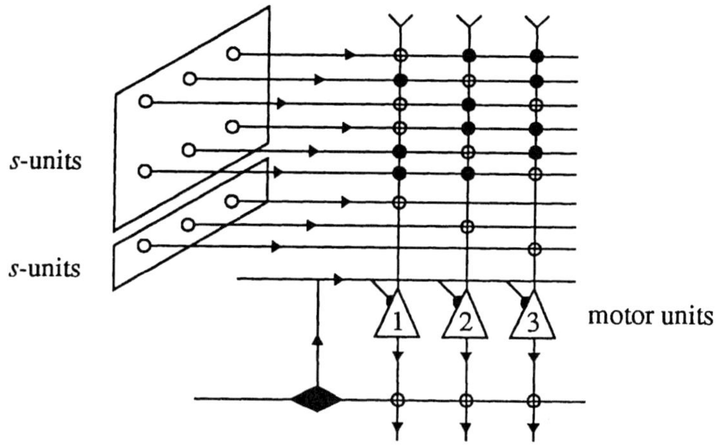

FIGURE 18. Taylor associative network. The network is trained in the style of the Perceptron on a subset of $s - m$ pairs. Other s patterns become "associated" with these pairs in a Pavlovian manner. The motor units are coupled together via an inhibitory neuron so that the unit with maximum activity shuts off all the others. Taylor called this a Maximum Amplitude Filter. It is currently termed a "winner-take-all" circuit.

more complex architecture incorporating hidden units with modifiable connections. Interestingly, Rosenblatt (1961) introduced such a scheme, which almost solved the problem.

1.1 Associative memory. Another notable feature of Perceptrons and Adalines is that their "memories" of the learned task are distributed over all the connections modified during the training phase, and are therefore less likely to be disrupted by damage. However, there is an important aspect of human memory, which Perceptrons and Adalines do not directly address; namely, human memory seems to be *associative* as well as distributed. Whatever is common to two differing memories binds them together—so that one may elicit the other if the two have sufficient overlap. Neural networks with associative memory have been studied extensively during the past thirty years, beginning with the work of W. K. Taylor (1956). Figure 18 shows the structure of one of Taylor's networks.

It is similar in structure to an elementary Perceptron with no hidden units, except that the units are not M-P neurons, but *analog* devices operating in the fashion described earlier, and the motor output units inhibit each other in what is now called a "winner-take-all" circuit. The training procedure also differs from that in Perceptrons and Adalines: it is simply Hebb's rule that activated synaptic weights increase if they activate their target unit (Hebb 1949, Brindley 1967). The network learns to associate differing sensory patterns through repeated presentation of pairs of patterns, one of which initially

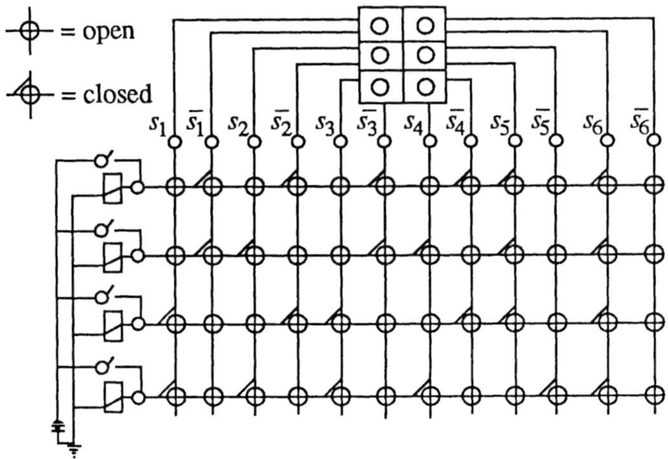

FIGURE 19. Structure of Steinbuch's learning matrix. It consists of a planar array of switches, each of which can be either open or closed. Each switch is connected to a receptor. However there are switches for both receptor ON and receptor OFF configurations. The switches control a set of relays, connected together in a winner-take-all circuit as in a Taylor network. The correspondence between the elements of a learning matrix—receptors, switches and relays, and those of Perceptrons and Taylor networks is evident.

elicits a motor response. Eventually the other pattern triggers the response. Thus Taylor networks exhibit simple Pavlovian conditioning (Rescorla and Wagner 1972), and the associated memory is stored in a distributed fashion in the pattern of weights.

Taylor suggested that the association areas of cerebral cortex and thalamus contained such networks. Shortly before this a very similar network was introduced by K. Steinbuch (1961), the *Learning Matrix* (see Figure 19.), comprising a planar network of switches interposed between arrays of "sensory" receptors and "motor" effectors. As in Taylor's scheme, the network learned to associate sensory with motor patterns. The associated memory is again stored in the pattern of opened switches. Learning matrices have a particularly simple mathematical structure, and their performance can be readily analyzed. Following Steinbuch, but in most cases quite independently, many other workers devised similar associative networks; e.g., J. A. Anderson (1968), D. Gabor (1969), D. J. Willshaw, O. P. Buneman and H. C. Longuet-Higgins (1969), D. Marr (1969, 1971), and T. Kohonen (1977), all of whom discovered that associative networks are also *content addressable*, i.e., stimulating such a network with some fragment of an associated memory will elicit the complete response. Thus the network can be "addressed" using the (partial) *content* of a memory, rather than just its location. Because of this property associative networks are now generally referred to as *associative content addressable memories* (ACAMs). Marr's work on this is particularly interesting in that it is formulated as a theory of how the *cerebellum* enables animals to make delicate and precise voluntary movements, and how memories may be temporarily stored in the *hippocampus.*

1.2. Hopfield networks. The current approach to associative memory uses analog neurons with symmetric connections, now called Hopfield networks. We have already described some of the properties of such networks. As is now well known, Hopfield showed that one could store "spin" configurations in the network by setting the w_{ij} equal to $\Phi_i^{(s)}\Phi_j^{(s)}$, where $\{\Phi_1^{(s)}, \Phi_2^{(s)}, \Phi_3^{(s)}, \cdots, \Phi_N^{(s)}\}$ is a configuration to be stored. Under such a constraint, each Φ_i converges to $\Phi_i^{(s)}$, and the network "stores" the configuration as a dynamical equilibrium. About $0.14N$ patterns can be stored in this manner, simply by setting s_{ij} equal to $\sum_s \Phi_i^{(s)}\Phi_j^{(s)}$. More exactly it can be shown (McEliece et al., 1987) that if error-free performance is required, then at most $N/4\log_2 N$ patterns can be stored and retrieved without error, whereas if a 5% error rate can be tolerated (Amit et al., 1985), about $0.14N$ patterns can be stored and retrieved. Table 2 shows numerically the relationships between N, $N/4\log_2 N$, and $0.14N$.

N	$N/4\log N$	$0.14N$
100	3	14
1 000	24	140
10 000	200	1 400
100 000	1 500	14 000
1 000 000	13 000	140 000

TABLE 2. Performance of Hopfield network as an ACAM. N denotes number of units in the network. (See text for details.)

Hopfield networks represent an important conceptual advance in the theory of neural networks. The principle they embody—storing information in dynamically stable configurations—is profound. Such a principle has its origins in the work of one of the early cyberneticians, W. R. Ashby (1950), who coined the term *ultrastability* to describe the way in which he thought brain activity patterns always tend towards dynamically stable configurations; and it is implicit in the work of Cragg and Temperley (1954) and Little and G. Shaw (1975), among others. However, it is in Hopfield's work that the principle is first made explicit.

How do Hopfield networks compare with Learning Matrices and ACAMs in their ability to store and retrieve information? This question has recently been answered in some detail by E. B. Baum, J. Moody, and F. Wilczek

(1988). They demonstrated that in the absence of malfunctions or failures, the most efficient way to store information is with "grandmother" cells (Barlow 1972), one for each memory to be stored. However, in the presence of malfunctions or failures, grandmother cells make too many mistakes, and information must be stored in a distributed fashion, as suggested in the early work of Winograd and Cowan (1963). It turns out that one very efficient scheme corresponds closely to Marr's model of the cerebellum as an ACAM (Marr 1969). In such a scheme suppose there are N inputs, N outputs, and G internal or "hidden" units (see Figure 17.). Let s be the fraction of inputs active at any instant, and c the corresponding number of active hidden units. (G is roughly equal to the number of ways of selecting c from N lines.) Baum et al. showed that the number of memories which can be stored and retrieved in such a store with an error rate of 5% is bounded by the function $0.05G/s(1-s)$. The fraction s is a measure of the *sparsity* of the input activity. Evidently the store is most efficient if the input lines are mostly inactivated or activated, just as in von Neumann's original scheme (von Neumann 1956). To facilitate comparison with Hopfield networks, we take $c = 3$. Table 3 shows the capacity of such a store for $s = 0.06$ and $s = 0.008$. It is evident that many more patterns can be stored and retrieved in such a system than in a Hopfield network comprising approximately G elements. It remains to be seen if the capacity of Hopfield networks can be improved to comparable levels.

G	$M(0.6)$	$M(.008)$
100	83	625
1 000	830	6 250
10 000	8 300	62 500
100 000	83 000	625 000
1 000 000	830 000	6 250 000

TABLE 3. Performance of Marr-like networks as an ACAM. G denotes the number of elements in the network, and corresponds to the number of hidden units in Marr's model of the cerebellum. Roughly speaking, G corresponds to the number of units in a Hopfield network. In general Marr networks outperform Hopfield networks as reliable ACAMs, by virtue of the extra units they possess for error control.

1.3. Neurocomputing. Hopfield networks have proved of interest in solving computational optimization problems. A well known example of this is the *Traveling Salesman Problem* in which a salesman needs to visit each of a number of cities once, on a schedule or tour which minimizes the total length of his journey. This is an example of what is called a *constrained optimization problem.* It was shown by S. Kirkpatrick, C. D. Gelatt, Jr., and M. P. Vecchi (1983) that the equilibrium configurations taken up by a spin-glass provide solutions to this problem. Following this Hopfield and D. W. Tank (1985) demonstrated that certain Hopfield networks also find good solutions. Although such networks do not find the shortest tour, they do find tours that compare quite favorably with those found by the Kernighan-Lin algorithm (Lin and Kernighan 1973), one of the better algorithms for solving constrained optimization problems. Recently R. Durbin and Willshaw (1987) have devised another neuron-like network which finds even shorter tours, and works better on big problems.

1.4. The Boltzmann machine. Hopfield networks suffer from one defect in their ability to find the best solution in constrained optimization problems. They can get trapped in local minima of the energy function E, essentially because they function in a deterministic fashion. To find the true global minimum requires that from time to time the network make random configurational changes that increase, rather than decrease E, thereby giving it the ability to escape from local minima. This is the essence of a well known version of the *Monte Carlo* algorithm, introduced by N. Metropolis et al. (1953), to find global minima. In this algorithm, the energy change ΔE produced by a random "flip" of one of the network spins is computed. If ΔE is negative, the new configuration is retained, since it has a lower energy than the previous one; otherwise the configuration is rejected; i.e., the flip is cancelled. Such a procedure, although slow, will eventually find the global minimum of E. G. E. Hinton and T. J. Sejnowski (1983) accordingly used the Monte Carlo algorithm to minimize the "energy" of a Hopfield network, in effect repeating S. J. Kirkpatrick et al.'s use of Monte Carlo methods in spin glass problems. In doing so they discovered a process by which the resulting networks, which they called *Boltzmann Machines* (Ackley et al. 1984) can modify their connectivity, in a way which solved the Perceptron credit assignment problem. Thus a Boltzmann machine is an adaptive Hopfield network with hidden units that implements a Monte Carlo algorithm for finding that configuration of active and inactive units which minimizes E. The training rule is as follows: Let p_{ij} be the average probability of two units (the ith and jth) being simultaneously active when the input-units are activated by a stimulus pattern which clamps the motor output-units into some activation pattern, and let p'_{ij} be the corresponding probability when the machine is free running in the absence of stimulation. Recall that w_{ij} is the weight of the ijth contact. The rule by which w_{ij} changes is very simple: if $p_{ij} > p'_{ij}$, add a small constant increment κ to w_{ij}; if $p_{ij} < p'_{ij}$, subtract κ from w_{ij}. {The reader should compare this with the Perceptron and Adaline training rules described earlier.} With such a rule in place Boltzmann machines are capable of solving a variety of constrained optimization problems.

1.5. Back-propagation. However, by far the most striking development in the theory and practice of neural networks was the successful implementation of the Rosenblatt-Widrow-Huff algorithms for training Perceptrons and

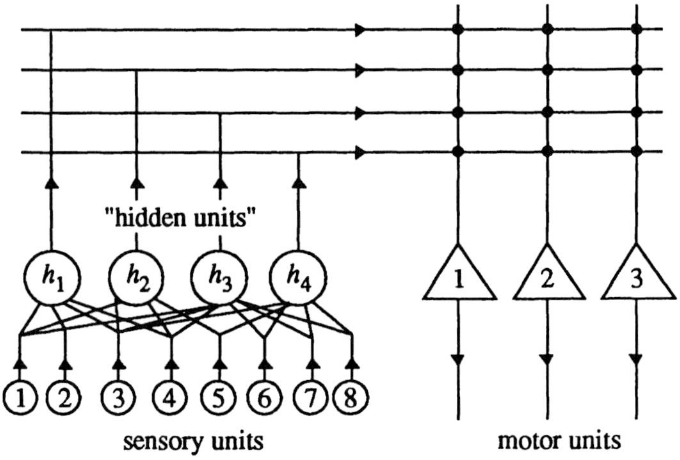

FIGURE 20. Structure of a multilayer Perceptron. In contrast to an elementary Perceptron, there is a layer of hidden units, the synaptic weights of which are also modifiable.

Adalines. This was first solved by D. E. Rumelhart, G. E. Hinton, and R. J. Williams (1986a) and others (Le Cun 1986, Parker 1986). Figure 20 shows the structure of the network used by Rumelhart et al. It is essentially a two-layer Perceptron, with sensory (s) and motor (m) units, and an intermediate layer of hidden units, $h_i (i = 1, 2, \ldots, n)$. The rules by which contacts are modified derive from the Adaline rules given earlier. Recall that in an Adaline, the rule by which weights change is $\Delta w_{ij} = s_i \varepsilon(s)$, where Δw_{ij} denotes the change in w_{ij}, s_i the state of the ith sensory unit (equal to ± 1), and $\varepsilon(s) = a(s) - e(s)$ the difference between the desired response pattern and motor unit excitation.

Rumelhart et al. showed that the following simple extension of the above rule solves the credit assignment problem. Consider two units somewhere in the network, say h_3 and m_2, connected with weight w_{32}. Let $\Delta w_{32} = h_3 \varepsilon_2(s) G(m_2)$, where $G(m_2)$ is the "gain" of the motor unit m_2, (obtained from its current-voltage characteristic (see Figure 14), as $(1/4)(m_2 + 1) \cdot (m_2 - 1)$). Apart from the gain factor, this is the Adaline rule. Now consider the units s_4 and h_3, connected with weight W_{43} and let $\Delta W_{43} = s_4 G(h_3) \cdot (w_{31}\varepsilon_1(s)G(m_1) + w_{32}\varepsilon_2(s)G(m_2) + w_{33}\varepsilon_3(s)G(m_3))$. This is the extended Adaline rule applied to all the m-units driven by h_3. (See Figure 21.) It can be shown that these weight changes are precisely those needed to minimize $\varepsilon(s)$. The actual computations are effected in two stages. In the first (forward) stage the network is stimulated by a pattern s, and the vectors $\varepsilon(s) = (\varepsilon 1(s), \varepsilon_2(s), \varepsilon_3(s))$ and $G(m(s)) = (G(m_1), G(m_2), G(m_3))$ are computed. In the second (backward) stage these are used first to adjust the weights w_{hm}, and then the weights W_{sh}, and so on—hence the description of the algorithm as *back-propagating error correction*. The Rumelhart et al. algorithm provides

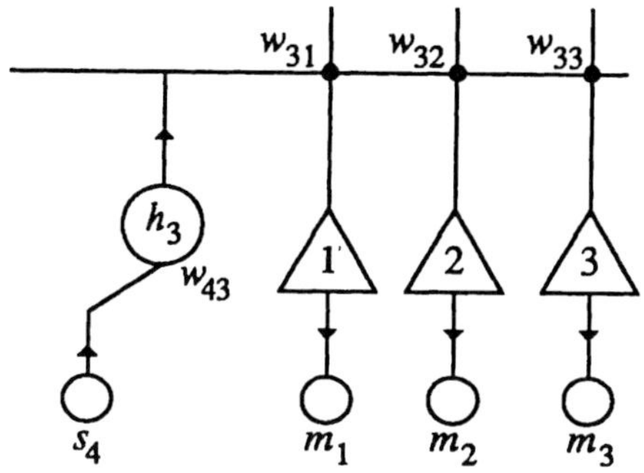

FIGURE 21. How back-propagation works. The network is first stimulated with a pattern of s activations—in this case s_4. The excitations $e_i (i = 1, 2, 3)$ are then obtained. The error signal $\varepsilon = \Sigma_i (a_i - e_i)^2$ is then used, first to adjust the weights w_{31}, w_{32}, and w_{33}, and then the weight W_{43}. (See text for details.)

a solution to the credit-assignment problem and makes multilayered analog Perceptrons and Adalines into powerful tools for investigating supervised learning in adaptive neural networks. In what follows a few applications of back-propagation are described.

1.6. The x OR ELSE y problem. As previously noted, elementary Perceptrons (and Adalines) fail to compute the logical function x OR ELSE y. This function is true if and only if x is true and y is false or vice versa, and is false otherwise. The *truth table* corresponding to x OR ELSE y is shown in Table 4. Using back-propagation, the network shown in Figure 22 is obtained after training comprising 558 sweeps through the four patterns shown in Table 4. The weights shown in the figure include threshold biases of -2.2 and -6.3 which are also determined by the algorithm, since thresholds are just weights from hidden units that are always active. In this case both the hidden units and the motor unit have negative thresholds, and are active $(+1)$ unless sufficiently inhibited. The hidden unit h_2 is active if neither s-unit is active, and when it is active it turns off the motor unit. The motor unit is also inactivated if both s-units are active. This network differs somewhat from that shown

x	y	x OR ELSE y
$+1$	$+1$	-1
$+1$	-1	$+1$
-1	$+1$	$+1$
-1	-1	-1

TABLE 4. "Truth table" for x OR ELSE y. $+1$ denotes "true" and -1 "false." Thus x OR ELSE y is false when x and y are both true or both false, otherwise x OR ELSE y is true.

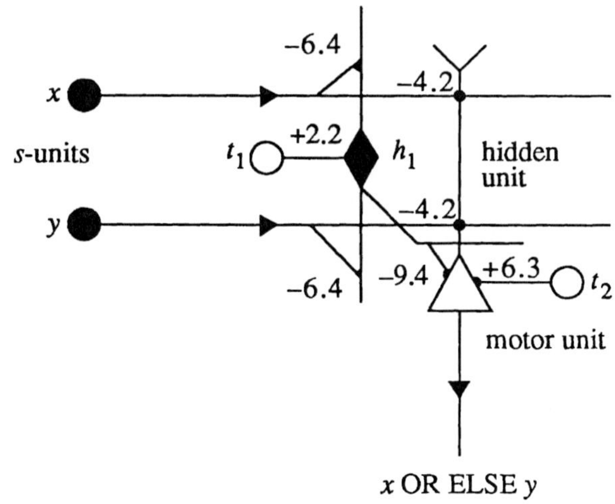

FIGURE 22. Network obtained via back-propagation which computes correctly the logical function x OR ELSE y. The hidden units h_1 and h_3 have 0 threshold and are always active. (See text for details.) [Redrawn from Rumelhart et al. (1986a).]

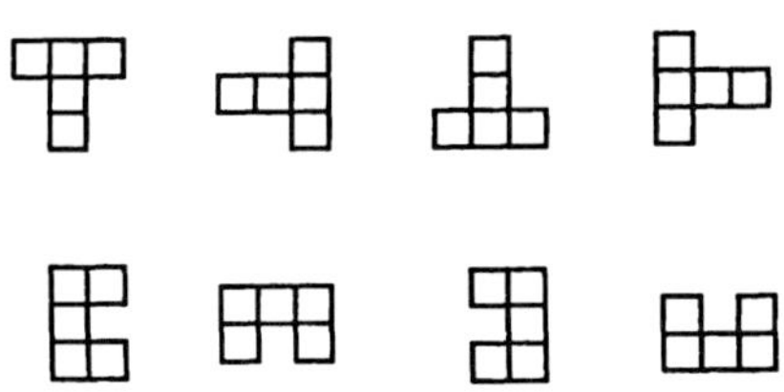

Figure 23. **T** and **C** patterns to be distinguished by a three-layer back-propagating Perceptron. [Redrawn from Rumelhart et al. (1986a).]

in Figure 11 in which all threshold biases are positive, nevertheless it also computes the logical function x OR ELSE y.

1.7. The T/C problem. Another problem solved by back-propagation is the geometric one of distinguishing between the letters **T** and **C**, independently of translation and rotation. Figure 23 shows the patterns, and Figure 24 the architect of the network used by Rumelhart et al. (1986a) to solve the problem. This structure is more or less that used by Rosenblatt and many others (Uhr 1966) on similar problems. It is noteworthy in that the hidden units are connected only with small regions of the sheet of s-units, called *receptive fields*, whereas the motor unit is connected with many widely separated hidden units. This mimics to some extent the architecture of the visual brain. After some 5,000 to 10,000 presentations of the **T** and **C** patterns, together

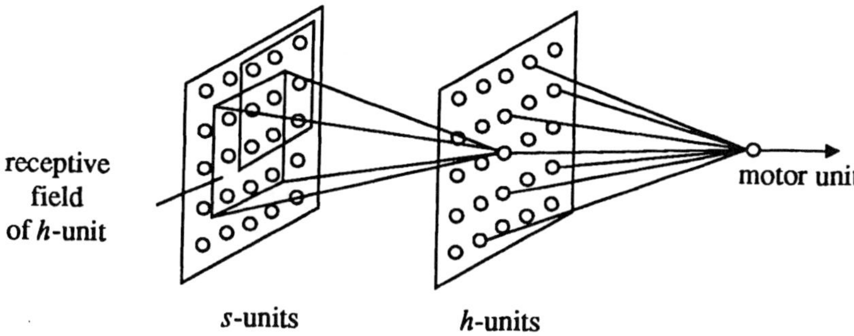

FIGURE 24. Structure of a three-layer Perceptron which can distinguish **T**s from **C**s. The input layer of *s*-units is a two-dimensional sheet, as is the layer of hidden units. The receptive field of each *h*-unit comprises a 3×3 square of *s*-units; i.e., it is area-limited (see Minsky and Papert 1969). Hidden unit weights are adjusted via back-propagation during the course of about 10,000 presentations of **T** and **C** patterns, and converge to a set which distinguishes **T** from **C** patterns. (See text for details.) [Redrawn from Rumelhart et al. (1986a).]

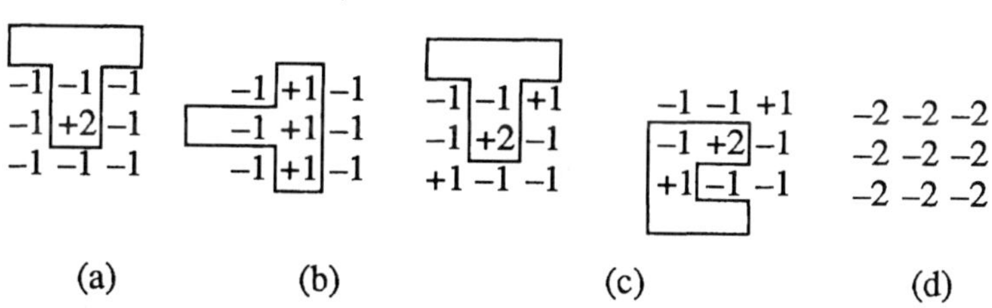

Figure 25. Receptive fields of some of the hidden units which solve the **T**/**C** problem. Each field comprises 9 *s*-units. The contact weights from each *s*-unit in the field of a given *h*-unit are shown as ± 1, and ± 2. (See text for details.) [Redrawn from Rumelhart et al. (1986a).]

with the appropriate responses, the network learns the task. In doing so the hidden unit receptive fields become adapted to the task in several differing ways. Figure 25 shows the weight patterns of some of these fields. Figure 25a shows a receptive field like that of a retinal ganglion cell with an ON center and an OFF surround. It turns out to be an excellent **T**-detector. Figure 25b shows a receptive field like that of a visual cortical edge or bar detector with vertical selectivity. This is also an excellent **T**-detector. Figure 25c is also reminiscent of a cortical edge or bar detector, but with oblique selectivity. Hidden units with such a receptive field can be activated by a **T** in five differing positions, but can be activated by a **C** in only three differing positions. Finally, hidden units with the field shown in Figure 25d are deactivated by

Ts or Cs. However, **C** is a more *compact* figure than **T** and therefore deactivates fewer hidden units. The net effect of all these fields is clearly such as to facilitate distinguishing **Ts** from **Cs** via a final simple threshold biasing of the motor unit.

1.8. NETtalk. Perhaps the most dramatic application of back-propagation is that of Sejnowski and C. R. Rosenberg (1986) who trained a network to read and *speak* English text. The network comprises 203 s-units arranged in 7 groups of 29, 80 h-units, and 26 m-units. In each group of s-units 26 encode one letter of the English alphabet, and the remaining 3 units encode punctuation and word boundaries. A stimulus pattern is thus a string of seven characters. The m-units encode speech sounds or *phonemes*, and also stress and syllable boundaries. The network was trained using back-propagation on a number of texts. One experiment used phonetic transcriptions of the informal continuous speech of a child. Approximately 1,000 words from this corpus were used and, after some 50,000 presentations, the network was able to read and speak with an accuracy of about 95%. Figure 26 shows the receptive field of one of the hidden units. It is evident that this unit has a distributed representation of many attributes of input strings. The network was then presented with a 439-word continuation of text from the same child (containing many novel words), which it read and spoke with an accuracy of 78%. This is an example of *generalization*, and the Sejnowski-Rosenberg network, which they called NETtalk, generalizes very well. In fact multilayer Perceptrons and Adalines generalize quite well over a variety of tasks, just as Rosenblatt claimed many years ago. Another property exhibited by NETtalk is resistance to damage. A substantially damaged NETtalk can still read and speak with an accuracy of some 40% and recovers quickly with retraining. Such properties are to be expected in all networks with distributed representations, as suggested by Winograd and Cowan (1963). Overall NETtalk works surprisingly well, but is limited in its ability to deal with syntactic and semantic ambiguities. More complex, structured versions of NETtalk may be expected to perform better in this respect.

1.9. Receptive fields and neurobiology. The examples described above show clearly that hidden units learn about the stimulus patterns which are presented to them, subject to feedback driven by the intended response of the network to such patterns. In so doing hidden units develop specialized receptive fields (determined by their input weights) in a highly cooperative fashion—each field is affected by all the others in the network—to form a distributed representation of the stimulus class. This has important implications for experimental neurobiology, for although it is unlikely that anything like the back-propagation is employed in the brain, the end product—receptive fields adapted to a given task—can supply clues about how brain networks performing similar tasks might work. This has very practical consequences, since one of the few ways neurobiologists have of directly observing brain operations is

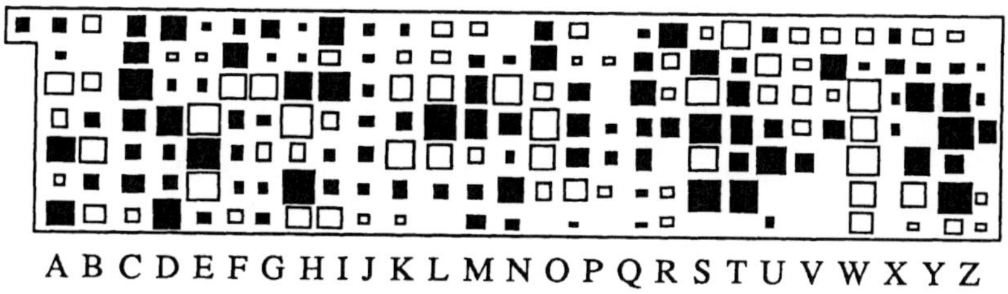

FIGURE 26. Receptive field of a hidden unit in NETtalk. It comprises 203 s-units plus 1 h-unit to set its threshold. The s-units are arranged in 7 groups of 29, 26 of which encode letters of the alphabet, and 3 encode punctuation and spaces. Thus each hidden unit responds to a string of 7 characters, in a specific fashion defined by its weights. (See text for details.) [Redrawn from Sejnowski and Rosenberg (1986).]

by determining which stimulus patterns activate the neuron nearest to a microelectrode implanted in the brain. A first application of back-propagation along these lines has recently been reported by R. Andersen and D. Zipser (1987) in connection with the receptive fields of neurons in the monkey posterior parietal cortex. In such neurons information about eye position relative to the head and to the position of the image on the retina is combined to locate objects with respect to the position of the head in space. Neurons in this area have very large overlapping receptive fields, and their responses are tuned so that they fire at rates proportional to the difference between prefered and actual eye position. Andersen and Zipser modeled this situation with a three-layer back-propagating Perceptron trained to learn spatial positions. After less than 1,000 trials R-unit responses accurately encode object locations in head centered coordinates, given eye and retinal image coordinates. The network has learned to associate eye and retinal image positions, and of course the hidden unit receptive fields reflect this. More important, these receptive fields resemble those in the monkey parietal cortex. It is evident that multilayer Perceptrons can be used in many similar situations to *predict* receptive field properties.

2.0. Concluding remarks. It is now almost fifty years since the publication of McCulloch and Pitts's first paper, and some forty years since von Neumann's work on the reliability problem. Following a period of slow but steady growth neural networks have now entered one of explosive growth: neural networks are now beginning to rival Artificial Intelligence studies in their potential for providing insights into how the brain works, and in providing practical machinery for a variety of nontrivial information processing tasks. Von Neumann's work, if not as seminal as that of McCulloch and Pitts, nevertheless triggered a number of investigations that ultimately led to the modern era of parallel distributed processors (PDP), sometimes referred to as "Non–von Neumann Architectures"!

REFERENCES

Ackley, D. H., Hinton, G. E. and Sejnowski, T. J. 1985. *A learning algorithm for Boltzmann machines*, Cognitive Sci. **9**:147–160.

Amit, D. J., Gutfreund, H. and Sompolinski, H. 1985. *Storing infinite numbers of patterns in a spin-glass model of neural networks*, Phys. Rev. Lett. **55** 14:1530–1533.

Andersen, R. and Zipser, D. 1987. "A neural network model of posterior parietal cortex," *Neurobiology of neocortex*, Dahlem Conferenz. **50**, Berlin.

Anderson, J. A. 1968. *A memory storage model utilizing spatial correlation functions*, Kybernetik, **5**:113–119.

Ashby, W. R. 1950. *The stability of a randomly assembled nerve-network*, EEG Clin. Neurophysiol. **2**:471–482.

Barlow, H. B. 1972. *Single units and sensation: a neuron doctrine for perceptual psychology*? Perception **1**:371–394.

Baum, E. B., Moody, J. and Wilczek, F. 1988. *Internal representations for associative memory*, Biol. Cybern. **59**, 4/5:217–228.

Beurle, R. L. 1956. *Properties of a mass of cells capable of regenerating pulses*, Phil. Trans. Roy. Soc. Lond. B **240**, 669:55–94.

Brindley, G. S. 1967. *The classification of modifiable synapses and their use in models for conditioning*, Proc. Roy. Soc. Lond. B **168**:361–376.

Caianiello, E. R. 1961. *Outline of a theory of thought processes and thinking machines*, J. Theor. Biol. **1**:204–235.

Cowan, J. D. 1967. *A mathematical theory of central nervous activity, Thesis*, University of London.

———. 1968. "Statistical mechanics of nervous nets," In ed. E. R. Caianiello, *Neural Networks*, Springer-Verlag, Berlin, 181–188.

———. 1971. "A statistical mechanics of nervous activity," in ed. M. Gerstenhaber, Proc. 2nd Amer. Math. Soc. Symp. *Mathematical Questions in Biology*, American Mathematical Society, Providence, R.I.

Cragg, B. G. and Temperley, H. N. V. 1954. *The organisation of neurones: a cooperative analogy*, EEG Clin. Neurophysiol. **6**:85–92.

———. 1955. *Memory: the analogy with ferromagnetic hysteresis*, Brain **78**, II:304–316.

Craik, K. J. W. 1943. *The nature of explanation*, Cambridge Univ. Press, Cambridge, MA.

Durbin, R. and Willshaw, D. J. 1987. *An analogue approach to the travelling salesman problem using an elastic net method*, Nature **326**:689–691.

Eccles, J. C. 1957. *The physiology of nerve cells*, Johns Hopkins University Press, Baltimore, MD.

———. 1964. *The physiology of synapses*, Springer-Verlag, Berlin.

Ermentrout, G. B. and Cowan, J. D. 1979a. *Temporal oscillations in neural networks*, J. Math. Biol. **7**:265–280.

———. 1979b. *A mathematical theory of visual hallucination patterns*, Biol. Cybernet. **34**:137–150.

Gabor, D. 1969. *Associative holographic memories*, IBM J. Res. Dev. **13**, 2:156–159.

Hamming, R. W. 1950. *Error detecting and error correcting codes*, Bell Syst. Tech. J. **29**:147–160.

Hebb, D. O. 1949. *The organization of behavior*, Wiley, New York.

Hinton, G. E. and Sejnowski, T. J. 1983. *Optimal perceptual inference*, Proc. IEEE Computer Soc. Conf. Computer Vision and Pattern Recognition, Washington D.C., 448–453.

Hopfield, J. J. 1982. *Neural networks and physical systems with emergent collective computational abilities*, PNAS **79**:2554–2558.

Hopfield, J. J. and Tank, D. W. 1985. *'Neural' computation and constraint satisfaction problems and the traveling salesman*, Biol. Cybern. **55**:141–152.

Kirkpatrick, S., Gelatt, C. D. Jr., and Vecchi, M. P. 1983. *Optimization by simulated annealing*, Science **229** 4598:671–679.

Kohonen, T. 1977. *Associative memory—a system-theoretical approach*, Springer-Verlag, Berlin.

———. 1984. *Self-organization and associative memory*, Springer-Verlag, Berlin.

Lasalle, J. and Lefschetz, S. 1961. *Stability by Liapunov's direct method with applications*, Academic Press, New York.

Le Cun, Y. 1985. *A learning scheme for asymmetric threshold networks*, Proc. Cognitiva **85**:599–607.

Lin, S. and Kernighan, B. W. 1973. *An algorithm for the TSP problem*, Oper. Res. **21**:498–522.

Linsker, R. 1986a. *From basic network principles to neural architecture: emergence of spatial-opponent cells*, PNAS **83**:7508–7512.

———. 1986b. *From basic network principles to neural architecture: emergence of orientation-selective cells*, PNAS **83**:8390–8394.

———. 1986c. *From basic network principles to neural architecture: emergence of orientation columns*, PNAS **83**:8779–8783.

Little, W. A. and Shaw, G. L. 1975. *A statistical theory of short- and long-term memory*, Beh. Biol. **14**:115–133.

Marr, D. 1969. *A theory of cerebellar cortex*, J.Physiol. Lond. **202**:437–470.

———. 1971. *Simple memory: a theory for archicortex*, Phil. Trans. Roy. Soc. Lond. B **262** 841:23–81.

———. 1974. *The computation of lightness by the primate retina*, Vision Res. **14**:1377–1388.

Marr, D. and Poggio, T. 1976. *Cooperative computation of stereo disparity*, Science **194**:283–287.

Metropolis, N., Rosenbluth, A., Rosenbluth, M., Teller, M., and Teller, E. 1953. *Equations of state calculation by fast computing machines*, J. Chem. Phys. **21**:1087–1092.

McCulloch, W. S. 1961. *What is a number, that a man may know it, and a man, that he may know a number?*, General Semantics Bull., 26–27, 7–18.

———. 1967. *Embodiments of Mind*, MIT Press, Cambridge, MA.

McCulloch W. S. and Pitts, W. 1943. *A logical calculus of the ideas immanent in nervous activity*, Bull. Math. Biophys. **5**:115–133.

McEliece, R. J., Posner, E. C., Rodemich, E. R., and Venkatesh, S. S. 1987. *The capacity of the Hopfield associative memory*, IEEE Trans. Inf. Theory **1**:33–45.

Minsky, M. and Papert, S. 1969. *Perceptrons: an introduction to computational geometry*, MIT Press, Cambridge, MA.

Neumann, J. von 1951. "The general and logical theory of automata," In ed. Jeffress, L. A., *Cerebral mechanisms in behavior*, the Hixon Symposium, Wiley, New York.

———. 1956. "Probabilistic logics and the synthesis of reliable organisms from unreliable components," In eds. Shannon, C. E. and McCarthy, J. *Automata studies*, Princeton University Press, Princeton, N.J., 43–98.

———. 1958. *The computer and the brain*, Silliman Lectures, Princeton University Press, Princeton, N.J.

Novikoff, A. 1963. "On convergence proofs for Perceptrons," In ed. Fox, J., *Symp. on Mathematical Theory of Automata*, Polytechnic Press, New York, 615–622.

Parker, D. B. 1986. "A comparison of algorithms for neuron-like cells," Proc. AIP Conf. **151**, In ed., Denker, J. S., *Neural networks for computing*, Amer. Inst. Physics, New York.

Rescorla, R. A. and Wagner, A. R. 1972. "A theory of Pavlovian conditioning: variations in the effectiveness of reinforcement and nonreinforcement." In eds. Black, A. H. and Prokasy, W. F. *Classical conditioning* II, Appleton-Century-Crofts, New York, 64–99.

Rosenblatt, F. 1958. *The Perceptron, a probabilistic model for information storage and organization in the brain*, Psych. Rev. **62**:386–408.

———. 1961. *Principles of neurodynamics: Perceptrons and the theory of brain mechanisms*, Spartan Books, Washington, D.C.

Rosenbleuth, A., Wiener, N. and Bigelow, J. 1943. *Behavior, purpose and teleology*, Philosophy of Science **10**:18–24.

Rumelhart, D. E., Hinton, G. E., and Williams, R. J. 1986a. "Learning internal representations by error propagation," In eds., Rumelhart, D. E. and McClelland, J. L., *Parallel distributed processing: explorations in the microstructure of cognition*, 1, *Foundations*, MIT Press, Cambridge, MA.

———. 1986b. *Learning representations by back-propagating errors*, Nature **323**: 533–536.

Sbitnev, V. I. 1982. *The origin and evolution of a strange attractor in a system of coupled oscillators*, USSR Acad. Sci., Leningrad Nucl. Phys. Inst., preprint.

Sejnowski, T. and Rosenberg, C. R. 1986. *NETtalk, a parallel network that learns to read aloud*, Tech. Rep. JHU/EECS-86/01 (Johns Hopkins Univ. EE and Comp. Sci.).

Shannon, C. E. 1948. *A mathematical theory of communication*, Bell Syst. Tech. J. **27**:379–423; 623–656.

Sheffer, H. M. 1913. *A set of five independent postulates for Boolean algebras, with application to logical constants*, Trans. Amer. Math. Soc. **14**:481–488.

Sherrington, D. and Kirkpatrick, S. 1975. *Spin glasses*, Phys. Rev. Lett. **35**:1972.

Steinbuch, K. 1961. *Die lernmatrix*, Kybernetik **1**:36–45.

Swindale, N. V. 1980. *A model for the formation of ocular dominance stripes*, Proc. Roy. Soc. Lond. B **208**:243–264.

——. 1982. *A model for the formation of orientation columns*, Proc. Roy. Soc. Lond. B **215**:211–230.

Taylor, W. K. 1956. *Electrical simulation of some nervous system functional activities*, Information Theory **3**, ed. Cherry, E. C., Butterworths, London, 314–328.

Turing, A. M. 1936. *On computable numbers with an application to the Entscheidüngsproblem*, Proc. London Math. Soc. **XLII**:230–265.

——. 1937. Proc. London Math. Soc. **XLIII**:544–546.

——. 1952. The chemical basis of morphogenesis, Phil. Trans. Roy. Soc. Lond. B **237**:32–72.

Uhr, L. 1966. ed., *Pattern recognition*, Wiley, New York.

Whitelaw, V. A. and Cowan, J. D. 1981. *Specificity and plasticity of retinotectal connections: a computational model*, J. Neurosci. **1** 12:1369–1387.

Widrow, B. and Hoff, M. E. 1960. *Adaptive switching circuits*, WESCON convention record **IV**:96–104.

Willshaw, D. J., Buneman, O. P., and Longuet-Higgins, H. C. 1969. *Non-holographic associative memory*, Nature, Lond **222**:960–962.

Willshaw, D. J. and v. d. Malsburg, Ch. 1976. *How patterned neural connections can be set up by self-organization*, Proc. Roy. Soc. Lond. B. **194**:431–445.

——. 1979. *A marker induction mechanism for the establishment of ordered neural mappings: Its application to the retinotectal problem*, Phil. Trans. Roy. Soc. Lond. B **287**:203–243.

Wilson, H. R. and Cowan, J. D. 1972. *Excitatory and inhibitory interactions in localized populations of model neurons*, Biophys. J. **12**:1–24.

——. 1973. *A mathematical theory of the functional dynamics of cortical and thalamic nervous tissue*, Kybernetik **13**:55–80.

Winograd, S. and Cowan, J. D. 1963. *Reliable computation in the presence of noise*, MIT Press, Cambridge, MA.

THE UNIVERSITY OF CHICAGO

Proceedings of Symposia in Pure Mathematics
Volume **50** (1990)

Two-person, Perfect-information GAMES

ELWYN BERLEKAMP

I want to talk to you today about GAMES, particularly mathematical games, a subject that was invented by John von Neumann. More precisely, I want to talk to you about

two-person, perfect-information games

which is a subset of game theory in which there has been enormous progress in the past twenty years.

Although this topic might be viewed as trivial from the abstract perspective of von Neumann's minimax theory, I believe that John von Neumann would have been very interested in the modern theory of two-person, perfect-information games because it deals so intimately with

- recursions,
- algebras,
- fundamental axioms,

and a wide range of applications, and because it provides mathematical solutions to certain problems for which no one suspected a priori that any such solutions would exist.

And, just as von Neumann believed that his general theory of games had something to say about economics, I believe that the modern combinatorial theory of two-person, perfect-information games may have something to say about management, where one is often trying to identify which program or subdepartment in which to make the next move.

Actually, the topic about which the theory has the most to say is

SUMS of Two-person, Perfect-information games

although that title would be considered overly restrictive.

One of the many games about which the theory has a lot to say is called DOMINEERING. This game is played on some subset of a (possibly enlarged) checkboard. Each player has a bountiful supply of 2×1 dominoes.

1980 *Mathematics Subject Classification* (1985 *Revision*). Primary 90D052, 00A08.

This paper is in final form. It overlaps the paper "Introduction to Blockbusting and Domineering", but no other version will be submitted for publication elsewhere..

FIGURE 1

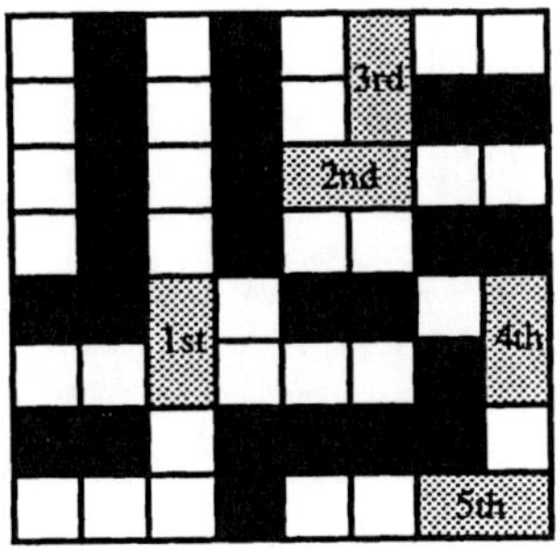

FIGURE 2

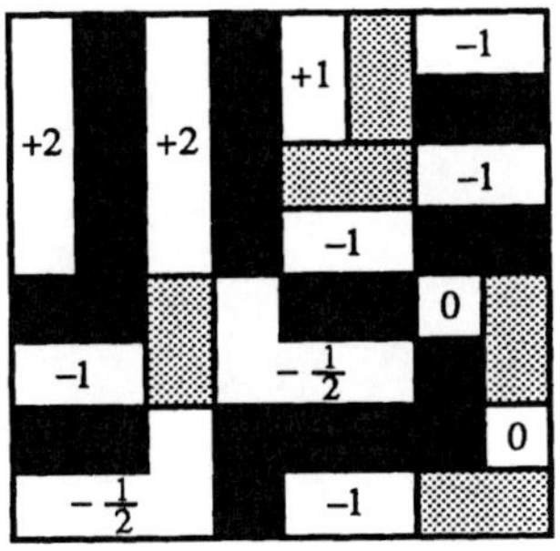

FIGURE 3

At each turn, Left places one of his dominoes onto any available space on the board in a verticaL position; right then places one of his dominoes onto the board in a hoRizontal position. The game continues until one player is unable to move and that player is then the loser.

Figure 1 shows a domineering game in progress, with Left's turn to play. The continuation of the game is shown in Figure 2. Left decides to move at location 1; Right, at location 2; Left, at location 3; then Right, at location 4; then Left, at location 5, after which the position is shown in Figure 3.

The position shown in Figure 3 is easily decomposed into several regions. Each of these regions can be assigned a value according to the number of moves advantage that it provides to Left. For example, a 2×1 region provides a place where Left can place one domino, but Right cannot place any. Hence, we say that the value of a 2×1 region is $+1$; it constitutes a one-move advantage for Left. Similarly, a 4×1 region provides a place where Left can place two dominoes while Right cannot move at all. Evidently, a 4×1 region has value $+2$. Similarly, a 1×2 region has a value of -1 because it offers a one-move advantage to Right. A 1×1 region has value 0 because it offers no moves to either player.

Regions of more complicated shapes have more complicated values. For example, each of the two regions of Figure 3 that is shaped like an L lying on its side has a value of $-1/2$. We shall justify this assertion later. At the moment, however, we merely wish to show that we can compute the value of the entire board simply by summing the values of the regions therein. Thus, the value of the game position shown in Figure 3 is $+2+2+1-1-1-1-1-1/2-1/2-1=-1$. Since this is negative, the game can be won by Right.

In general, we say that a game is

- Positive iff Left can win (no matter who plays first),
- Negative iff Right can win (no matter who plays first),
- Null iff second player can win (no matter whether he is Left or Right), or
- Fuzzy iff first player can win (no matter whether he is Left or Right).

Every Domineering region has a value, and the value of a sum of regions is equal to the sum of the corresponding values. Sometimes, as in Figure

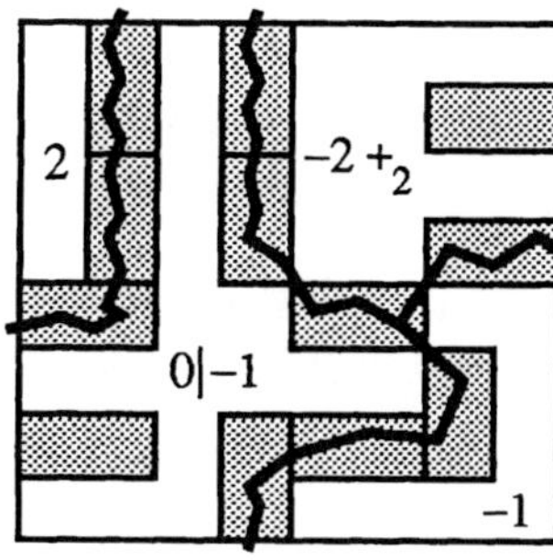

FIGURE 4

3, the values are familiar numbers. However, other positions, such as those which appear in Figure 1, have the more complicated values shown in Figure 4. We shall return to describe the properties of these values later.

In general, we define a game, recursively, as a pair of sets of games. Symbolically, these two sets are separated by a vertical slash:

$$G = \{G^L | G^R\}.$$

In words, G is defined as the game whose left followers comprise the set G^L and whose Right followers comprise the set G^R. We usually simply denote these sets by listing their elements, separated by commas.

The simplest game is the one whose sets of Left and Right followers are both empty. This game is denoted by "0":

$$0 = \{\ |\ \}.$$

In general, we may define the BIRTHDAY of a game as one more than the birthday of any of its followers, with the starting proviso that the game 0 has birthday 0. There are then three different games born on Day 1:

$$1 = \{0|\ \}, \qquad -1 = \{\ |0\}, \qquad * = \{0|0\}.$$

The game "$*$" is called "star". It appears in Domineering as

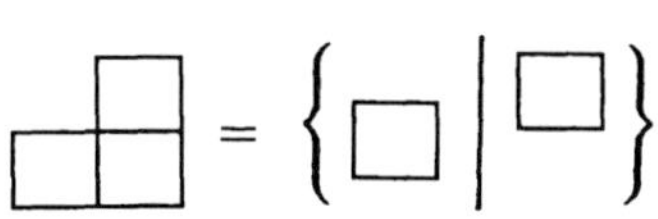

$= \{0|0\} = *.$

Altogether, there are four games born on Day 0 or Day 1. A game born on or before Day 2 can have any subset of these four games as its Left followers and any subset of these four games as its Right followers. This gives $2^4 = 16$ possible sets of Left followers and 16 possible sets of Right followers, for 256 choices altogether. Among these 256 different choices we find such games as these:

$$\{-1|1\}, \qquad \{-1|*\}, \qquad \{-1, *|\ \}.$$

Each of these three games has the property that the second player can win. It is not hard to show that adding such a game into a sum has no effect on the outcome. More specifically, if I can win a game H, and if G is a game in which the second player can win, then I can also win $G+H$. My strategy is simple: I never play on G except when my opponent does, and then I always respond immediately on G according to the winning strategy for the second player there.

This argument justifies calling ANY game which the second player can win a NULL game, and denoting its value by zero:

$$\{-1|1\} = 0, \qquad \{-1|*\} = 0, \qquad \{-1, *|\ \} = 0.$$

Similarly, we may define the NEGATIVE of a game G as

$$-G = \{G^L|G^R\}.$$

It is not hard to show that $G + (-G)$ is a game which the second player can win, i.e., zero. It is then reasonable to say that two games, G and H, are EQUAL iff $G - H$ is a game which the second player can win. This definition of equality partitions the 256 games born on or before Day 2 into 22 equivalence classes, such as the following:

$$\begin{aligned} 2 &= \{0, 1|\ \} = \{1|\ \}, \\ 1/2 &= \{0|1\} = \{-1, 0|1\}, \\ \uparrow = \text{“up”} &= \{0|*\} = \{0, *|*\}, \\ *2 &= \{0, *|0, *\}, \\ \pm 1 &= \{1|-1\}, \\ -1/2 &= \{-1|0\} = \{-1|0, 1\}. \end{aligned}$$

The game $-1/2$ appeared in Figure 3:

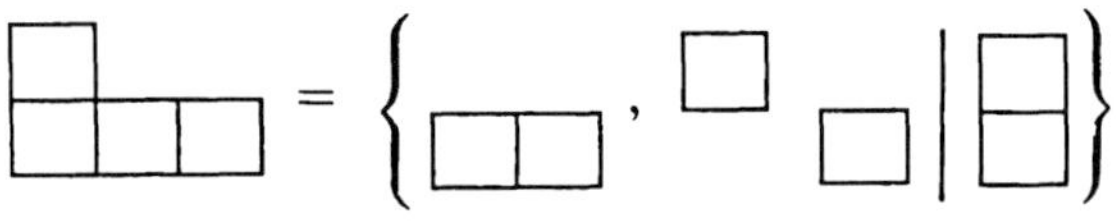

$$= \{-1|0+0, 1\} = \{-1|0, 1\}$$
$$= \{-1|0\}.$$

It is now also possible to justify the reasonableness of our choices of names. To justify calling $\{0|1\}$ by the name “1/2”, we should show that

$$1/2 + 1/2 - 1 = 0$$

in the sense that the second player can win from

$$1/2 + 1/2 - 1.$$

We may also define inequality in the obvious manner: $G < H$ means that Left can win on $G - H$ no matter who goes first, while $G \leq H$ means that Left, going second, can win on $G - H$.

These basics, which I have just described intuitively, can also be stated more formally as the following axioms of combinatorial game theory. This formulation of the axioms is due to John Conway.

Conway's axioms

1. Game: $G = \{G^L | G^R\}$.
2. Sum: $G + H = \{G + H^L, G^L + H | G + H^R, G^R + H\}$.
3. Minus: $-G = \{G^L | G^R\}$.
4. Partial Order: $G \geq H$ unlesss (unless and only unless) $H \geq$ some G^R or some $H^L \geq G$.

To see the correspondence between Axiom 4 and the intuitive view that we discussed above, we might restate Axiom 4 as

$$G - H \geq 0 \text{ unlesss,}$$
$$G^R - H \leq 0 \text{ for some } G^R, \text{ or}$$
$$G - H^L \leq 0 \text{ for some } H^L.$$

This assertion can now be directly interpreted as saying that Left, going second on $G - H$, can win unless and only unless Right, going first, can win on $G - H$. For, if Right has a winning opening move on $G - H$, then it must be to a position of the form $G^R - H$ or $G - H^L$ and, in either case, such a position must then be one from which Right can win going second.

Intuitively, these axioms apply directly to games such as Domineering, in which the winner is the last player to move. Remarkably, the theorems and results derived from these axioms turn out to be sufficiently broad in scope that they also give results which apply to many games with scores. Among the games to which the theory is particularly relevant are the following:

- Go, a popular Japanese board game;
- Dots & Boxes, a pencil and paper game popular among schoolchildren in some sections of the U.S.A.;
- Domineering;
- Toads & Frogs;
- Col;
- Hackenbush Hotchpotch;
- Blue-Red Hackenbush, Green Hackenbush

and dozens of others, many of which were introduced in the book *Winning Ways* to illustrate various salient aspects of this general theory.

The types of results that the theory provides include the following:

- Complete, perfect strategies for arbitrary sums of "solved" games.
- Equivalent positions occur in many different games.
- Some more difficult games are solved to within infinitesimals.
- Others solved to within close approximation.
- Many more solved "Asymptotically" via "Best Numerical Approximation" or "Mean Value".
- Some classes are proved NP-Hard.

In some games, such as the Japanese game of "Go", the theory offers explicit solutions to a variety of endgame puzzles which appear very difficult even to experienced Go players. However, the sorts of positions that appear in these puzzles don't seem to arise in play between expert Go players. The computations needed to apply the theory to more realistic Go positions appear to be so cumbersome that the theory offers little or no help to experienced Go players, who play excellent Go without using any serious mathematics.

In other games, such as "Dots and Boxes", the computations suggested by the theory are tractable only in limited types of situations, but these situations occur so frequently that anyone who has studied the book *Winning Ways* seems to enjoy an enormous advantage over any "Dots and Boxes" opponent who has not.

The axioms define an Abelian group of objects called games. One interesting subgroup of games, called NUMBERS, can be defined as those games which satisfy two additional restrictions:

- all G^L and all G^R are numbers,
- no $G^R \geq$ any G^L.

If G^L is empty, then the value of the game is the least integer greater than all G^R; if G^R is empty the value of the game is the greatest integer less than all G^L. If both G^L and G^R are nonempty, then it turns out that the value of such a game is the SIMPLEST number in the interval between G^L and G^R.

Simplicity can be defined in several different ways which yield equivalent results. One approach is to say that among integers, the smaller the magnitude, the simpler the integer; among nonintegers, the smaller the power of two in the denominator, the simpler the number. Another approach is to define simplicity explicitly in terms of the game's birthday: the integers n and $-n$ are born on Day n; any fraction between 0 and 1 which is represented by an odd number divided by 2^{-n} is also born on Day n. The earlier the birthday, the simpler the game.

Abstractly, examples of numbers include the following:

$$\tfrac{1}{8} = \{0|\tfrac{1}{4}\}$$

$$\dots$$

$$2^{-(n+1)} = \{0|2^{-n}\}$$

$$\tfrac{3}{4} = \{\tfrac{1}{2}|1\}$$

$$\tfrac{7}{8} = \{\tfrac{3}{4}|1\}$$

$$\dots$$

$$3 = \{2|\ \}$$

$$4 = \{3|\ \}$$

$$\dots$$

$$1\tfrac{1}{2} = \{1|2\}$$

$$2\tfrac{3}{4} = \{2\tfrac{1}{2}|3\}$$

$$\dots$$

And here are some numbers that are not in canonical form:

$$\tfrac{1}{2} = \{0|\tfrac{3}{4}\}, \qquad 1 = \{0|\{1|1\}\}, \qquad \tfrac{3}{4} = \{\tfrac{5}{8}|1\}.$$

Domineering positions are numbers:

$\frac{3}{4} =$ 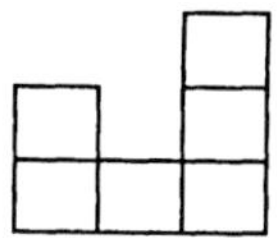

$\frac{7}{8} =$ 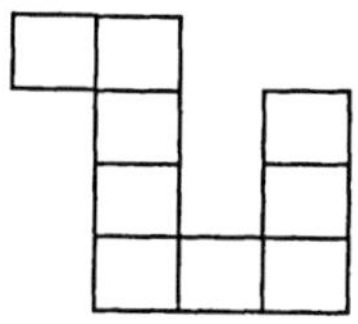

$1\frac{1}{2} =$ 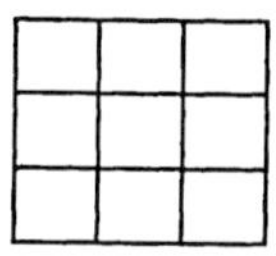

$0 =$ 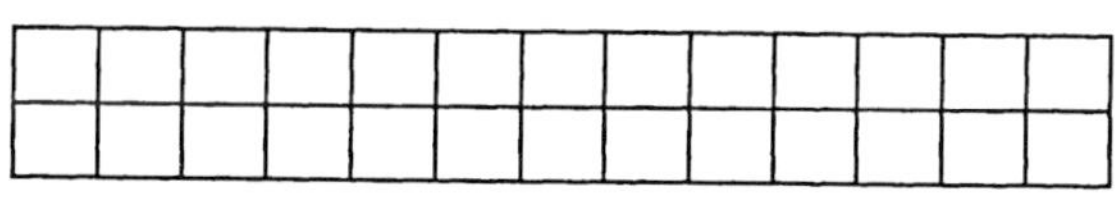

These assertions are by no means obvious. The proof that the 2×13 board has value 0 is well beyond the scope of the present exposition; it appears in *Blockbusting and Domineering*.

It is instructive to examine the ordering of sums of some of the games born on or before Day 2. For example, we find that

$$* + * = 0.$$

This means that, in some sense, $*$ has characteristic 2. The game $\uparrow$ (pronounced "up") and its negative $\downarrow$ (pronounced "down"), have several interesting properties. These games occur in Domineering:

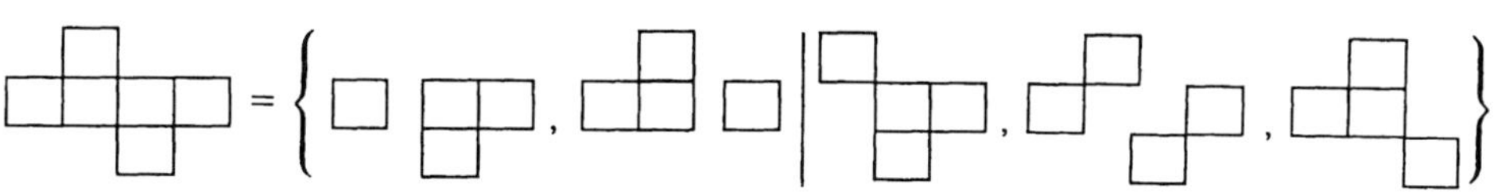

$$= \{0 + *, **+)|0 + *, 0 + 0 + 0 + 0, * + 0\}$$
$$= \{*|0, *\} = \downarrow .$$

It is easily seen that Left can win the game $\uparrow$, no matter who goes first. Hence,

$$\uparrow > 0.$$

Therefore,

$$0 < \uparrow < \uparrow + \uparrow < \uparrow + \uparrow + \uparrow < \cdots .$$

It is convenient to denote

$$2 \cdot \uparrow = \uparrow + \uparrow, 3 \cdot \uparrow = \uparrow + \uparrow + \uparrow, \cdots .$$

If n is any positive integer, then we still find

$$n \cdot \uparrow < 1.$$

In other words, $\uparrow$ is a positive infinitesimal and we write

$$0 < \uparrow \ll 1.$$

$*$ is also an infinitesimal. In fact,

$$2 \cdot \downarrow < * < 2 \cdot \uparrow,$$

although each of the games $*$, $* + \uparrow$, and $* + \downarrow$ is fuzzy.

The 4×2 rectangular Domineering game has the value $\{0|\{0|-2\}\}$. We christen this game "tiny-2". More generally, if x is any nonnegative number we define the game

$$\text{"tiny-}x\text{"} = \{0|\{0|-x\}\}.$$

In particular, $\uparrow =$ "tiny-0".

It can be shown that, $0 < x < y$, then

$$0 < \text{"tiny-}y\text{"} \ll \text{"tiny-}x\text{"}.$$

In particular, "tiny-2" is a higher order infinitesimal than $\uparrow$.

It turns out that there are many other types of infinitesimals other than the tinys. However, $\uparrow$ has the lowest order of them all, in the sense that given any such infinitesimal, one can find an integer multiple of $\uparrow$ that is larger. The position in the upper right corner of Figure 4 has value equal to -2 +"tiny-2" which we write as $-2+_2$. Just as elementary school teachers tell us to omit the implied "plus" sign in numbers like $2\frac{1}{4}$, game theorists find it convenient to omit plus signs between numbers and infinitesimals. This differs from the common convention in physics or engineering, where omitted arithmetic operators are assumed to be "times" In game theory, we assume omitted arithmetic operators are "plus".

One of the games born on Day 2, $*2 = \{0, *|0, *\}$, is part of an obvious sequence of games born on successive days:

$$*3 = \{0, *, *2|0, *, *2\}, \ *4 = \{0, *, *2, *3|0, *, *2, *3\},$$
$$*5 = \{0, *, *2, *3, *4|0, *, *2, *3, *4\}, \cdots .$$

As far as I know, 0 and $*$ are the only games in this sequence which occur in Domineering. However, these games all appear in the classical game of NIM, which enjoyed popularity in the late 1800s. This game is played on several heaps of counters. At each turn, a player can remove as many counters as he wishes, subject to the constraint that all removed counters must come from the same heap. The popular game of Nim was solved by Bouton (a Harvard professor) in 1901, with his observation that

$$*n + *m = *(n \overset{*}{+} m)$$

where $\overset{*}{+}$ is the exclusive-or operation performed on the binary representation of the numbers. Thus, for example, from the Nim position consisting of four heaps of sizes 7, 5, 3, 1 (which occurs in the popular movie called "Last Year at Marianbad"), there is no winning move because

$$*7 + *5 + *3 + * = 0$$

as can be verified by this calculation:

$$\begin{array}{c} 111 \\ 101 \\ 011 \\ 001 \\ --- \\ 000 \end{array}$$

The games 0, $*$, $*2$, $*3$, are called NIMBERS. Within the additive Abelian group of all games, the nimbers form an elementary Abelian 2-group. All of the nimbers are infinitesimals. They are all smaller than $\uparrow + \uparrow$.

The finite games which are numbers form a dense set in the real line and any finite game can be compared with the set of numbers. A typical game will be less than all sufficiently large positive numbers, greater than all sufficiently small negative numbers, and "confused" with all numbers in some intermediate range (in the sense that the difference between the game under consideration and any number in that range is a "fuzzy" game which can be won by the first player). This range is called the "confusion interval". For some games, like $\uparrow$, the confusion interval can be empty: $\uparrow$ is less than all positive numbers but greater than zero and all negative numbers. For other games, the confusion interval may consist of a single number: $*$ is confused only with 0; it is greater than all negative numbers and less than all positive numbers. However, some other games have bigger confusion intervals. Several such games appear on Day 2, including the following:

$$\{1|-1\}, \qquad \{0|-1\}, \qquad \{1|*\}.$$

The confusion intervals of these games are $[-1, 1]$, $[0, -1]$, and $[1, 0)$, respectively. Games whose confusion intervals consist of the single number 0 are infinitesimals. Games whose confusion intervals contain more than one number are said to be "hot".

Many interesting hot games appear in Domineering, including those shown in Figure 5.

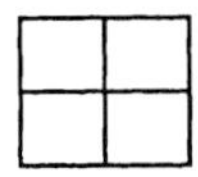

$$= \{1 | -1\} = \pm 1$$

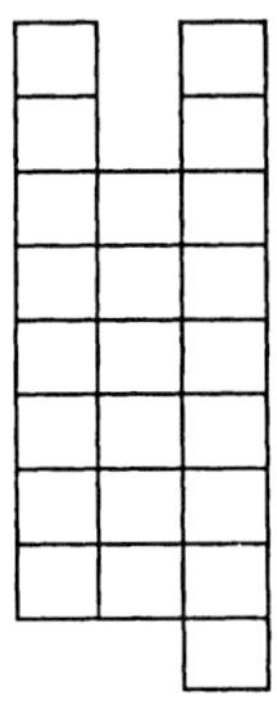

$$= 5\tfrac{3}{4} ||| 5\tfrac{1}{2} | 3\tfrac{1}{4} || 3 | \tfrac{1}{2}$$

$$= \{5\tfrac{3}{4} | \{\{5\tfrac{1}{2} | 3\tfrac{1}{4}\} | \{3 | \tfrac{1}{2}\}\}\}$$

FIGURE 5

(As seen in the second example in Figure 5, instead of writing multiple levels of brackets, we often find it convenient to omit all brackets entirely and to indicate the precedence of slashes by using multiple slashes. Double slash takes precedence over single slash; triple slash takes precedence over double slash, etc.)

We now have this complicated-looking expression:

$$5\tfrac{3}{4} ||| 5\tfrac{1}{2} | 3\tfrac{1}{4} || 3 | \tfrac{1}{2}$$

This expression becomes tractable when it is viewed as an OVERHEATED NUMBER. Abstractly, such games are defined by applying "heating" and "overheating" operators to numbers. It turns out that, under appropriate restrictions too complicated to detail here, such operators are linear and order-preserving. To emphasize the linearity of these operators, we denote them by integral signs. Since combinatorial game theory is totally discrete, there is no risk of confusion with the conventional meanings of integrals.

Heating and overheating are discussed in Chapter 6 of *Winning Ways*. In *Blockbusting and Domineering*, those concepts were refined and extended to the following:

Heating and overheating. Let G be a game whose canonical form is:

$$G = \{G^L | G^R\}.$$

Then we define "G heated by t" as:

$$\int^t G = \begin{cases} G & \text{if } G \text{ is a number,} \\ \left\{ t + \int^t G^L \middle| -t + \int^t G^R \right\}, & \text{otherwise,} \end{cases}$$

and for $s > 0$, we define "G overheated from s to t" as:

$$\int_s^t G = \begin{cases} Gs = \overbrace{s + s + \cdots + s}^{G \text{ times}} & \text{if } G \text{ is an integer,} \\ \left\{ t + \int_s^t G^L \middle| -t + \int_s^t G^R \right\} & \text{otherwise.} \end{cases}$$

For example,

$$\begin{aligned} \int_1^{1*} \frac{3}{4} &= \int_1^{1*} \left\{ \frac{1}{2} \middle| 1 \right\} \\ &= \left\{ 1* + \int_1^{1*} \frac{1}{2} \middle\| -1* + \int_1^{1*} 1 \right\} \\ &= \left\{ 1* + \int_1^{1*} \{0|1\} \middle\| -1* + 1 \right\} \\ &= \left\{ 1* + \left\{ 1* + \int 0 \middle| -1* + \int 1 \right\} \middle\| * \right\} \\ &= \{ 1* + \{ 1* | * \} \| * \} = 2|1\| * . \end{aligned}$$

The heating and overheating operators allow us to solve many Domineering positions explicitly. For example:

$$5\tfrac{3}{4}|||5\tfrac{1}{2}|3\tfrac{1}{4}||3|\tfrac{1}{2} = 4\tfrac{5}{8} + \int^{9/8} * - \int^{9/8} \int_{1/4}^{1/4*} \tfrac{7}{8}.$$

Similar expressions provide explicit and usable solutions to Domineering boards of sizes $2 \times 2n + 1$ and $3 \times 2n + 1$, with one extra square attached, and many positions that occur in such games. These results may be found in my recent paper, *Blockbusting and Domineering.*

Let us now step back from specific games, such as Domineering, and take a more global view of the current status of this subject. Here is an outline of some of the more interesting classes of games:

- Hot games
    - Overheated Numbers
- Infinitesimals
    - Higher-order Infinitesimals
    - Entire games

- Ups & downs
- Nimbers
- Numbers
  - Surreal Numbers
    - Transfinite Numbers
  - Dyadic Rationals
  - Integers
    - Null Games

As we continue to investigate more games, in more detail, we encounter two types of obstacles:

- Mathematical: Structures may appear intractable.
- Computer Science: Computations may be NP hard or P-space hard.

However, it is more productive to view these obstacles as the corresponding challenges:

- Find better, more tractable structure.
- Find large, interesting, tractable proper subsets of the hard games.

We have achieved a quantitative understanding of notions like

- On which project do we move next?
- Should we seize the initiative?
- Does a proposed move keep sente?

I also personally believe that this understanding offers a small amount of additional insight into a wide variety of real-life negotiating situations which are game-like only in a weak or superficial sense.

References

1. E. R. Berlekamp, J. H. Conway, and R. K. Guy, *Winning ways*, Academic Press, New York, 1982.

2. E. R. Berlekamp, *Introduction to blockbusting and domineering*, paper presented at Strens Memorial Conference, Univ. of Calgary, July 1986.

3. ——, *Blockbusting and domineering*, J. Combin. Theory Ser. A. **49** (1988), 67–116.

Titles of Viewgraphs Used at the von Neumann Conference Talk

01 Title GAMES, with refs
02 2-person perfect-information
03 sums of
04 Sum of Chess & Checkers
05 Gallimaufary: T& F, col, hackenbush
06–09 Games to which the theory is relevant
07 Management
08 Types of Results
09 Pointer
10 (Destructable Cover)
11 Dots and Boxes
12 Partitioning
13X (Blank overlay)
14 Domineering (make 5 moves)
15 Regional display
16 Earlier Regional Display (Remove Pointer)
17 Axioms Postulate last player wins
18 4 outcome classes
19 Conway's Axioms
20 Left Going First...
21 Genesis of Games
22 Positive, Negative...
23 Number, nimber
24 Domineering $0 = 1 \times 1 = 2 \times 13$
25 $1/2 = L$; up $= \cdots 3/4 =$
26 Append 1/4, 1/8 numbers
27 Append nimbers and retract tiny-two
28 5.75|||5.5|3.25||31.5
29 HOT GAMES
30 Finite Games
31 Heating and Overheating
32 Integral 1 to $1^*3/4 =$
33? Basis Positions
34 Nice Boards
35 Uncovered long rectangle with 2 ends
36 $2n \times 3$
36.5 $2n \times 3$?
37 Obstacles
38 /Challenges

Kodak/Cyclotomics and University of California, Berkeley

Proceedings of Symposia in Pure Mathematics
Volume **50** (1990)

The Origins of John von Neumann's Theory of Automata

WILLIAM ASPRAY

The contributions of John von Neumann (1903–1957) to the development of modern computing have received considerable attention, particularly over the last decade, as historians and computer pioneers have attempted to reconstruct the history of the electronic, digital, stored-program computer. Two of his contributions are widely hailed as critical events in the development of the computer: the report he wrote in 1945 offering the first account of the stored-program computer, and the computer project he directed at the Institute for Advanced Study beginning in 1946. The focus on these two (undoubtedly important) events has overshadowed von Neumann's main contributions to computing. The current historiography treats von Neumann principally as a computer designer and compares him with J. Presper Eckert, Jay Forrester, and other computer designers of the vacuum tube era. However, von Neumann's main interest in computers was not in machine design, *per se*, but in having a scientific instrument, comparable to the cyclotron or the 100-inch telescope, that would open up new areas of scientific research. If he participated in the design of computing machinery, it was because no workable design able to meet his scientific needs yet existed. In fact, his contributions to computing encompassed much more than hardware and software design; they also included numerical analysis, scientific computation, and the theory of automata.

If the current picture of von Neumann were accurate, it would be hard to understand why the 1945 EDVAC report [**35**], in which he first described the stored-program computer, was written as it was. The machine he described there was not to be built from vacuum tubes, electromechanical relays, or even mechanical switches, but instead from idealized neurons described in

---

1980 *Mathematics Subject Classification* (1985 *Revision*). Primary 01A70; Secondary 68Q70.
This work was supported by the Charles Babbage Institute and National Science Foundation grant NSF/SES-8609543.

This subject will be covered in greater detail in my forthcoming book, *John von Neumann and the origins of modern computing.*

0082-0717/90 $1.00 + $.25 per page

a 1943 paper by Warren McCulloch and Walter Pitts [**16**]. Von Neumann followed the description with a comparison of the computer and the human nervous system: making an analogy of the associative, sensory, and motor neurons with the central processing, input, and output equipment of the computer; and contrasting the synchronous timing of the clocked ENIAC circuits with the asynchronous system in which neurons fire (seemingly) autonomously. It may seem anomalous to capture the design of a technology in terms of biological organs; Eckert and Mauchly, the principal designers of the EDVAC, the computer this report described, certainly thought so.[1]

Yet this report was not an aberration, for von Neumann continued this line of research throughout the remainder of his career—towards a general theory of information processing automata that would apply to both the biological and technological realms. He hoped his work would give scientific coherence to the study of information processing and guide the future design of computing technology. Briefly, this research falls into six segments. His first project was to construct a logical theory of automata that incorporated the work of McCulloch and Pitts on neural networks and of Alan Turing on computability as its foundations. Von Neumann next suggested how the logical theory could be extended to a probabilistic theory of automata, based in part upon the work of Claude Shannon on communication theory and of Norbert Wiener on the statistical mechanics underlying cybernetics. The purpose of the probabilistic theory was to explore the complexity of biological and technological systems able to process information, and the mechanisms by which such complex systems could function reliably with unreliable switching components (vacuum tubes and neurons). Studies of complexity led him to investigate the mechanisms by which complex organisms are able to self-replicate and to attempt to build theoretical automata with this ability. In turn, these questions led him to the notion of a cellular automaton, and to show how self-reproductive faculties could be built into one. Overarching all of these studies were his efforts to compare the computer and the brain as information processing systems, and to learn lessons from the human nervous system which could be applied to the construction of more powerful computing systems.

This paper does not focus on these results, which are already widely known, but instead upon their origins. It recounts von Neumann's early experiences with scientific aspects of information, traces his abiding interest in biological information processing, and suggests how these shaped his work in computing technology and science. The story reaches back to a time long before the development of the computer and delves into many branches of science and technology.

[1]See Stern [**22**] for a more thorough discussion of this issue.

Von Neumann's early experiences with information as a scientific concept. Von Neumann's earliest experience with information as a scientific concept came as a young man, in the late 1920s, as part of his work on the mathematical foundations of quantum mechanics. In 1871 James Clerk Maxwell had first put forth his famous Demon paradox:

> ...a being whose faculties are so sharpened that he can follow every molecule in his course, and would be able to do what is at present impossible to us.... Let us suppose that a vessel is divided into two portions, A and B by a division in which there is a small hole, and that a being who *can see the individual molecules* opens and closes this hole, so as to allow only the swifter molecules to pass from A to B, and only the slower ones to pass from B to A. He will, thus, without expenditure of work raise the temperature of B and lower that of A, in contradiction to the second law of thermodynamics.[2]

Several generations of physicists had puzzled over the paradox until it was resolved in 1929 by von Neumann's Hungarian colleague, Leo Szilard [**26**].[3] Szilard showed that the answer lay in the relationship between entropy and information. The entropy that was lost in the gas on account of the Demon's separation of low and high energy particles was compensated for by the information the Demon gained about the motion of the gas particles. In order to obtain this information the Demon had to introduce negative entropy into the system, explaining the paradoxical net loss of entropy.

Szilard's explanation was accepted by the physics community, and information was accepted as a scientific concept, defined by its statistical-mechanical properties as a kind of negative energy that introduced order into a system. Von Neumann adopted Szilard's solution in his own account of quantum thermodynamics in 1932 [**38**, Chapter 5].[4] In answer to a thought experiment similar to Maxwell's Demon, von Neumann proposed "the intervention of an intelligent being in thermodynamical systems" [**38**, p. 369].

Von Neumann returned to the statistical definition of information, and its connection with entropy, in the 1940s in his considerations of Claude Shannon's work on communication engineering. Building on work carried out in the 1920s and 1930s by Harry Nyquist, R. V. Hartley, and other communication engineers, in 1948 Shannon published his mathematical theory of communication. He presented precise definitions of the components of a communication system (source, transmitter, channel, receiver, destination)

---

[2]Maxwell [**13**, p. 328] as quoted in Brillouin [**3**, p. 162].

[3]See the discussions of Szilard's 1929 paper and the connection between entropy and information in Brillouin [**3**], Cherry [**6**], and Singh [**24**].

[4]It is likely that von Neumann knew of Szilard's result and its significance for quantum mechanics in the late 1920s.

and gave general theorems about the theoretical limits for the capacity of information flow on a channel with noise. Shannon's work was the first to differentiate carefully between the concepts of information and meaning. As his collaborator and popularizer, Warren Weaver, wrote [**23**, pp. 8–9]:

> The word *information*, in this theory, is used in a special sense that must not be confused with its ordinary usage. In particular, *information* must not be confused with meaning.
>
> In fact, two messages, one of which is heavily loaded with meaning and the other of which is pure nonsense, can be exactly equivalent, from the previous viewpoint, as regards information....
>
> To be sure, this work *information* in communication theory relates not so much to what you do say, as to what you could say. That is, information is a measure of one's freedom of choice when one selects a message. If one is confronted with a very elementary situation where he has to choose one of two alternative messages, then it is arbitrarily said that the information, associated with this situation, is unit. Note that it is misleading (although often convenient) to say that one or the other messages conveys unit information. The concept of information applies not to the individual messages (as the concept of meaning would), but rather to the situation as a whole, the unit information indicating that in this situation one has an amount of freedom of choice, in selecting a message, which is convenient to regard as a standard or unit amount.

Shannon knew that his definition of information both in conception, as a measure of the orderliness of the message set, and in its mathematical formulation

$$H = -(p_1 \log_2 p_1 + \cdots + p_n \log_2 p_n),$$

where the possible messages have probabilities $p_1, \dots, p_n$, strongly resembled the entropy concept in statistical mechanics [**23**]:

> Quantities of the form $H = -P_i \log P_i$ (the constant K merely amounts to a choice of unit of measure) play a central role... as measures of information, choice and uncertainty. The form of H will be recognized as that of entropy as defined in certain formulations of statistical mechanics where P_i is the probability of a system being in cell i of its phase space. H is then, for example, the H in Boltzmann's famous H theorem.

It was von Neumann who had made Shannon aware of the historical antecedents of his ideas [**23**, p. 3, footnote]:

> Dr. Shannon's work roots back, as von Neumann has pointed out, to Boltzmann's work on statistical physics (1894), that entropy is related to "missing information," inasmuch as it is related to the number of alternatives which remain possible to a physical system after all the macroscopically observable information concerning it has been recorded. L. Szilard (Z. Phys. **53** (1925)) extended this idea to a general discussion of information in physics, and von Neumann (*Math. Foundation of Quantum Mechanics*, Berlin, 1932, Chapter V) treated information in quantum mechanics and particle physics.

These ideas were developed further by Shannon, Robert Fano, Denis Gabor, and others into the rich modern engineering discipline of communication theory.

A second, early involvement with information processing came to von Neumann through his contact with Alan Turing.[5] In the spring of 1935, as an undergraduate at King's College, Cambridge, Turing enrolled in a course in the foundations of mathematics taught by M. H. A. Newman. The course covered current research problems in the logical foundations of mathematics, in particular David Hilbert's formalist program for the foundations of mathematics and Kurt Gödel's incompleteness results, which destroyed Hilbert's program. Newman showed how Gödel's results left open one important question formulated by Hilbert, known as the decision problem (*Entscheidungsproblem*): whether there is a fixed procedure which, for every formally statable mathematical assertion, will determine whether the assertion is true.[6]

Turing was attracted to this question and in 1936 wrote a paper that answered it in the negative. He framed his answer in terms of theoretical machines, known later as "Turing machines", able to carry out the decision procedures. Turing believed, as do most computer scientists today, that the numbers that can be calculated by any purely mechanical procedure in mathematics are exactly those that can be calculated by one of his Turing machines.[7] Moreover, Turing constructed a universal Turing machine, which can have instructions encoded in it so as to make it function like any other Turing machine. In effect, Turing had given a mathematically precise, theoretical model of the stored-program computer a decade before the first computer came under construction.

Turing explicitly designed his machines to model, down to specific functional details, the human computational process. His machines were each

[5]The contact between Turing and von Neumann is discussed in Davis [7] and Hodges [**10**].

[6]This problem had been stated by Hilbert at the 1928 International Congress of Mathematicians held in Bologna. A statement is considered here as formally statable if it can be expressed in the language of a logical theory being used to develop a foundation for mathematics. One system commonly used then was that developed by Bertrand Russell and Alfred North Whitehead in *Principia Mathematica*.

[7]This is one way of stating what is commonly known as Church's Thesis.

supplied with "a tape (the analogue of paper)" divided into squares that were "scanned" for symbols. The square being scanned at a given time he wrote of as the only one of which the machine is "directly aware". By altering the internal configuration the machines could "remember" some of the symbols they had previously "seen". Turing concluded his anthropomorphic description with the following statement, using the word "computer" in the standard 1930s sense of a human carrying out a calculation [**27**, pp. 231–232]:

> We may now construct a machine to do the work of this computer. To each state of mind of the computer corresponds an "m-configuration" of the machine. The machine scans B squares corresponding to the B squares observed by the computer.... The move which is done, and the succeeding configuration, are determined by the scanned symbol and the m-configuration.... A computing machine can be constructed to compute...the sequence computed by the computer.

In 1936, Turing received a fellowship to study mathematics at Princeton University for a year with Alonzo Church. Church was pursuing these same questions of mathematical logic, formulated in terms of his lambda-calculus. Turing refined his computability results in Princeton, and eventually decided to return for a second year to complete a doctoral dissertation under Church's direction.

Turing and von Neumann probably first met in April 1935, when von Neumann came to Cambridge to lecture on almost periodic functions at Newman's invitation. Only several weeks earlier Turing had submitted for publication a minor improvement on a result of von Neumann on this very subject; and although there is no firm evidence of their meeting, it is very likely to have happened through Newman. Turing and von Neumann came into contact again in Princeton, for at the time of Turing's graduate study the Institute for Advanced Study was housed in the building of the Princeton University mathematics department. Their earliest contacts apparently involved their mutual interests in almost periodic functions and the theory of continuous groups, though at the latest by 1938 von Neumann knew of Turing's work on computable numbers. Von Neumann regarded Turing's abilities highly, and his imagination was captured by Turing's machines. In the spring of 1938 von Neumann offered Turing a position as his assistant at the Institute, but Turing declined it in order to return to his Cambridge fellowship and his war-threatened homeland. Von Neumann's friend, Stanislaw Ulam, reports of von Neumann playing with Turing-like mechanical descriptions of numbers in the summer of 1938, and speaking highly of Turing's work on several occasions in 1939.[8] Von Neumann was later to have his employees on the Institute computer project read Turing's paper as part of their education in the theory of computing,[9] and he used it himself as one base of his logical theory of automata.

Another formative influence was von Neumann's correspondence during the years 1939–1941 with the Hungarian physicist Rudolf Ortvay.[10] Ortvay was an important figure in the organization of Hungarian science and a friend of the von Neumann family. He became interested in what the results of quantum mechanics could tell us about the brain. He contacted von Neumann, who responded with alacrity, noting that he had already been thinking "extensively, mainly from last year [1938] about the nature of the 'observer' in quantum mechanics as a kind of "quasi-physiological concept".[11]

In the course of their correspondence Ortvay tutors von Neumann on the anatomy of the brain and suggests a program of brain study that closely resembles the one he eventually followed. In a letter written on Christmas day, 1939, Ortvay returned to a topic "I have already brought forward many times before, namely to the theory of the operation of the brain". In the letter he predicts that science is on the verge of a rapid advancement in the study of the brain as a "switching system" and argues that the fundamental breakthrough will not be made by physicians or physiologists, but by mathematicians or physicists who are better able "to see the simple and theoretical structure of a complicated complex". In his letter of 30 March 1940 Ortvay first makes the comparison between the brain and electronic calculating equipment, if only to deny the similarity.

> Of course I don't think that the mechanism of the brain is very similar to a computing equipment consisting of electronic valves but I do believe it is a complicated and very

[8]This information is given in notes prepared by Ulam for his aborted biography of von Neumann (Ulam Papers, American Philosophical Society Library).

[9]Julian Bigelow described how von Neumann first assigned him to read Turing's computability paper when he came to work on the Institute computer project in 1946. (Private communication, 6 November 1987.)

[10]Von Neumann discussed questions of the computer and the brain with a number of Hungarian scientists. Eugene Wigner read drafts of his works on automata. There is later correspondence between von Neumann and fellow Hungarians Zoltan Bay and Albert Szent-Gyorgyi. See, for example, von Neumann's letter to Szent-Gyorgyi of 22 June 1949 (Von Neumann Papers, Library of Congress).

[11]See, for example, von Neumann's letter to Ortvay of 29 March 1939. There are approximately sixty letters in their correspondence, all written in Hungarian. Some are in the von Neumann Papers at the Library of Congress, others in the Library of the Hungarian Academy of Science, Manuscript Division, in Budapest. Eight of these letters, chosen for what they contribute to the understanding of the development of von Neumann's ideas on the computer and the brain, are translated into English in [**17**]. All of the letters cited here are from this collection, with occasional slight improvements in the English translation. Denes Nagy has informed me that there is little of substance in the part of the correspondence they did not translate, and that many of the letters written by von Neumann merely ask Ortvay for additional information about the brain. (Private communication, 9 March 1988.)

> specialized equipment, which may be very simple in terms of basic properties!... I am ready to believe that a less detailed treatment, that with the mechanism for the brain, whereby the cell would be the unit and some idea for the interconnection, is perhaps even now timely.

This is startlingly early for the discussion of any electronic calculating equipment; what Ortvay may have had in mind were the cosmic ray counters internationally available in the late 1930s. Von Neumann agreed about the importance of having a simple theory, which conforms with his previous efforts to provide simple mathematical regularity in the complicated domains of set theory, quantum mechanics, and the theory of games.

> I cannot accept that a theory of prime importance, which describes processes that everybody believes to be elementary, can be right if it is too complicated, i.e. if it describes these elementary processes as being horribly complex and sophisticated ones. Thus I feel instinctively horror of the present day quantum chemistry and especially biochemistry.[12]

Ortvay writes repeatedly of being on the eve of a breakthrough in the understanding of the brain. (See, for example, his letter of 30 May 1940). In his letter of 29 January 1941 he goes one step further and suggests that von Neumann is the appropriate person to carry forward this research program. Ortvay also amplifies on the suggestion in his 30 March 1940 letter about the mathematical approach to be taken; and what he suggests resembles very closely the program that von Neumann developed after the war:

> I liked your paper [on the theory of games] very much at the time and it gave birth to the hope that, if I could succeed in drawing your attention to the problem of brain cells, you may succeed in formulating the problem.
>
> The problem seems to be this: the train can be conceived as a network, having in its nodes the brain cells. These are interconnected in a way that each cell can receive from and transmit to several other cell impulses. As to which of these impulses are taken in from or passed on to the other cells may depend on the state of the cell, which in turn is dependent on the effects previously affecting the cell in question. For the cell it may perhaps be enough to have a limited number of potential states. (Although it is genetics which shows us how differentiated a cell can be since we know many hundreds of genes in the chromosomes.) The actual state of cells conceived as numbered ones could characterize the state of the

[12]Letter, von Neumann to Ortvay, 13 May 1940.

> brain. There would be such a distribution corresponding to every spiritual state and the state would be relevant to every reaction, e.g. that which is the way of transmitting a stimulus to a nerve. It may recall an automatic telephone center where there is a change however after every communication in the connection. Perhaps the ever refining techniques in the switching equipment would prove a fertile analogy.
>
> The big trouble is that the physicians, who know the facts, are so hostile towards the more abstract way of thinking.

In a final letter of interest, dated 16 February 1941, Ortvay suggests a program of study similar to that outlined in Norbert Wiener's *Cybernetics* seven years later:

> Now I wish to mention another matter: these days everybody is talking about organization, totality. Today's computing machines, automatic telephone exchanges, high-voltage equipment like cascade transformers as well as radio transmitter and receiver equipment, but also an industrial plant or an office are all technical examples of such organizations. I think there is a common element in all these which is capable of being axiomatized. I don't know if there has been an attempt in this direction. I am interested in knowing this because I believe that if it is possible to sharply accentuate the essential elements relevant to the organization, as such, this would give an overview of the alternatives and would facilitate the understanding of such systems as, for instance, the brain.

The event that triggered von Neumann's active interest in the theory of information processing was the paper on neural networks by McCulloch and Pitts referred to above. At the recommendation of Norbert Wiener and Julian Bigelow, von Neumann read **[16]** soon after its publication in 1943 and was enormously impressed by its potential for bringing mathematical regularity to the ill-understood phenomena of information processing in the human brain.[13] The paper presented a simple logical calculus of neural nets, using as axioms rules McCulloch had developed in an earlier research project to describe psychons, and as logical framework an amalgam developed by Pitts from Rudolf Carnap's logical calculus and Bertrand Russell and Alfred North Whitehead's *Principia Mathematica*.[14]

---

[13]Wiener and Bigelow's role in introducing von Neumann to this work was related to the author in a private communication from Julian Bigelow 6 November 1987. According to Bigelow, von Neumann read widely in Nicholas Rashevsky's Bulletin of Mathematical Biophysics, but complained about the scientific depth of the papers.

[14]This work is described in McCulloch [**15**, pp. 9–10] and presented in historical context in Aspray [**1**]. A psychon is a least psychic event, akin to a Leibnizian monad.

The McCulloch-Pitts model does not describe the structure or internal operation of the individual neurons. The neurons are characterized as "black boxes", about which is known only the mathematical rules governing their input and output of signals. Networks of neurons cooperate to calculate, learn, store information, and perform other information processing tasks. The neutral networks have a power similar to Turing's machines; and the authors claim to have "proved, in substance, the equivalence of all general Turing machines—man-made or begotten" [**15**]. (McCulloch and Pitts cite Turing machines as a model for their work.) McCulloch, sometimes in collaboration with Pitts, continued these investigations. (For example, in a 1945 paper he uses a related model to explain how humans believe universal ("for all") statements. See McCulloch [**15**], a collection of his writings, for his work carrying out this research program.) Pitts accepted a position at the MIT Research Laboratory of Electronics in 1943, and McCulloch joined him there in 1952. At MIT their work came under the influence of Norbert Wiener, who regarded it as contributing to his research in cybernetics.[15] This fundamental relationship between the biological and technological worlds implicit in the work of Turing, and of McCulloch and Pitts also, attracted von Neumann's attention. While McCulloch and Pitts continued to develop their ideas, von Neumann began independently to build on them in his own direction, making them a centerpiece of his logical theory of automata.[16]

Contacts with the biomedical community. Following his reading of the McCulloch-Pitts paper and his first involvement with high-speed calculating equipment, von Neumann took a more active interest in the scientific work of what might be called biological information processing. He began to read the published literature on neurophysiology, establish contacts in the biomedical community, and participate in interdisciplinary conferences on the subject.[17]

[15]See, for example, McCulloch [**14**] which describes the applications of cybernetic techniques to understanding the functioning of the central nervous system. An application of cybernetic principles developed by McCulloch and Pitts was a 1947 project on prosthetic devices designed to enable the blind to read by ear (Wiener [**43**, pp. 31–32]; de Latil [**8**, pp. 12–13]).

[16]The psychological and physiological communities were not enthusiastic about McCulloch and Pitts' axiomatic model, objecting that it was too simplistic and did not model many of the important physiological features of biological neurons. See McCulloch [**15**, p. xvii] and von Neumann [**42**, pp. 422–431]. Von Neumann was aware of the primitive character of the model:

> Following W. Pitts and W. S. MacCulloch [sic]... we ignore the more complicated aspects of neuron functioning: Thresholds, temporal summation, relative inhibition, changes of the threshold by after effects of stimulation beyond the synaptic delay, etc. (von Neumann [**35**, p. 24]).

His use of their work had the salutary effect of bringing McCulloch and Pitts' work to the attention of the biology community. See McCulloch [**15**, p. 9].

[17]This point is discussed by Klara von Neumann in the preface to *The computer and the brain*. There is an indication that during the war years, or possibly earlier, von Neumann had discussions on the literature of neurophysiology with Princeton University biologist Newton Harvey. See, for example, the letter von Neumann wrote to the Princeton mathematics librarian, Miss Blake, 29 November 1944 (Von Neumann Papers, Library of Congress).

Von Neumann's scientific interchanges on biological information fall into two categories. There were his close personal and intellectual associates, with whom he discussed his ideals regularly over a period of years. They include Julian Bigelow, Herman Goldstine, Stanislaw Ulam, Norbert Wiener, and Eugene Wigner, and perhaps also Warren McCulloch, Walter Pitts, Arturo Rosenblueth, and Abraham Taub. Unfortunately, this discussion occurred in informal visits and did not leave any documentation for the historical record. There were also scientists with whom von Neumann had brief, but occasionally intensive contact, usually involving inquiries he made about biological phenomena or discussion of ideas he had presented in public lectures. These contacts are more representatively documented.

Contacts with the biomedical community began at about the same time as the IAS computer project. A sample from a ten-year span indicates their nature and suggests their importance to the formation of von Neumann's ideas on information. The earliest occurs in December 1946, when von Neumann corresponded with Washington University, St. Louis biochemist Sol Spiegelman, following a meeting that Spiegelman attended in Washington, D.C., at which von Neumann lectured on the possible role of servomechanisms in chromosome pairing.[18] Von Neumann provides Spiegelman with information on the literature of servomechanisms, while Spiegelman sends information on the biophysics literature and discusses his ideas for "self-duplication".[19]

Von Neumann continued to discuss his ideas on biological information processing with other scientists in the late 1940s. Around this time he corresponded with the population biologist A. J. Lotka of Metropolitan Life Insurance Company in New York about the similarities between Lotka's results on reproduction and multiplication in biological populations and war-time work at Los Alamos on neutron multiplication in near critical states.[20] In 1948 or 1949 von Neumann met twice in Princeton with Heinz von Foerster, later the director of the Biological Computer Laboratory at the University of Illinois, about von Foerster's theory of memory.[21] There was also an interesting correspondence with the distinguished German physical chemist K. F. Bonhoeffer.[22] Through a mutual scientific contact, Max Delbruck, Bonhoeffer sent von Neumann a copy of his 1948 paper presenting a theory of nervous

[18]See letters from Speigelman on 3, 14, and 31 December 1946 and replies by von Neumann on 10 and 16 (two letters) December (Von Neumann Papers, Library of Congress).

[19]In the first 16 December letter von Neumann informs Spiegelman that his encouragement has convinced him to try to write up his own ideas on self-reproduction.

[20]See the letter from von Neumann to Mortimer Spiegelman (no known relation to Sol Spiegelman) of Metropolitan Life Insurance Company, 13 February 1950 (Ulam Papers, American Philosophical Society Library).

[21]See the letter of von Neumann to Warren McCulloch, 16 March 1949 (McCulloch Papers, American Philosophical Society Library). Von Neumann found von Foerster to be "really very brilliant" but had serious reservations about his theory of memory.

[22]See letters from von Neumann dated 17 February and 1 April 1949, and from Bonhoeffer dated 21 January, 10 March, and 7 May 1949 (Von Neumann Papers, Library of Congress).

excitation and conduction based on an investigation of an iron-wire analog of the nerve developed in the 1930s by Ostwald, Heathcote, and Lillie.[23] Von Neumann answered with a lengthy analysis of McCulloch and Pitt's work, an explanation of his ideas for building counters and learning automata out of idealized neurons and for developing a probabilistic theory of automata, and a statement of his reasons for not studying humans, rats, or even army ants to learn about biological information processing.[24] That same year K. S. Lashley of the Yerkes Laboratory of Primate Biology wrote to von Neumann with suggestions for a physiological research program that would test the general and logical theory of automata.[25]

These contacts continued into the 1950s, until von Neumann was no longer healthy enough to pursue his scientific interests. In 1950, he agreed to participate in a study by Alejandre Arrellane of Massachusetts General Hospital, who was collecting electroencephalographs of "privileged brains".[26] In 1951 the Rockefeller Institute physiologist R. Lorente de No encouraged von Neumann to continue with his work on reliability of automata with unreliable organisms, as a means to improve the general understanding of the cerebral cortex.[27] In August 1955 von Neumann corresponded with the University of Wisconsin geneticist Joshua Lederberg regarding the adequacy of von Neumann's model of self-reproduction in explaining biological reproduction.[28] Later the same year von Neumann was in contact with the physiolo-

[23]See von Neumann's letter to Warren McCulloch, 17 March 1949 (McCulloch Papers, American Philosophical Society Library). Von Neumann compares Bonhoeffer's model favorably with one developed by Nicholas Rashevsky. (Letter, von Neuman to Bonhoeffer, 17 March 1949. Von Neumann Papers, Library of Congress.)

[24]About the work of McCulloch and Pitts, von Neumann wrote: "I think that they contain very good and important ideas and results, but are unfortunately burdened with philosophical terminology that is not entirely clear and a hyper-formalized way of designating things. The latter pains me especially, as an ex-logician. In my opinion, it is the cumbersome and inappropriate logical apparatus of R. Carnap. In spite of this the real ideas and significant results are clear and effective." (Letter, von Neumann to Bonhoeffer, 1 April 1949, Von Neumann Papers, Library of Congress.)

[25]Letter, 12 December 1949 (Von Neumann Papers, Library of Congress). Lashley was responding to the lecture von Neumann had given in 1948 at the Hixon Symposium. Lashley suggested work in four areas: direct study of neurological phenomena, like the work of Wolfgang Kohler at Swarthmore College, who was trying to establish neurological activity as fundamentally analog in nature; research on the relationship between switching elements and simultaneous input impulses, as might be obtained from examination of the rat's visual cortex; a study of precision and the kinds of errors made by the human nervous system, mainly to show the limitations of the McCulloch-Pitts model; and studies like those of Nicholas Rashevsky on the brain and a complex network of circuits, but modified to use statistical tools and assume the circuits to be randomly distributed.

[26]Letter, Arrellane to von Neumann (Von Neumann Papers, Library of Congress). Wiener had apparently interceded on behalf of Arrellane, who also hoped to get Albert Einstein's participation.

[27]Letter, Lorente de No to von Neumann, 31 October 1951 (Von Neumann Papers, Library of Congress).

[28]There are letters from Lederberg on 4 and 10 August 1955, and replies by von Neumann of 8 and 15 August (Von Neumann Papers, Library of Congress). They disputed the role of information in the process of self-reproduction. It is clear that their letters talk at cross-purposes, partly because of semantic difficulties. No resolution is clear in the correspondence.

gist C. M. Williams, who jointly with Graham Weddell carried out a research project at Oxford in the years 1952–1954 on the innervation of ear hairs in a rabbit. Their work uncovered an information mechanism similar to the multiplexing technique von Neumann had developed to improved reliability in computing machinery.[29] These and the many other contacts with scientists from the biomedical community provide strong evidence of von Neumann's abiding interest in biological information processing.

This interest is also apparent in von Neumann's association with two interdisciplinary organizations. In late 1944 von Neumann, Harvard Computation Laboratory director Howard Aiken, and Norbert Wiener decided that there was sufficient interest in questions of information processing to bring together a small group to discuss the future development of this field of effort, which they pointed out "as yet is not even named".[30] It was decided that an open meeting was not then appropriate because so many of the projects involving information processing were connected with the war effort. People were invited on the basis of their interest in "communications engineering, the engineering of computing machines, the engineering of control devices, the mathematics of time series in statistics, and the communication and control aspects of the nervous system".[31] Aiken and Wiener proposed that the group be known as the Teleological Society, for, as it was explained to Herman Goldstine:

> Teleology is the study of purpose of conduct, and it seems that a large part of our interests are devoted on the one hand to the study of how purpose is realized in human and animal conduct and on the other hand how purpose can be imitated by mechanical and electrical means.[32]

They convened a meeting January 6–7, 1945 in Princeton. In attendance were Leland Cunningham (Aberdeen Proving Ground), W. E. Deming (Census Bureau), Goldstine (University of Pennsylvania), R. Lorente de No (Rockefeller Institute), McCulloch (University of Illinois Medical School), Pitts (Kellex Corporation), E. H. Vestine (Carnegie Institute), and Samuel Wilks (Princeton University), together with Aiken, von Neumann, and Wiener.[33] The first day von Neumann lectured on computing machines

[29]See the letter from Williams of 7 September 1955 and von Neumann's reply of 3 October. Von Neumann encouraged them to publish their paper, *Error control in a biological information system*.

[30]Letter, Aiken, von Neumann, and Wiener to Herman Goldstine, 22 December 1944 (Herman Goldstine Papers, American Philosophical Society Library). The original invitation went out from these three organizers 4 December 1944, but I have found no copies of it.

[31]Letter, Aiken, von Neumann, Wiener to Goldstine, 22 December 1944 (Goldstine Papers, American Philosophical Society Library).

[32]Letter, Aiken, von Neumann, and Wiener to Goldstein [sic], 28 December 44 (Goldstine Papers, Americal Philosophical Society Library).

[33]Memorandum to the participants from von Neumann, dated 12 January 1945 (Von Neumann Papers, Library of Congress). Von Neumann suggested adding to the group J. Presper Eckert, S. Chandrasekhar, and George Stibitz. (Letter, von Neumann to Wiener, 1 February 1945.)

and Wiener on communication engineering. The second day Lorente de No and McCulloch made presentations on the organization of the brain.[34] The agenda called for the group to discuss publication of a journal, steps to form and support a research center, a patent policy for the group, measures to bring the work to public attention (e.g., through the American Association for the Advancement of Science), and "how to protect the researches of the group from dangerous and sensational publicity".[35] The outcome was the formation of four study groups, reflecting common interests of the participants: filtering and prediction problems (Pitts, Wiener), applications of fast computing methods to statistical problems (Deming, Vestine, Wilks, von Neumann), applications to differential equations that arise in astronomy, hydrodynamics, and ballistics (Aiken, Cunningham, Goldstine, von Neumann), and neurology (No, McCulloch, Pitts). The study groups agreed to prepare within three months memoranda about research they would be interested in pursuing, with the understanding that another meeting would be called within six months to formulate a final memorandum on a research program. Correspondence about the Teleological Society rapidly tailed off, no second meeting was convened, and no research agenda was written.[36] Julian Bigelow attributes this in part to a clash of personalities among the organizers.[37]

Although plans for the Teleological Society had fallen apart, von Neumann continued to think about the question of biological information processing and to discuss his ideas with other scientists. A good picture of his thinking in late 1946 is given in a long letter to Norbert Wiener[38] describing topics for a meeting they had planned for December.

The first part of the letter is highly pessimistic, expressing doubts that von Neumann says he has felt "for the better part of a year". He argues that the results of Turing and McCulloch and Pitts are so general as not to inform us about the working of the neural mechanism:

> What seems worth emphasizing to me is, however, that after the great positive contribution of Turing-cum-Pitts-and-

---

[34]Letter, Wiener to Arturo Rosenblueth, 24 January 45 (Wiener Papers, MIT Archives).

[35]Letter, Aiken, von Neumann, and Wiener to Goldstein [sic], 28 December 1944 (Goldstine Papers, American Philosophical Society Library).

[36]One informative letter on the topic exists from Wiener to von Neumann, dated 24 January 1945. Wiener reports that he and Pitts are working on their report, and that they have added to their task "the problem of transition from the computing machine to the control machine." Wiener also addresses the prospects for MIT becoming the home for this new organization: "The attitude at Tech towards making it the home of such a project is as far as I can see friendly but not decisive as yet. One disadvantage of our community is Harvard Medical to which we could naturally look for backing on the physiological side has become completely clinical and offers little prospect as a home for the physiological side of the undertaking. I strongly expect that if we can get outside funds available it will not be too difficult to persuade the Institute to embark to a limited extent on a physiological program tied to our plans, but I have no definite commitment in this direction. I think that Tech should be put very high up in the list of our possibilities as a home for the undertaking, but the time is not yet come to assume that it represents the only solution." Von Neumann replies on 1 February 1945 with his encouragement to pursue the matter at an opportune time (Norbert Wiener Papers, MIT Archives).

> McCulloch is assimilated, the situation is rather worse than before. Indeed, these authors have demonstrated in absolute and hopeless generality, that anything and everything Brouwerian can be done by an appropriate mechanism, and specifically by a neural mechanism—and that even one, definite mechanism can be "universal". Inverting the argument: Nothing that we may know or learn about the functioning of the organism can give, without "microscopic", cytological work any clues regarding the further details of the neural mechanism.... I think you will feel with me the type of frustration that I am trying to express.

The problem he foresees with microscopic work is that the complexity of the brain is "overawing", as he explains by analogy:

> To understand the brain with neurological methods seems to me about as hopeful as to want to understand the ENIAC with no instrument at one's disposal that is smaller than about 2 feet across its critical organs, with no methods of intervention more delicate than playing with a fire hose (although one might fill it with kerosene or nitroglycerine instead of water) or dropping cobblestones into the circuit.... And our intellectual possibilities relatively to it are about as good as some bodies [sic] vis-a-vis the ENIAC, if he has never heard of any part of arithmetic.

Von Neumann discusses several alternative approaches, such as studying the nervous systems of frogs or employing Gestalt theory, but dismisses them all as being unlikely to lead to any direct knowledge of the human nervous system. He also discounts the axiomatic study of multi-cellular organisms

[37]Personal communication to the author, 6 November 1987. Wiener explained one reason for the delay in forming a society: "While we are going to meet again in the Spring, we have not organized in a formal permanent society. The reasons are that owing to the control of different and by no means unified government departments over parts of our program, it is best not to stir up a fuss until the military situation makes matters of classification less important. However, we definitely do have the intention of organizing a society and a journal after the war, and founding either at Tech or elsewhere in the country a center of research in our new field. In this matter Moe is giving his good will and expects to be able to help us with fellowships. We are also getting a good backing from Warren Weaver, and he has said to me that this is just the sort of thing that the Rockefeller should consider pushing. In addition, McCulloch and von Neumann are very slick organizers, and I have heard from von Neumann mysterious words concerning a some thirty megabucks which is likely to be available for scientific research. von Neumann is quite confident that he can siphon some of it off." (Letter to Arturo Rosenblueth, 24 January 1945, Wiener Papers, MIT Archives.)

[38]29 November 1946 (von Neumann Papers, Library of Congress). Von Neumann was concerned enough about this letter that he asked Julian Bigelow for comments before mailing it to Wiener. Bigelow's comments are also found in the von Neumann Papers.

axiomatically, in which cells appear as the undefined basic elements, arguing by analogy that the cell is too refined an object to be regarded as elementary:

> To be more *par terre*: Consider, in any field of technology, the state of affairs which is characterized by the development of highly complex "standard components", which are at the same time individualized, well suited to mass production, and (in spite of their "standard" character) well suited to purposive differentiation. This is clearly a late, highly developed style, and not the ideal one for a first approach of an outsider to the subject, for an effort toward understanding. For the purpose of understanding the subject, it is much better to study an earlier phase of its evolution, preceding the development of this high standardization-with-differentiation. I.e. to study a phase in which these "elegant" components do not yet appear. This is especially true, if there is reason to suspect already in that archaic stage mechanisms (or organisms) which exhibit the most specific traits of the simplest representatives of the above mentioned "late" stage.

The second half of the letter is more optimistic and suggests a program of empirical research studying the bacteriophage, which he hopes "may be the first relevant step forward and possibly the greatest step that may at all be required".[39] His optimism is based on the fact that these organisms are simple enough to study, yet complex enough to self-reproduce, orient themselves in an unorganized milieu, and locate and appropriate food. He suggests high precision x-ray analysis or advanced electron microscopy as the most promising methods of investigation.

These ideas had arisen through von Neumann's discussions with Irving Langmuir, Associate Director of General Electric's Research Laboratory in Schenectady, New York. In October 1946 they met to discuss the study of proteins by chemical and crystallographic analyses, and Langmuir had introduced him to the work of the crystallographer Dorothy Wrinch.[40] Von Neumann found x-ray crystallography unsuitable because of problems with data analysis. He estimated that some proteins would require 1000-term Fourier analyses, which involve millions of multiplications.[41] Thus he began

---

[39]Von Neumann also mentions work he is carrying out on the theoretical front. He claims to have formulated the problem of self-reproduction rigorously, in the style of Turing, and promises to complete the details in the next two months. He anticipates that he will learn that it takes in the high 10,000s or 100,000s in number of basic components to produce an automaton capable of self-replication. He also mentions that he has been musing over aspects of the gene-enzyme relationship and mutation as they relate to self-reproduction.

[40]See Langmiur's letter of 11 November 1946 and von Neumann's of 12 November 1946 (Von Neumann Papers, Library of Congress).

[41]Letter, von Neumann to Mina Rees, Office of Naval Research, 20 January 1947 (Von Neumann Papers, Library of Congress).

to search for other procedures to use until computers then under development became available.

The plan that he settled on involved scaling up the existing "linear" and "cyclol" models of proteins by a factor of 10^8, and using similarly scaled-up electromagnetic centimeter waves to simulate x-rays, as a means to study medium- and large-sized proteins.[42] This model avoided the large amounts of calculation required by x-ray crystallography.[43] He discussed the engineering aspects of the plan with RCA scientist Vladimir Zworykin and then submitted a funding proposal to Mina Rees of ONR.[44] Apparently, the funding did not materialize and the project was scratched.

The Macy Conferences on Feedback Mechanisms and Circular Causal Systems in Biology and the Social Sciences, or as they were later known the Conferences on Cybernetics, provided a substitute for the Teleological Society. The Josiah Macy, Jr. Foundation in New York regularly organized conferences, especially in the biomedical sciences, to bring together groups of about 25 distinguished scientists to discuss focused, interdisciplinary topics. Frank Fremont-Smith of the Macy Foundation suggested a conference series in which mathematicians, neurophysiologists, psychiatrists, psychologists, and sociologists would discuss "mechanisms underlying purposive behavior", and recruited Warren McCulloch to chair the meetings.[45] The work of Wiener and others on cybernetics was a major stimulus for calling the meetings, as McCulloch explained (see [**34**, p. 70]):

> Our meetings began chiefly because Norbert Wiener and his friends in mathematics, communication engineering, and physiology, had shown the applicability of the notions of inverse feedback to all problems of regulation, homeostasis, and goal-directed activity from steam engines to human societies. Our early sessions were largely devoted to getting these notions clear in our heads, and to discovering how to employ them in our dissimilar fields.

---

[42]See, for example, the letter of von Neumann to D. Harker of General Electric Research Laboratory, 16 December 1946; and the 20 January 1947 letter to Mina Rees cited above (Von Neumann Papers, Library of Congress).

[43]In this scaled-up model intro-atomic distances where 1 cm and atoms were modeled by metal balls. The scale was such that the measurements were greatly simplified.

[44]See the interdepartmental correspondence, Edward G. Ramberg to V. K. Zworykin, 27 December 1946 estimating the equipment and personnel needs to realize von Neumann's plans (Von Neumann Papers, Library of Congress). Von Neumann argued that the plan called for (a) an engineering group with radar experience, (b) an x-ray crystallographer, (c) a group of physical chemists, and (d) a theoretical coordinator. He mentioned that Zworykin and the RCA engineers had agreed to do (a), H. S. Taylor and J. Turkevitch of the Princeton University Chemistry Department to do (b), the trained British crystallographer A. D. Booth who was coming to the Institute as a visitor in 1947 could handle (c), and he would assume duty (d). He asked for initial funding of \$50,000 and annual continuing support of \$15,000.

[45]Letter, McCulloch to Walter Pitts, 2 February 1946 (McCulloch Papers, American Philosophical Society Library).

The first meeting was held in New York in March 1946, with subsequent meetings scheduled at six-month intervals. After the fifth meeting it was decided that a full year was needed between meetings in order that the participants could have adequate time to make observations and conduct experiments to test ideas originating at the conferences. The tenth and last meeting was held in April 1953.[46] A core of people participated from the beginning: mathematicians Pitts, von Neumann, and Wiener; electrical engineer Julian Bigelow; physiologists Freemont-Smith and Ralph Gerard; anthropologists Gregory Bateson and Margaret Mead; psychologist Hienrich Kluever, and psychiatrist McCulloch. Several others, including the biophysicist Heinz von Foerster and the philosopher F. S. C. Northrop, were regulars for a few meetings; and there were also visitors invited to attend a single meeting on a topic close to their special interests.

Von Neumann was a central figure and a regular lecturer, especially at the early meetings. At the March 1946 meeting he lectured on "Computing Machines: Their Formal Behavior including Memory, Learning, and Recording". At the third meeting, in March 1947, he presented his ideas on the connection between entropy and information, and compared feedback and memory systems in biological and artificial information processing systems.[47] At the fourth meeting, in October 1947, he lectured on "Circuits for Typical Logical Tasks". After this, the number of formal talks was reduced and von Neumann gave no more. Outside obligations caused him to miss several of the later meetings and eventually to resign at the ninth meeting to make room for other participants. The resignations of von Neumann and Wiener may have been a primary reason for terminating the conferences after the tenth meeting.

Two examples will illustrate the interdisciplinary nature of the conferences and suggest the context in which von Neumann developed his ideas about automata. As McCulloch describes part of the March 1947 meeting:

> The second day, at von Neumann's request, began with a discussion of consciousness, as to whether there is an anatomical substrate for what everybody has [called] "knowing things." Von Bonin held it was probably not [the] cerebral cortex that determined awareness but more likely something in the vicinity of the midbrain, the cortex determining what we know

[46]Proceedings of the last five meetings were published by the Macy Foundation and are still readily available. There is little information available about the first five meetings, other than an occasional letter, program, or unpublished summary in Warren McCulloch's papers at the American Philosophical Society Library. The Josiah Macy, Jr. Foundation has destroyed its administrative records of the conference. It is possible that some documentation can be found among the papers of other participants.

[47]Donald Marquis summarized the discussion following von Neumann's presentation. A copy of this summary is to be found in the McCulloch Papers at the American Philosophical Society Library.

> rather than that we know; and I suggested the importance of the thalamus as suggested by experiments with strychnine. Fremont-Smith went into the different degrees of consciousness from full attention through twilight states to sleep and hysterical amnesias. Molly Harrower described the stimulation of the cortex as not disturbing patients even when it induced all sorts of sensations, but throwing them into a panic when it caused vocalization resembling speech. Fitch sharpened the distinction between consciousness and awareness, and I tried vainly to help him with Spinoza's definition of consciousness as the idea of ideas, or else awareness of awareness, but was unable to tie it to the mid-brain. I then presented my scheme of how we know universals, including diagrams that had the circuit action I expected in the auditory and visual cortex. Their likeness to the known anatomy was self-apparent, but I had difficulty in formulating the process of averaging over a group of transformations.... Von Neumann was disturbed by these notions for there was too much chaos instead of exact connections in the cortex and I failed to make clear, but Pitts succeeded, that the beauty of these circuit actions was that they only required statistical regularity, which is all I thought I had a right to expect of the tissue.[48]

The second example is from the eighth meeting, held in March 1951, on the topic of communication. K. von Frisch spoke on communication among bees, the literary scholar I. A. Richards on communications among men, Claude Shannon on the theory of communication, and the philosopher Suzanne Langer on the problem of symbols. Von Neumann was scheduled to talk about communication among machines, but was called away at the last moment to meet government obligations.

What are we to make of all these experiences? Contrary to what the current literature about von Neumann suggests, he was not merely interested in computer design, or even only in computer science and technology as it is practiced today. Wiener's cybernetics of the 1940s, or today's cognitive science, may be a better fit; but even they do not encompass all of his interests in information processing, nor provide the organizing principle that seems to run through all his research.

Von Neumann's scientific interests were not confined by traditional disciplinary boundaries. His interest in information came from early experiences

[48] "To the Members of the Conference on Teleological Mechanisms, Oct. 23 & 24, 1947", a summary of the previous meetings prepared by McCulloch (McCulloch Papers, American Philosophical Society Library).

in physics, mathematical logic, and biology; and his proclivity towards interdisciplinary study, already present in his early work on quantum mechanics and game theory, was undoubtedly reinforced by his interdisciplinary activities during the second world war. The interest in biological information processing that evolved from these early experiences was strong, and remained with him throughout his career. Very likely he would have continued to pursue these subjects if his health had improved and his government duties diminished; and he would have been very interested in the work on DNA that came to fruition only shortly after his death. His contacts with the biomedical community were an important stimulus to his thinking about computing, one which he hoped and believed would contribute to the improvement of computer technology. Only by examining von Neumann in this larger scientific context do we gain a good measure of his plans and accomplishments.

Bibliography

1. William Aspray, *The scientific conceptualization of information: A survey*, Ann. Hist. Comput. **7** (1985), 117–140.
2. K. F. Bonhoeffer, *Activation of passive iron as a model for the excitation of nerve*, J. General Phys. **32** (1948), 69–91.
3. Leon Brillouin, *Science and information theory*, Academic Press, New York, 1956.
4. A. Burks, *Essays on cellular automata*, Univ. of Illinois Press, Urbana, Ill., 1970.
5. E. C. Cherry, *The communication of information (an historical review)*, Amer. Sci. **40** (1952), 640–664.
6. ——, *On human communication*, MIT Press, Cambridge, 1957.
7. Martin Davis, *Mathematical logic and the origin of modern computers*, in Studies in the History of Mathematics (Esther R. Phillips, ed.), vol. 26, Math. Assoc. of America Studies in Math., The Mathematical Association of America, Washington, D.C., 1987.
8. P. de Latil, *Thinking by machine*, Houghton Mifflin, Boston, 1956.
9. R. V. Hartley, *Transmission of information*, Bell system Tech. J. **7** (1928), 535–553.
10. A. Hodges, *Alan Turing: The Enigma*, Simon & Schuster, New York, 1983.
11. Lloyd A. Jeffress, *Cerebral mechanisms in behavior*, The Hixon Symposium, Wiley, New York, 1951.
12. John G. Kemeny, *Man viewed as a machine*, Scientific American, vol. 192, no. 4, 1955, pp. 58–67.
13. J. C. Maxwell, *Theory of heat*, Longmans, Green, London, 1871.
14. W. S. McCulloch, *Brain and behavior*, in W. C. Halstead (Editor), Comparative Psychology, Univ. of California Press, Berkeley, 1950, pp. 39–50.
15. ——, *Embodiments of mind*, MIT Press, Cambridge, 1965.
16. W. S. McCulloch and W. Pitts, *A logical calculus of the ideas immanent in nervous activity*, Bull. Math. Biophys. **5** (1943), 115–133.
17. Denes Nagy, Ferenc Nagy, and Peter Norvath, *The computer, the brain, and the missing link* (in Memoriam John von Neumann), Pamphlet, National Technical Information Centre and Library, Budapest and Tempe, 1987.
18. H. Nyquist, *Certain factors affecting telegraph speed*, Bell System Tech. J. **3** (1924), 324–346.
19. A. Rosenblueth, N. Wiener, and J. Bigelow, *Behavior, purpose, teleology*, Philos. Sci. **10** (1943), 18–24.
20. Bertrand Russell and Alfred North Whitehead, 3 volumes, *Principia Mathematica*, Cambridge Univ. Press, Cambridge, 1910–1913.
21. Claude E. Shannon, *Von Neumann's contributions to automata theory*, Bull. Amer. Math. Soc. **64** (1958), 123–129.

22. C. Shannon, *Information theory*, Encyclopaedia Britannica, 14th edition, 1971.
23. C. Shannon and W. Weaver, *The mathematical theory of communication*, Univ. of Illinois Press, Urbana, Ill., 1949.
24. Jagjit Singh, *Great ideas in information theory, language, and cybernetics*, Dover, New York, 1966.
25. Nancy Stern, *Form ENIAC to UNIVAC*, Digital Press, Bedford, MA, 1981.
26. Leo Szilard, *Uber die Entropieverminderung in einem Thermodynamischen System bei Eingriffen Intelligenter Wesen*, Z. Phys. **53** (1929), 840.
27. A. Turing, *On computable numbers, with an application to the Entscheidungsproblem*, Proc. London Math. Soc. (2) **42** (1937), 230–265.
28. S. Ulam, *The late John von Neumann on computers and the brain*, Scientific American, vol. 198, no. 6, 1959, pp. 127–130.
29. Stanislaw M. Ulam, *Science, computers, and people*, Birkhauser, Boston, 1986.
30. Heinz von Foerster, *Cybernetics: Circular causal, and feedback mechanisms in biological and social systems*, Transactions of the Sixth Conference (March 24–25, 1949, New York), Josiah Macy, Jr. Foundation, New York, 1950.
31. ——, *Cybernetics: Circular causal and feedback mechanisms in biological and social systems*, Transactions of the Seventh Conference (March 23–24, 1950, New York), Josiah Macy, Jr. Foundation, New York, 1951.
32. ——, *Cybernetics: Circular causal and feedback mechanisms in biological and social systems*, Transactions of the Eighth Conference (March 15–16, 1951, New York), Josiah Macy, Jr. Foundation, New York, 1952.
33. ——, *Cybernetics: Circular causal and feedback mechanisms in biological and social systems*, Transactions of the Ninth Conference (March 20–21, 1952, New York), Josiah Macy, Jr. Foundation, New York, 1953.
34. ——, *Cybernetics: Circular causal and feedback mechanisms in biological and social systems*, Transactions of the Tenth Conference (April 22, 23, and 24, 1953, Princeton, N.J.), Josiah Macy, Jr. Foundation, New York, 1955.
35. J. von Neumann, *First draft of a report on the EDVAC*, Moore School of Electrical Engineering, University of Pennsylvania, Philadelphia, 1945. (Published in N. Stern, *From ENIAC to UNIVAC*, Digital Press, Bedford, MA, 1981, pp. 177–246.
36. ——, Review of Norbert Wiener, *Cybernetics*, Physics Today **2** (1949), 33–34.
37. ——, *General and logical theory of automata*, in Cerebral Mechanisms in Behavior—The Hixon Symposium, Reprinted in A. H. Taub (ed.), *Von Neumann's collected works*, vol. V, 1951, pp. 288–328.
38. ——, *Mathematical foundations of quantum mechanics*, Translated from the 1932 German edition by Robert T. Beyer, Princeton Univ. Press, Princeton, 1955.
39. ——, *The computer and the brain*, Yale Univ. Press, New Haven, 1958.
40. ——, *Collected works*, Six volumes (A. H. Taub, ed.), Pergamon Press, New York, 1961–1963.
41. ——, *Theory of self-reproducing automata* (A. Burks, ed.), Univ. of Illinois Press, Urbana, Ill., 1966.
42. ——, *Papers of John von Neumann on computing and computer theory* (William Aspray and Arthur W. Burks, eds.), Charles Babbage Institute Reprint Series, vol. 12, MIT Press and Tomash Publishers, Cambridge, 1987.
43. N. Wiener, *Cybernetics: or control and communication in the animal and the machine*, Technology Press, Cambridge, 1948.

CHARLES BABBAGE INSTITUTE, UNIVERSITY OF MINNESOTA

Proceedings of Symposia in Pure Mathematics
Volume **50** (1990)

Developments in "The Synthesis of Reliable Organisms from Unreliable Components"

NICHOLAS PIPPENGER

Abstract. In a series of lectures given at the California Institute of Technology in 1952, John von Neumann laid the foundations of the theory of reliable computation by machines built from unreliable components. This paper surveys the developments that have taken place in the theory since that time.

1. Introduction. In January 1952, John von Neumann gave a series of five lectures at the California Institute of Technology. Notes of these lectures were revised by von Neumann and published in 1956 [**vN56**]. It is fair to say that von Neumann's paper founded the theory of reliable computation with unreliable components. The goal of this paper is to give an account of the developments that have taken place in that theory.

Von Neumann's paper, like his earlier *The general and logical theory of automata* [**vN51a**], his later *The computer and the brain* [**vN58**], and the posthumously published *Theory of self-reproducing automata* [**vN66**], was written in heady times, the fragrance of which can still be savored in Wiener's *Cybernetics* [**Wie48**]. It was a time of heightened optimism about the prospects for understanding communication, computation, observation, control, self-correction, and self-reproduction, both in natural and artificial systems, and a time of growing awareness of the interrelationships among them. Von Neumann's bibliography in [**vN56**] cites papers by Kleene [**Kl56**], McCulloch and Pitts [**McCP43**], Shannon [**Sh48**], Szilard [**Sz29**], and Turing [**Tu36**], and emphasizes the connections among mathematics, physics, and biology.

2. Von Neumann's model. To study reliable computation with unreliable components presupposes a satisfactory model for computation with reliable

1980 *Mathematics Subject Classification* (1985 *Revision*). Primary 68M15, 94C10; Secondary 03B48, 68Q80, 82A25, 82A60, 94B30.

This paper is in final form and no version of it will be submitted for publication elsewhere.

components. The model used by von Neumann is roughly speaking the "neural net" of McCulloch and Pitts [**McCP43**], with the adjunction of a finite probability space to model the failures of components. Since the neural nets he discusses are all in fact "combinational" in nature, most later workers have used a model that is inherently combinational, such as the "network of gates" of Muller [**Mul56**], and we shall follow that practice here.

A precise discussion of errors in networks of gates requires a distinction between two types of errors, which (following a suggestion of Peter Gács) we shall call "failures" and "deviations". If the output of a gate is not what we expect from the inputs to that gate, we shall say that the gate "fails". If the output of a gate is not what we expect from the inputs to the network, we shall say that the output "deviates"; this may be due to a failure of that gate or to the propagation of failures from antecedent gates.

The occurrence of deviations may depend on the inputs to the network as well as on the locations of failures. We shall always be interested in the reliability of the network for the least favorable choice of the inputs. Thus we assume that an adversary to the network, with knowledge of the network, chooses the inputs to the network; then the failures are determined by a process to be described below.

The question of how to model failures is not completely straightforward; consider the following two quotations from [**vN56**].

> It is the author's conviction, voiced over many years, that error should be treated by thermodynamical methods, and be the subject of a thermodynamical theory, as information has been, by the work of L. Szilard and C. E. Shannon [...].

> The simplest assumption concerning errors is this: With every basic organ is associated a positive number ε such that in any operation, the organ will fail to function correctly with the (precise) probability ε. This malfunctioning is assumed to occur statistically independently of the general state of the network and of the occurrence of other malfunctions. A more general assumption, which is a good deal more realistic, is this: The malfunctions are statistically dependent on the general state of the network and on each other. In any particular state, however, a malfunction of the basic organ in question has a probability [...] which is $\leq \varepsilon$.

Here we have a statement of philosophy together with two concrete models. The philosophy is certainly consistent with that adopted by Wiener [**Wie49**] and Shannon [**Sh48**], in which "noise", the source of unreliability, is modeled as a random process with known parameters. Von Neumann's first model, with its parenthetical "precisely", adheres to this philosophy: the gates behave unreliably, but they can be relied upon to behave unreliably! If the value of ε is sufficiently small, then von Neumann shows that such gates can be assembled to form a network that simulates (except with a modest probability of deviation) a given network of perfectly reliable gates. But gates with probability of failure precisely ε can also be assembled to form "random number generators" with known statistical properties. Now random number generators can in many cases be eliminated from networks of perfectly reliable gates (see Adleman **[Ad78]**), but it seems unlikely that this elimination can be accomplished without substantial increase in size. In any event, in this model it is certainly not true that a network that computes reliably in the presence of failures would necessarily continue to do so in their absence. Thus, in this first model, the failures add to as well as detract from the computational resources.

The second, more "realistic" model, is ambiguous because of the phrase "general state of the network". When this ambiguity is resolved in what seems the most natural way, the counterintuitive features of the first model disappear, and some new attractive features (which will be mentioned later) replace them. (For a more complete discussion of the relationships among these and other models, see Pippenger **[Pi88b]**.) This second model is not completely consistent with the "thermodynamic" philosophy, however; the arbitrary dependence of probabilities on the general state of the network has an adversarial aspect akin to that of "Maxwell's demon".

Fortunately, it is not necessary to resolve the disparity between these two models. All of the negative results we shall discuss can be formulated for the first model, and all of the positive results for the second, so each result is automatically applicable to the other model.

An attractive feature of von Neumann's model is that it assumes the failure probability of the components to be fixed while contemplating arbitrarily large networks; thus, larger and larger networks must tolerate more and more failures. This may be contrasted with studies of networks that tolerate a fixed number of failures, independent of the size of the network, or which tolerate a large number of failures only if they are well dispersed. Such schemes include the well-known "triple-modular redundancy" (Lyons and Vanderkulk **[LV62]**) and "quadded logic" (Tryon **[Tr60]**, **[Tr62]**), as well as some schemes related to error-correcting codes (see Armstrong **[Ar61]** and Ray-Chaudhuri **[R61]**, for example). Von Neumann's model also assumes that all gates are unreliable; this contrasts with theories in which some reliable gates may also be used in critical portions of the network (see Muchnik and Gindikin **[MucG62]**, Kirienko **[Ki64, Ki70]**, Ulig **[U74]**, and Ortyukov **[Or78]** for results along these lines).

Another feature of von Neumann's model is that it considers the evaluation of a Boolean function on one set of Boolean arguments, rather than considering the evaluation of one function on many disjoint sets of arguments. This feature is best understood by consulting works that take a contrary stand, such as Winograd and Cowan **[WinC63]**, in the light of subsequent results concerning the evaluation of functions on many disjoint sets of arguments (see Ulig **[U74]** and Paul **[Pa76]**).

3. Computability and complexity. The main result that is proved in von Neumann's paper is the following. Fix a basis B of Boolean functions (we shall always assume B to be finite, and "complete" in the sense of Post **[Po41]**). For all sufficiently small $\varepsilon > 0$, there exists a $\delta < 1/2$ with the following property. Every Boolean function is computed by a network over B that deviates with probability at most δ when its gates fail with probability at most ε. The supremum of the ε for which there exists such a $\delta < 1/2$ will be denoted $\varepsilon_0(B)$. For $0 < \varepsilon < \varepsilon_0(B)$, the infimum of the δ with the cited property will be denoted $\delta_0(B, \varepsilon)$.

Three prominent features of von Neumann's argument are (1) that $\varepsilon_0(B) < 1/2$ for all finite bases B, (2) that $\delta_0(B, \varepsilon) > \varepsilon$ for all $0 < \varepsilon < \varepsilon_0(B)$, and (3) the network of unreliable gates may have greater depth (or "delay"), by a constant factor depending on B, ε, and δ, than a network of reliable gates computing the same function. This situation is in marked contrast with that of reliable communication over an unreliable channel. There, a channel (say, a binary symmetric channel with error probability ε) has a positive capacity $C(\varepsilon) = 1 + \varepsilon \log_2 \varepsilon + (1 - \varepsilon) \log_2(1 - \varepsilon)$ for all $\varepsilon < 1/2$, and any probability of incorrect decoding $\delta > 0$ is achievable (see Shannon **[Sh48]**). This contrast is evidently due to the fact that in the case of communication, it is assumed that coding and decoding can be performed without error, while in the case of computation, any coding and decoding must be done by components that are themselves subject to failure.

It is natural to ask whether this contrast is inherent, or whether it might be avoided by some more sophisticated argument. Pippenger **[Pi88a]** showed that for any finite, complete basis B, $\varepsilon_0(B)$ is bounded below $1/2$, at least if the computation is done by a formula with unreliable gates, rather than a network (a formula is a network in which the output of any gate serves as an input to at most one other gate). Furthermore, the ratio of the depth of the unreliable formula to that of a reliable formula is bounded above 1. Feder **[F88]** has extended these results to networks, and also shown that $\delta_0(B, \varepsilon)$ is bounded above ε in this case. Unfortunately, all of these negative results differ quantitatively from the positive results that emerge from von Neumann's analysis, even when this analysis is carried to its natural limit (see Pippenger **[Pi87]**). Indeed, while the negative results give bounds that are analytic functions of ε (like Shannon's capacity), the positive result gives a ratio of unreliable depth to reliable depth that is a "devil's staircase" function: a function that is monotonic and continuous, but constant throughout each of a countably infinite collection of intervals.

After establishing the possibility of reliable computation with unreliable components and ascertaining the effect on depth, von Neumann turned to assessing the effect on the size (or "cost") of the network. He came to the conclusion that a function computed by a network of l reliable gates could be computed by a network of $O(l \log l)$ unreliable gates (where the constant implicit in the O-notation depends on B, ε, and δ). He drew this conclusion by considering a method of converting a network of reliable gates into one of unreliable gates by (1) replacing each reliable gate by an "executive organ" comprising $O(\log l)$ unreliable gates; (2) replacing each wire by a "bundle" of $O(\log l)$ wires; and (3) introducing "restoring organs" (comprising $O(\log l)$ unreliable gates) that correct the deviations caused by the executive organs. The analysis he gave is not entirely satisfactory, however; it considers the operation of the network for a single choice of the inputs, and assumes that certain interconnections are made "at random" in the restoring organ. Now there must be a particular pattern of interconnection that performs as well as the average pattern for a particular input, but there is no guarantee that the same pattern of interconnection will perform well for all inputs. Thus this argument cannot deal with a situation where the input to the network is chosen by an adversary with knowledge of the network, as we have assumed in our models.

A rigorous proof for these models was given by Dobrushin and Ortyukov **[DO77b]**. They also give an example of a function of n arguments (namely, the sum modulo 2 of n arguments) computed by a network of $O(n)$ reliable gates but requiring $\Omega(n \log n)$ unreliable gates. (As is customary in computer science, we use $\Omega(\dots)$ to denote a lower bound that holds "eventually", rather than merely "frequently".) Pippenger **[Pi85]**, on the other hand, gives an example of a function requiring $O(n)$ gates, whether reliable or unreliable. It seems to be difficult to give any simple criterion for when the extra logarithm appears.

It should be mentioned that the method of **[DO77a]** cannot give a lower bound larger than $\Omega(n \log n)$, and in particular cannot give $\Omega(l \log l)$ unless $l = O(n)$. It would be of interest to develop techniques that give larger lower bounds; since many such techniques are available for networks of reliable "monotone" gates (see Wegener **[We87]**, for example), and since monotone gates suffice for the correction of deviations, it would be natural to try first to find such lower bounds for networks of unreliable monotone gates.

In discussing these results on the sizes of networks, we have not explicitly indicated the dependence of the results on B, ε, and δ. In the theory of networks of reliable gates, it is an important principle that changing B from one finite and complete basis to another can affect the size and depth of optimal networks by at most constant factors (see Muller **[Mul56]**). It is not at all obvious that this principle carries over to networks of unreliable gates; indeed, for the model in which gates fail with probability precisely ε it seems unlikely that it holds, since there appears to be no way to construct a network over a basis that imitates the behavior of a gate not in that basis. For the second model, however, it can be shown that changing B, ε, and δ (within obvious limits) can affect the size and depth of optimal networks by at most constant factors (see Pippenger [**Pi88b**]). This justifies the imprecision of the O- and Ω-notation that we have used above and will use often throughout the rest of the paper.

4. Explicit constructions. As we have seen, both von Neumann's original plan for the restoring organ and Dobrushin and Ortyukov's implementation of it depend on random interconnections. This method of proof is familiar from Shannon's communication theory; it provides an existence proof but no explicit construction. Von Neumann was interested in the deterministic simulation of randomness from both a mathematical and a philosophical point of view, as is evident from the following well-known passage from [**vN51b**].

> Any one who considers arithmetical methods of producing random digits is, of course, in a state of sin. For, as has been pointed out several times, there is no such thing as a random number—there are only methods to produce random numbers, and a strict arithmetic procedure of course is not such a method. (It is true that a problem that we suspect of being solvable by random methods may be solvable by some rigorously defined sequence, but this is a deeper mathematical question than we can now go into.)

It is natural then, that he tried to supply an explicit construction for the restoring organ, and conjectured that a certain permutation (based on the reversal of binary digits) would serve as the basis for such a construction. From today's perspective, von Neumann's suggestion seems rather naive; it is vague enough to evade definitive refutation, but there is no evidence in its favor. In recent years, however, several provably correct explicit constructions have been given.

The first such explicit construction was given by Pippenger [**Pi85**]; it depends on the explicit construction of graphs with certain special spectral properties. These graphs (essentially "expanding graphs") have themselves been the subject of a long quest for explicit constructions. Contributions by Margulis [**Mar75**] and by Gabber and Galil [**GabG81**] were refined by Jimbo and Maruoka [**JM87**]; the use of Jimbo and Maruoka's graph in Pippenger's construction yields restoring organs with about 2^{57} gates per wire. Fortunately, subsequent work by Lubotzky, Phillips, and Sarnak [**LubPS88**] on "Ramanujan graphs" allows this constant to be deflated to about 2^9, still too large to be practical, but at least no longer ridiculous. Peter Gács has pointed out that another explicit restoring organ can be fashioned from the "approximate halvers" that constitute the fundamental building block of the celebrated Ajtai-Komlós-Szemerédi sorting network [**AKS83a**, **AKS83b**]; this also depends on explicit constructions for expanding graphs, and leads to similar constants to the ones cited above.

Before leaving the subject of explicit restoring organs, it should be mentioned that the explicit constructions for expanding graphs are all based on the action of a discrete group on a finite set, where the discrete group contains at least a free group on two generators. The proofs given by Gabber and Galil and by Jimbo and Maruoka both have a certain resemblance to von Neumann's proof that a group containing a free group on two generators cannot have an invariant mean [**vN29**]; so it may not be too arrogant to believe that von Neumann would have been delighted by this resolution of his problem.

5. Generic and linear functions. One of the central features of Shannon's theory of communication is that it associates with an unreliable channel a real number (called the "capacity" of the channel) that summarizes the information carrying ability of the channel in an absolute and quantitative way. It has been the dream of von Neumann's successors to associate with an unreliable gate a real number that summarizes the information processing ability of the gate in a similar way. This can clearly be done if we look at the depths of networks, as has been mentioned in §3. (The resulting number depends on some of the details of the model, but the situation is not substantially different in this respect from that of communication.) The $O(l \log l)$ estimate of von Neumann, together with the $\Omega(l \log l)$ bound of Dobrushin and Ortyukov [**DO77a**], would seem to dash all hopes of establishing similar results for the sizes of networks, since it suggests that reliable gates are stronger in a "non-Archimedian" way than unreliable ones. Recent results have, however, given rebirth to some of these hopes.

Muller [**Mul56**] showed that "almost all" Boolean functions of n Boolean arguments are computed by networks the minimum possible size of which is $\Theta(2^n/n)$. (Here $\Theta(\dots)$ denotes both $O(\dots)$ and $\Omega(\dots)$.) This result (like its precursor for networks of relay contacts, due to Shannon [**Sh49**]) is proved in two steps: an upper bound of $O(2^n/n)$ is proved by explicit construction that applies to all (rather than almost all) functions of n arguments; then a lower bound of $\Omega(2^n/n)$ is proved by a counting argument that compares the number of networks of a given size with the number of functions of a given number of arguments. The estimates given by Muller in this way differ by substantial constant factors; it was Lupanov [**Lup58**] who showed that more delicate methods give constant factors that asymptotically coincide, thus providing an exquisitely precise theory of the complexity of almost all functions. Lupanov's result is that almost all functions of n arguments are computed by networks the minimum possible size of which is asymptotic to $\rho(B)2^n/n$, where $\rho(B)$ is a constant that depends on the basis B in a known and easily determinable way.

Pippenger [**Pi85**] showed that all functions of n arguments are computed by networks with $O(2^n/n)$ unreliable gates. Comparing this with the lower bound of Muller shows that for almost all functions, the minimum possible numbers of reliable and unreliable gates differ at most by a constant factor, depending on the basis B, the gate failure probability ε, and the network deviation probability δ. The methods used by Pippenger are comparable to those of Muller in their crudity; but Uhlig [**U87a**, **U87b**] has used the more delicate methods of Lupanov to derive an upper bound asymptotic to $\sigma(B,\varepsilon,\delta)2^n/n$, where $\sigma(B,\varepsilon,\delta) \to \rho(B)$ as $\varepsilon \to 0$. It would be of great interest to obtain an asymptotically matching lower bound of this form, for it would justify considering $\rho(B)/\sigma(B,\varepsilon,\delta)$ as a "capacity" for unreliable gates in the basis B.

Von Neumann's theory deals with the computation of an arbitrary Boolean function, rather than with particular functions (such as arithmetic operations) to which special methods may apply. Nevertheless, the computation of linear functions occupies a special place. The first hint that this is so appears in a paper of Elias [**E58**], in which he analyzes two-argument Boolean functions and points out an essential difference between "exclusive-or" (or addition modulo 2) and "inclusive-or". Specifically, certain forms of coding that are possible for exclusive-or (or its complement) are impossible for inclusive-or (and the functions obtained from it by complementing the function and its arguments in some combination). The negative result is developed in subsequent papers by Peterson and Rabin [**PeteR59**], Winograd [**Win62**, **Win63**], Pradhan and Reddy [**PrR72**], and Ahlswede [**Ah84**]. Winograd's result, for example, can be formulated as follows: let ϕ: $\{0,1\}^k \to \{0,1\}^n$ be a code that preserves componentwise inclusive-or (that is, $\phi(x \vee y) = \phi(x) \vee \phi(y)$) and has minimum distance d. Then d/n and k/n cannot both be bounded away from zero as $n \to \infty$.

If inclusive-or is replaced by exclusive-or, the situation is quite different; indeed, most error-correcting codes are linear (over the field $GF(2)$), and thus preserve componentwise exclusive-or. A class of linear codes that is particularly well suited to computation with unreliable gates is the class of low-density parity-check codes, introduced by Gallager [**Gal63**]. This suitability was exploited by Taylor [**Ta68a**] for reliable storage of information with unreliable gates and memory cells, and in [**Ta68b**] for reliable componentwise exclusive-or with unreliable gates (the results in [**Ta68b**] concerning other operations are incorrect).

The result of Taylor on storage of information has been refined by Kuznetsov [**Kuz73**] to a form that is (within constant factors) optimal as regards both the amount of information stored and (the logarithm of) the length of time for which it is preserved. The only drawback of Kuznetsov's result is that, while explicit constructions have now been found for low-density parity-check codes (see Margulis [**Mar82**] and Imrich [**Im84**]), Kuznetsov's result still depends upon a random-coding argument. It would be of interest to replace this by an explicit construction. It would also be of interest to determine the complexity of the initial coding and final decoding required by this scheme.

Let us now consider the simultaneous computation of a set of m functions, each of which is the sum modulo 2 of some of their n arguments. If $m = 1$ and the function is the sum of all n arguments, then (as we have seen in §3), the minimum possible size with reliable components is $\Theta(n)$, but with unreliable components it is $\Theta(n \log n)$. If $n = m$, the minimum size depends of course on the particular functions. Lupanov **[Lup56]** has shown that for all such sets of functions, the minimum possible number of reliable gates is $O(n^2/\log n)$, and for "almost all" such sets it is $\Omega(n^2/\log n)$. Pippenger **[Pi85]** has shown that for all such sets, the minimum possible number of unreliable components is also $O(n^2/\log n)$. This provides yet another example of a situation in which the minimum possible numbers of reliable and unreliable components differ by at most a constant factor.

6. Variations. This section is devoted to a discussion of two other approaches to the problem of reliable computation with unreliable components, approaches that are based on models for computation and unreliability that are different from those von Neumann used.

The neural nets used by von Neumann have come to be viewed as a member of a class of models in which "gates" are interconnected to form networks. There is a similar but distinct class of models in which "contacts" are the basic components, and it was such a model that Shannon **[Sh38]** used in the paper that first applied Boolean algebra to logical networks. Such models were used by Shannon in other papers on logical networks (in particular, **[Sh49]**), including one with Moore **[MoSh56]** entitled *Reliable circuits using less reliable relays.* A great deal of additional work has been done for such models; most of this work assumes a fixed or slowly growing number of failures (see Andreev **[An86]** for references), but Petri **[Petr69]** retains the assumption used by Moore and Shannon of independent failures with fixed probability.

With contact networks, it is possible to tolerate component failure probabilities arbitrarily close to 1/2; and for any such fixed component failure probability, it is possible to achieve network failure probabilities arbitrarily close to 0. This striking difference from gate networks may be attributed to the fact that in contract networks, more of the computation is done by the interconnection pattern (which is not subject to failure), rather than by the components. This is clearly a weakness of the failure model, for in practice there are certainly "open circuits" and "short circuits" that do not correspond to the failure of any single contact.

There is an analogous weakness in the failure model of von Neumann, in that "wires" are assumed to carry signals from gate to gate in a way that is not subject to failure. For wires of bounded length, this presents no problem, since a failure of the wire can be ascribed to the gate that drives it or the gate that senses it. But if gates occupy a fixed volume, large networks embedded in three-dimensional space will have large distances between most pairs of gates, and unless special care is taken many wires will traverse such distances. (Random constructions, such as those proposed by von Neumann, and explicit constructions, such as those proposed by Pippenger and Gács, inevitably have long wires.) It may reasonably be argued that the probability of a signal successfully traversing a long wire decreases exponentially with its length, if the cross-section of the wire remains fixed as its length increases.

The prospect of formulating and justifying a model accounting for the physical disposition of gates in space and for the failures of communication among them may appear daunting, but there is a simple model that easily meets objections based on lengths of wires, and which can overcome a number of objections related to power supply and heat removal as well. This model is the "cellular automaton", or "iterative array", which was used by von Neumann in the context of self-reproduction [**vN66**]. In this model, computation is performed by an infinite collection of finite automata positioned periodically in one- or more-dimensional space. The state of each automaton at a given time step depends on its state and those of its neighbors at the immediately preceding time step; this dependence is the same for all automata and all time steps. This model requires no direct communication beyond neighbors and, once furnished with conventions for input and output, provides a satisfactory computing medium. The automata can, of course, be made unreliable in the same way as are gates.

This model bears a strong resemblance to many that are used in physics to study magnetization, condensation, percolation, and other phenomena that exhibit a "phase transition". (Most such physical models are "static", and are used to study equilibria, but many have "dynamic" counterparts; it is these that are analogous to cellular automata.) The distinguishing feature of a phase transition is that a qualitative change in the overall organization of the system occurs when a parameter passes a critical threshold. After consideration of many specific examples, the following principle became part of the folklore of statistical mechanics: phase transitions can occur in systems with two or more dimensions, but not in one-dimensional systems. In the context of cellular automata with unreliable components, this principle takes the form of the following conjecture: a one-dimensional cellular automaton with unreliable components is "ergodic", that is, tends to an equilibrium distribution that is independent of the initial state. Put simply, a one-dimensional unreliable cellular automaton cannot "remember".

This conjecture was disproved by Gács [**Gác86**]; the proof makes use of ideas from a rather vague attempt at disproof due to Kurdjumov [**Kur78**]. The essential feature of these constructions is that they simultaneously solve the problems of storage and computation with unreliable components; they do this recursively and depend upon "universality" and "fixed-point" arguments similar to those used by von Neumann in his discussion of self-reproduction [**vN51a, vN66**]. (If the requirement that the state transition rule be independent of time and position is dropped, reliable storage by a one-dimensional unreliable cellular automaton can be achieved by a much simpler scheme due to Cirelson **[C78]**, which does not involve universal computation.)

In two or more dimensions, the situation is much simpler. Despite an early misstatement by Ising **[Is25]**, it has been known since 1936 that two-dimensional static magnetic systems can exhibit a phase transition (see Peierls **[Pei36]**, Kramers and Wannier **[KrW41]**, and Onsager **[On44]** for a beautiful example). It remained to formulate an analog with suitable dynamics to achieve reliable storage and computation with unreliable components. This was done by Toom **[To74]** in 1974; in **[To80]** Toom gave a remarkably simple transition rule that is nonergodic: at each step, take a majority vote among your state, that of your northern neighbor, and that of your eastern neighbor. The use of Toom's rule for reliable computation by two- and three-dimensional unreliable cellular automata has been discussed by Gács **[Gác87]** and by Gács and Reif **[GácR88]**, who have given an independent proof of the nonergodicity of Toom's rule based on a "renormalization" argument. Recently, Berman and Simon **[BS88]** have given a simplification of Toom's proof. This development leads to a most interesting open problem (formulated by Gács **[Gác86]**) that concerns "self-organization" with unreliable components. The two-dimensional nonergodic media constructed by Toom have distinct translation-invariant equilibrium distributions; is it possible for a one-dimensional medium to have this property?

This survey began with remarks on von Neumann's interest in the connections among computation, physics, and biology. It is fitting that it has concluded with examples of how ideas connected with phase transitions and self-reproduction have played a role in solving some of the problems arising from his legacy.

References

[Ad78] L. Adleman, *Two theorems on random polynomial time*, IEEE Sympos. on Foundations of Computer Science, vol. 19, IEEE Press, New York, 1978 pp. 75–83.

[Ah84] R. Ahlswede, *Improvements of Winograd's result on computation in the presence of noise*, IEEE Trans. Inform. Theory **30** (1984), 872–877.

[AKS83a] M. Ajtai, J. Komlós, and E. Szemerédi, *Sorting in $c \log n$ parallel steps*, Combinatorica **3** (1983), 1–19.

[AKS83b] ——, *An $O(n \log n)$ sorting network*, ACM Sympos. on Theory of Computing, vol. 15, ACM, New York, 1983, pp. 1–9.

[An86] A. E. Andreev, *A universal principle of self-correction*, Math. USSR-Sb. **55** (1986), 145–169.

[Ar61] D. B. Armstrong, *A general method of applying error correction to synchronous digital systems*, Bell System Tech. J. **40** (1961), 577–593.

[BS88] P. Berman and J. Simon, *Investigations of fault-tolerant networks of computers*, ACM Sympos. on Theory of Computing, vol. 20, ACM, New York, 1988, pp. 66-77.

[C78] B. S. Cirelson, *Reliable storage of information in a system of unreliable components with local interactions*, in Locally Interacting Systems and Their Applications in Biology (R. L. Dobrushin, V. I. Kryukov, and A. L. Toom, eds.), Springer-Verlag, New York, 1978.

[DO77a] R. L. Dobrushin and S. I. Ortyukov, *Lower bound for the redundancy of self-correcting arrangements of unreliable functional elements*, Problems Inform. Transmission **13** (1977), 59–65.

[DO77b] ——, *Upper bound for the redundancy of self-correcting arrangements of unreliable functional elements*, Problems Inform. Transmission, **13** (1977), 203–218.

[E58] P. Elias, *Computation in the presence of noise*, IBM J. Res. Develop. **2** (1958), 346–353.

[F88] T. Feder, *Reliable computation by networks in the presence of noise*, IEEE Trans. Info. Theory, **35** (1989), 569–571.

[GabG81] O. Gabber and Z. Galil, *Explicit constructions of linear-sized superconcentrators*, J. Comput. System Sci. **22** (1981), 407–420.

[Gác86] P. Gács, *Reliable computation with cellular automata*, J. Comput. System Sci. **32** (1986), 15–78.

[Gác87] ——, *Error-correction in two-dimensions: A trade-off in complexity* (to appear).

[Gác88] P. Gács and J. H. Reif, *A simple three-dimensional real-time reliable cellular array*, J. Comput. System Sci. **36** (1988), 125–147.

[Gal63] R. G. Gallager, *Low-density parity-check codes*, MIT Press, Cambridge, MA, 1963.

[Im84] W. Imrich, *Explicit construction of regular graphs without small cycles*, Combinatorica **4** (1984), 53–59.

[Is25] E. Ising, *Beitrag zur Theorie des Ferromagnetismus*, Z. Phys. **31** (1925), 253.

[JM87] S. Jimbo and A. Maruoka, *Expanders obtained from affine transformations*, Combinatorica **7** (1987), 343–355.

[Ki64] G. I. Kirienko, *O Samokorrektiruyushchikhsya Skhemakh iz Funktsionalnykh Elementov*, Problemy Kibernet. **12** (1964), 29–37.

[Ki70] ——, *Sintez Samokorrektiruyushchikhsya Skhem iz Funktsionalnykh Elementov dlya Sluchaya Rastushchego Chisla Oshibok v Skheme*, Diskret. Anal. **16** (1970), 38–43.

[Kl56] S. C. Kleene, *Representation of events in nerve nets and finite automata*, in Automata Studies (C. E. Shannon and J. McCarthy, eds.), Princeton Univ. Press, Princeton, N.J., 1956.

[KrW41] H. A. Kramers and G. H. Wannier, *Statistics of the two-dimensional ferromagnet*, Phys. Rev. **60** (1941), 252, 263.

[Kur78] G. L. Kurdjumov, *An example of a nonergodic one-dimensional homogeneous random medium with positive transition probabilities*, Soviet Math. Dokl. **19** (1978), 211–214.

[Kuz73] A. V. Kuznetsov, *Information storage in a memory assembled from unreliable components*, Problems Inform. Transmission **9** (1973), 254–264.

[LubPS88] A. Lubotzky, R. Phillips, and P. Sarnak, *Ramanujan graphs*, Combinatorica **8** (1988), 261–277.

[Lup56] O. B. Lupanov, *O Ventilnykh i Kontaktno-Ventilnykh Skhemakh*, Dokl. Akad. Nauk SSSR **111** (1956), 1171–1174.

[Lup58] ——, *Ob Odnom Metode Sinteza Skhem*, Izv. Vyssh. Uchebn. Zaved. (Radiofizika) **1** (1958), 120–140.

[LV62] R. E. Lyons and W. Vanderkulk, *The use of triple-modular redundancy to improve computer reliability*, IBM J. Res. Develop. **6** (1962), 200–209.

[McCP43] W. S. McCulloch and W. Pitts, *A logical calculus of the ideas immanent in nervous activity*, Bull. Math. Biophys. **5** (1943), 115–133.

[Mar75] G. A. Margulis, *Explicit construction of concentrators*, Problems Inform. Transmission **9** (1975), 325–332.

[Mar82] ——, *Explicit constructions of graphs without short cycles and low density codes*, Combinatorica **2** (1982), 71–78.

[MoSh56] E. F. Moore and C. E. Shannon, *Reliable circuits using less reliable relays*, J. Franklin Inst. **262** (1956), 191–208, 281–297.

[MucG62] A. A. Muchnik and S. G. Gindikin, *The completeness of a system made up of non-reliable elements realizing a function of algebraic logic*, Soviet Phys. Dokl. **7** (1962), 477–479.

[Mul56] D. E. Muller, *Complexity in electronic switching circuits*, IRE Trans. Electr. Comput. **5** (1956), 15–19.

[vN29] J. von Neumann, *Zur allgemeinen Theorie des Masses*, Fund. Math. **13** (1929), 73–116.

[vN51a] ——, *The general and logical theory of automata*, in Cerebral Mechanisms in Behavior (L. A. Jeffress, ed.), Wiley , New York, 1951.

[vN51b] ——, *Various techniques used in connection with random digits*, J. Res. Nat. Bur. Standards (Appl. Math) **3** (1951), 36–38.

[vN56] ——, *Probabilistic logics and the synthesis of reliable organisms from unreliable components*, in Automata Studies (C. E. Shannon and J. McCarthy, eds.), Princeton Univ. Press, Princeton, N.J. 1956.

[vN58] ——, *The computer and the brain*, Yale Univ. Press, New Haven, CT, 1958.

[vN66] ——, *Theory of self-reproducing automata*, (A. W. Burks, ed.), Univ. of Illinois Press, Urbana, IL, 1966.

[On44] L. Onsager, *Crystal statistics*, I. *A two-dimensional model with an order-disorder transition*, Phys. Rev. **65** (1944), 117–149.

[Or78] S. I. Ortyukov, *Synthesis of asymptotically nonredundant self-correcting arrangements of unreliable functional elements*, Problems Inform. Transmission **13** (1978), 247–251.

[Pa76] W. J. Paul, *Realizing Boolean functions on disjoint sets of variables*, Theoret. Comput. Sci. **2** (1976), 383–396.

[Pei36] R. Peierls, *Ising's model of ferromagnetism*, Proc. Cambridge Philos. Soc. **32** (1936), 477.

[PeteR59] W. W. Peterson and M. O. Rabin, *On codes for checking logical operations*, IBM J. Res. Develop. **3** (1959), 163–168.

[Petr69] N. V. Petri, *The complexity of realizing a function of algebraic logic with contact circuit using unreliable contacts when high reliability is required*, Problemy Kibernet. **21** (1969), 159–169.

[Pi85] N. Pippenger, *On networks of noisy gates*, IEEE Sympos. on Foundations of Computer Science, vol. 26, IEEE Press, New York, 1985, pp. 30–38.

[Pi87] ——, *Analysis of error-correction by majority voting*, (to appear).

[Pi88a] ——, *Reliable computation by formulae in the presence of noise*, IEEE Trans. Inform. Theory **34** (1988), 194–197.

[Pi88b] ——, *Invariance of complexity measure for networks with unreliable gates*, J. ACM **36** (1989), 531–539.

[Po41] E. L. Post, *Two-valued iterative systems of mathematical logic*, Princeton Univ. Press, Princeton, N.J., 1941.

[PrR72] D. K. Pradhan and S. M. Reddy, *Error-control techniques for logic processors*, IEEE Trans. Comput. **21** (1972), 1331–1336.

[R61] D. K. Ray-Chaudhuri, *On the construction of minimally redundant reliable system designs*, Bell System Tech. J. **40** (1961), 595–611.

[Sh38] C. E. Shannon, *A symbolic analysis of relay and switching circuits*, Trans. AIEE **57** (1938), 713–723.

[Sh48] ——, *A mathematical theory of communication*, Bell System Tech. J. **27** (1948), 379–423, 623–656.

[Sh49] ——, *The synthesis of two-terminal switching circuits*, Bell System Tech. J. **28** (1949), 59.

[Sz29] L. Szilard, *Über die Entropieverminderung in einem thermodynamischen System bei Eingriffen intelligenter Wesen*, Z. Phys. **53** (1929), 840–856.

[Ta68a] M. G. Taylor, *Reliable information storage in memories designed from unreliable components*, Bell System Tech. J. **47** (1968), 2299–2337.

[Ta68b] ——, *Reliable computation in computing systems designed from unreliable components*, Bell System Tech. J. **47** (1968), 2339–2366.

[To74] A. L. Toom, *Nonergodic multidimensional systems of automata*, Problems Inform. Transmission **10** (1974), 239–246.

[To80] ——, *Stable and attractive trajectories in multicomponent systems*, in Multicomponent Random Systems (R. L. Dobrushin and Ya. G. Sinai, eds.), Marcel Dekker, New York, 1980.

[Tr60] J. G. Tryon, *Redundant logic circuitry*, U.S. Patent 2,942,193, June 21, 1960.

[Tr62] ——, *Quadded logic*, in Redundancy Techniques for Computing Systems (R. H. Wilcox and W. C. Mann, eds.), Spartan Books, New York, 1962.

[Tu36] A. M. Turing, *On computable numbers*, Proc. London Math. Soc. (2) **42** (1936), 230–265.

[U74] D. Ulig (=Uhlig), *On the synthesis of self-correcting schemes from functional elements with a small number of reliable elements*, Math. Notes. Acad. Sci. USSR **15** (1974), 558–562.

[U87a] D. Uhlig, *On reliable networks from unreliable gates*, Lecture Notes in Comput. Sci., vol. 269, Springer-Verlag, New York, 1987, pp. 155–162.

[U87b] ——, *Reliable networks from unreliable gates with almost minimal complexity*, Lecture Notes in Comput. Sci., vol. 278, Springer-Verlag, New York, 1987 pp. 462–469.

[We87] I. Wegener, *The complexity of Boolean functions*, Wiley, New York, 1987.

[Wie48] N. Wiener, *Cybernetics: or control and communication in the animal and the machine*, MIT Press, Cambridge, MA, 1948.

[Wie49] ——, *Extrapolation, interpolation, and smoothing of stationary time series*, MIT Press, Cambridge, MA, 1949.

[Win62] S. Winograd, *Coding for logical operations*, IBM J. Res. Develop. **6** (1962), 430–436.

[Win63] ——, *Redundancy and complexity of logical elements*, Inform. and Control **5** (1963), 177–194.

[WinC63] S. Winograd and J. D. Cowan, *Reliable computation in the presence of noise*, MIT Press, Cambridge, MA, 1963.

UNIVERSITY OF BRITISH COLUMBIA, CANADA

Symposium Participants

Symposium Participants

Photographs donated to the American Mathematical Society by John Impagliazzo.

Marshall Stone, Marina von Neumann Whitman

Robert Whitman, Marina von Neumann Whitman,
Nicholas Vonneuman, Ann Vonneuman

Shizuo Kakutani, Dorothy Stone, Edwin Hewitt

Uffe Haagerup, Jack Cowan, James Glimm, George Mackey

Robert Bartle, Charles Ford, Marshall Stone, Huzihiro Araki

Francis Murray, Isadore Singer, Jack Cowan

Lee Dembart, Elwyn Berlekamp

Herman Goldstine, Ellen Goldstine, Shmuel Winograd

Eugene Wigner, Edward Teller

Abraham Taub, James Glimm

Mina Rees, Peter Lax, Françoise Ulam

Creighton Buck, Donald Ornstein, Richard Kadison, Harry Furstenberg

John Conway, Nicholas Pippenger

Barry Simon, Irving Segal

Israel Halperin, Dorothy Stone, Richard Kadison

Enrico Clementi, Irving Segal

Shmuel Winograd, Jack Cowan, John Conway,
John Impagliazzo, Daniel Hillis

Alain Connes, Arthur Jaffe

John Conway, William Aspray, Irving Segal